Manual of Water Supply Practices

M37

Coagulation, Clarification, and Filtration in Water Treatment

Fourth Edition

American Water Works Association

Manual of Water Supply Practices—M37, Fourth Edition

Coagulation, Clarification, and Filtration in Water Treatment

Sr. Editorial Manager, Books: Suzanne Snyder
Manager – Manuals: Molly Beach
Manager – Publishing Operations: Gillian Wink
Specialist – Copyright and Permissions: Peggy Tyler
Director of Publishing: John Fedor
Technical Editor: Wiley
Cover Composition: Michael Labruyere
Cover Image: Henrik A. Jonsson/Shutterstock.com
Production: Innodata

Library of Congress Cataloging-in-Publication Data
CIP data pending from the Library of Congress

ISBN 978-1-64717-232-9

ISBN, electronic 978-1-61300-755-6

http://doi.org/10.12999/AWWA.M37ed4

American Water Works Association

American Water Works Association
6666 West Quincy Avenue
Denver, CO 80235-3098
awwa.org

Contents

This page intentionally blank.

Figures

This page intentionally blank.

Tables

Preface

A group of volunteers from the Coagulation and Filtration Committee of AWWA, with support from members of several other committees and individual experts, prepared this manual of practice. This publication is the fourth edition of AWWA Manual M37, *Coagulation, Clarification, and Filtration in Water Treatment*. The previous editions were titled *Operational Control of Coagulation and Filtration Processes,* and it was originally published in 1992. Interested readers are encouraged to pursue additional information regarding coagulation, clarification, and filtration through many additional AWWA resources, including other publications and in-person engagement.

The purpose of this edition of this manual is focused on enabling operations staff at existing water treatment plants to get the most effective and optimized treatment reasonably possible from the systems and facilities they have at their disposal. This is intended to be not a design manual but rather a tool for operators and engineers who support operators. The contents of this manual include an overview of coagulation chemistry (Chapter 1); an in-depth look at rapid mixing and flocculation (Chapter 2), clarification (Chapter 3), and filtration (Chapter 4); an overview of how coagulation is used for membrane systems (Chapter 5) and for non–drinking water applications (Chapter 6); guidance for bench- (Chapter 7), pilot-, and full-scale testing (Chapter 8); a review of typical instruments used in these processes (Chapter 9); and several relevant case studies (Chapter 10). The authors strongly recommend that readers take the time to read Chapter 1 in its entirety before beginning the rest of the manual, as a clear understanding of coagulation chemistry is essential to understand or improve the operation of any other process discussed in this manual.

The Coagulation and Filtration Committee and AWWA welcome your comments and suggestions for improving future editions of this manual. Please e-mail AWWA staff at etssupport@awwa.org to provide feedback on the contents of this manual.

This page intentionally blank.

Acknowledgments

The AWWA Water Quality and Technology Division and the Coagulation and Filtration Committee gratefully acknowledge the contributions made by those volunteers who drafted, edited, and provided the significant and critical commentary essential to updating M37. The Steering Committee members dedicated many hours in the final stages of preparation of this edition to ensure the overall technical quality, consistency, and accuracy of the manual.

Chair

Enoch Nicholson, Jacobs, Bellevue, Wash.

Steering Committee Members

Brock Emerson, Intuitech, Salt Lake City, Utah
Ann Malinaro, Aurora Water, Aurora, Colo.
Enoch Nicholson, Jacobs, Bellevue, Wash.
David J. Pernitsky, Stantec Consulting, Pasadena, Calif.

Chapter Leads

John Civardi, Carollo Engineers, Canal Winchester, Ohio
Jeff Cochran, Birmingham Water Works Board, Birmingham, Ala.
Jason Curl, Hazen and Sawyer, Greenwood Village, Colo.
Ashley Kent, Arcadis, Austin, Tex.
Ali Leeds, Carollo Engineers, Seattle, Wash.
William Pennock, New Jersey Institute of Technology, Newark, N.J.
David J. Pernitsky, Stantec Consulting, Pasadena, Calif.
Robert Peters, Carollo Engineers, Hood River, Ore.
Damon Roth, Brown and Caldwell, Ritzville, Wash.
Ayman Shawwa, Stantec, Walnut Creek, Calif.
Rebecca Venot, Jacobs, Anchorage, Alaska

Contributors to the 4th edition

Chathura Abeyrathna, Central Contra Costa Sanitary District, Martinez, Calif.
Amir Alansari, Stantec Consulting, Winston-Salem, N.C.
James Amburgey, University of North Carolina at Charlotte, Charlotte, N.C.
Mark Beebe, HDR, Inc., Denver, Colo.
Allyssa Brauer, HDR, Inc., Denver, Colo.
George Budd, Retired, Richmond, Va.
Jesse Burch, Beaver Water District, Lowell, Ark.
Cassandra Butler, Central Contra Costa Sanitary District, Martinez, Calif.
William Carter, Jacobs, Denver, Colo.
John Clark, Chemtrac LLC, Norcross, Ga.

James DeWolfe, Hazen and Sawyer, State College, Pa.
Tracy Doane, Retired, Greenwood, Ind.
George Duval, Chesterfield County, Midlothian, Va.
Tom Elford, Shaw, Calgary, Alta.
Kim Ervin, Jacobs, Bellevue, Wash.
Josh Fortenbaugh, AWWA, Denver, Colo.
Martin Freudenberger, Retired, Ludwigsburg, Germany
Joseph Gemin, Retired Kitchener, Ont.
Mac Gifford, Portland Water Bureau, Portland, Ore.
Alexandra Griffin, Cleveland Water, Cleveland, Ohio
Paul Hargette, Hargette LLC, Greenville, S.C.
Catherine Hughes Keenan, HDR, Inc., Raleigh, N.C.
Stuart Humphries, Orthos Liquid Systems, Bluffton, S.C.
Tyson Ingels, Colorado Department of Public Health and Environment, Denver, Colo.
Frederick Kohlmann, Retired, Sussex, Wis.
Kerim Kollu, ABB, Inc., Whitby, Ont.
Nathan Kutil, HDR Inc., Missoula, Mont.
Eric Lawrence, Vessco Inc., Ames, Iowa
Kevin Linder, Aurora Water, Aurora, Colo.
Ann Malinaro, Aurora Water, Aurora, Colo.
Gregg McLeod, MarMac Water, Denver, Colo.
Jose Mendiola, A&S Engineers, Inc., Humble, Tex.
Craig Mizutani, Central Contra Costa Sanitary District, Martinez, Calif.
Ana Morfesis, Malvern Instruments, Pittsburgh, Pa.
Joseph Nattress, CDM Smith, Wilmington, Del.
Daniel Nix, Texas Section AWWA, Wichita Falls, Tex.
Christopher Bordlemay Padilla, Eastern Research Group, Inc., Upper Marlboro, Md.
Enric Palacios Donaque, Retired, Barcelona, Spain
Keith Riley, Endress+Hauser, Greenwood, Ind.
Jacqueline Rhoades, City of Arvada Public Works, Arvada, Colo.
Meric Selbes, Hazen and Sawyer, Fairfax, Va.
Holly Shorney-Darby, PWNT, Andijk, Netherlands
Robert Smith, Black & Veatch, Columbus, Ohio
Steven Smith, University of Colorado, Denver, Colo.
Darin St. Germain, WesTech Engineering, Ames, Iowa
Lynn Stephens, Brown and Caldwell, Seattle, Wash.
Supramaniam Sutha Suthaker, Associated Engineering, Edmonton, Alta.
Amanda Tyndall, Veolia Water Technologies & Solutions, Boulder, Colo.
Yamuna Srinivasan Vadasarukkai, Jacobs, Ottawa, Ont.
Adriano Vieira, Veolia Water Technologies & Solutions, Glen Allen, Va.
John Welsh, OI Analytical, College Station, Tex.
Brian Wilson, City of Shelby, Shelby, N.C.
Kristine Yung, East Bay Municipal Utility District, Oakland, Calif.
Ivan Zhu, Evoqua Water Technologies, Sewickley, Pa.

Chapter 1

Coagulation Chemistry

This chapter describes the fundamentals of coagulation chemistry and presents coagulation selection guidelines to aid operators in selecting the proper coagulant type, dose, and coagulation pH. Coagulation is essential for the proper functioning of clarification and granular media filtration processes and plays an important role in the removal of natural organic matter (NOM) for the control of disinfection byproducts (DBPs). Coagulation treatment chemistry is complex and involves the chemistry of the source water particles and NOM, the pH and alkalinity of the source water, and the chemistry of the coagulant itself. Two main groups of metal-salt coagulants are common in drinking water treatment: aluminum (Al)-based and iron (Fe)-based. Aluminum metal-salt coagulants include alum and various types of polyaluminum chloride (PACl). Iron-based coagulants include ferric chloride ($FeCl_3$); ferric sulfate ($Fe_2(SO_4)_3$); and to a much lesser extent, ferrous chloride and ferrous sulfate. All metal-salt coagulants depress the pH when added, so the selection of the appropriate coagulant type depends on the required coagulant dose, the source

Chapter Summary

- Natural organic matter (NOM) generally controls the coagulant dose, not turbidity.
- The pH of coagulation affects the charge of NOM, particles, and the coagulant itself. pH also affects the amount of floc formed by the coagulant.
- The coagulant type and dose must be matched to the source water alkalinity.
- Best results are generally achieved at coagulation pH of
  - 6.0–7.0 for alum,
  - 6.5–7.5 for polyaluminum chloride, and
  - 5.5–6.5 for iron coagulants.
- Acid or base addition may be desirable to maintain these pH ranges.
- Key performance indicators (KPIs) for coagulation effectiveness must include filter performance, not just clarifier performance. Unit filter run volumes are the best KPI.
- Each water is unique, and the guidelines in this chapter are general; testing is recommended in all cases.

water alkalinity, and whether any other acids or bases are added (like carbon dioxide [CO_2] or lime). This chapter will emphasize several times that the pH at which coagulation occurs—the coagulation pH—is one of the most important variables for optimizing coagulation performance.

Coagulation processes are used in water treatment to aid in the removal of both particulate and dissolved contaminants. Particulate material may include clays, sands, algae cells, bacteria, viruses, and protozoa, as well as particles generated in the water treatment process itself, such as coagulant floc particles or iron and manganese (Mn) precipitates. The removal of pathogenic microorganisms via coagulation, clarification, and filtration is one of the basic functions of water treatment. Chemical and/or physical disinfection is also dependent on the effective removal of particulate matter that may shield microorganisms from disinfectant contact and/or reduce the effectiveness of disinfection chemicals.

Most natural particles are negatively charged, and without treatment, they repel each other, settle slowly, and filter poorly. The addition of positively charged coagulants neutralizes this negative charge and allows particles to approach each other and flocculate together to form larger particles that are easier to remove. The coagulation process is essential for the proper operation of granular media filters. Without coagulation, particles present in the source water would simply pass through granular media filters as the negative surface charges of the particles cause them to be repelled by the negatively charged filter media. Once particle surface charges are neutralized, particles adhere to the surface of granular filter media, and particles that are much smaller than the space between the filter media grains can be efficiently removed. Membrane filters, on the other hand, remove particles by physical exclusion of the particles that are larger than the pore size of the membrane itself.

For many waters, coagulation is required to remove dissolved NOM, color, and DBP precursors. Coagulation reactions convert dissolved NOM to a solid phase that can be removed by clarification, granular media filters, or microfiltration (MF) and ultrafiltration (UF) membrane filters. Without coagulation, these dissolved NOM molecules would pass through granular media and MF–UF filters. In this case, only particulate NOM would be removed, which is usually a small fraction of the total NOM. Only nanofiltration and reverse osmosis (RO) membranes can remove NOM molecules without coagulation. This chapter will explain that the required coagulant dose is almost always controlled by the amount of NOM in the source water, not the source water turbidity.

Coagulation processes are also used to remove other compounds, remove NOM to reduce interference with iron and manganese removal processes and ultraviolet (UV) disinfection processes, reduce fouling in membrane treatment applications, and adsorb and precipitate various chemical contaminants, both regulated and nonregulated.

Coagulation selection is discussed first, as this practical information is why most operators use this manual. Additional explanation of the concepts used in the coagulant selection guidelines is provided afterward for readers who may be a bit rusty on water chemistry or those who want to develop a deeper understanding of coagulation chemistry. This chapter is meant as an overview of coagulation fundamentals; more detail can be found in AWWA's *Water Quality and Treatment: A Handbook on Drinking Water* (AWWA 2011). See Tables 1-5 through 1-7 at the end of the chapter for a coagulant unit conversion table, solids production factors, and alkalinity consumption for various coagulants.

COAGULANT SELECTION GUIDELINES

The selection of the dose and type of coagulant and pH of coagulation depends on the source water total organic carbon (TOC) concentration and turbidity, the source water alkalinity, and the water treatment processes used. The recommended steps for coagulant selection are outlined as follows.

Table 1-1 Estimated coagulant dose (mg/L coagulant per mg/L TOC) as a function of source water TOC and coagulation pH

	Coagulation pH					
Coagulant Dose	5.5	6.0	6.5	7.0	7.5	8.0
Al-based (mg/L as Al)	0.5	0.5	0.51	1.0	1.8	*
Fe-based (mg/L as Fe)	1.8	2.8	3.3	*	*	*

Source: Adapted from Letterman and Yiacoumi 2011.

* Reliable estimates are not available for this coagulation pH or greater.

Table 1-2 Effects of pH on particle, NOM, and coagulant charge

	Low pH	High pH
NOM charge	Less negative (easier to remove)	More negative (harder to remove)
Particle charge	Less negative (easier to neutralize)	More negative (harder to neutralize)
Coagulant charge	More positive (more powerful)	Less positive (less powerful)

Step 1: Estimate Coagulant Dose

As described previously, source water TOC generally controls the required coagulant dose, and a good starting point for estimating the coagulant dose is to use the values in Table 1-1 to estimate the coagulant (milligrams per liter) required per source water TOC (milligrams per liter). The estimated coagulant dose increases as the coagulation pH (the pH after coagulant addition) increases. For example, source water with 3 mg/L of TOC would require an alum dose of 1.53 mg/L as Al if the coagulation pH was controlled to pH 6.5.

Step 2: Match Coagulant Type to Source Water Alkalinity

The pH at which coagulation occurs has a strong influence on the charge of the particles and NOM in the source water and on the charge of the dissolved species and the floc particles formed by the addition of coagulant. Generally, low pH is better for coagulation performance, as summarized in Table 1-2.

Although most coagulants can be used over the broad range of pH 6.0–8.5 as well as higher pH values in softening applications, consideration of solubility and charge (discussed in detail in the later sections of this chapter) suggests narrower pH ranges at which they will work most efficiently (i.e., at lower doses). Alum is generally most effective when coagulation occurs between pH 6.0 and 7.0. PACls are most effective between pH 6.5 and 7.5, with higher pH values possible for higher-basicity products. Iron coagulants are generally most effective at pH 6.5 or lower. The coagulant basicity (i.e., alum, medium- or high-basicity PACl) and/or the use of supplemental acid or base should be selected such that the pH after coagulant addition is in these ranges. This is illustrated in a few examples:

- For a low-alkalinity water, if high doses of coagulant are needed because of high NOM concentrations, the use of acidic coagulants like alum or iron coagulants may result in the pH dropping to very low levels at which floc will not form. In this situation, a high-basicity PACl or a lower alum dose plus a cationic polymer may be a better choice, as less alkalinity will be consumed. Coagulant plus caustic or lime addition to maintain pH may also work well.
- For a high-alkalinity water, a very high-basicity PACl or aluminum chlorohydrate (ACH) may perform well, as these coagulants form positively charged floc particles even at high pH and their pH of minimum solubility is higher.

- An alternate approach for high-alkalinity waters is to use a more acidic coagulant such as alum, iron coagulants, or a low- to medium-basicity PACl. Supplementary acid addition to depress the pH may also be required.
- Sometimes for medium- to high-alkalinity waters, a utility cannot (or does not want to) drop the pH into the optimal range discussed previously. In this case, higher coagulant doses, use of polymers, and/or the use of conservative clarification and filtration loading rates can be combined to yield good overall treatment.

There may be several combinations of coagulants, polymers, and pH-adjustment chemicals that will result in good conditions. Desktop chemical modeling can be used to predict final coagulant pH to screen alternatives. This can then be followed up with jar testing for confirmation, with measurements including TOC, UV absorbance, turbidity, and particle charge, as outlined in Chapter 7.

Step 3: Water Treatment Process Considerations

In addition to producing near-neutral particles, there are some additional, process-specific aspects to coagulant selection. Some of these are outlined as follows:

- Some clarification processes, like sedimentation, benefit from the creation of large, heavy floc particles. High-molecular-weight polymers may be beneficial in these cases.
- Direct filtration processes require low concentrations of suspended solids to avoid high head loss and short filter runs. In high-NOM waters, the necessary metal-salt coagulant doses may result in high floc production that overwhelms the filters. Replacing some of the metal-salt coagulant dose with a low- to medium-molecular-weight cationic polymer may achieve near-neutral particle charge with lower floc concentrations. PACls have been shown to have lower head loss than alum in direct filtration applications (Pernitsky and Edzwald 2006).
- In some cases, media filters benefit from a filter-aid polymer if filter rates are high and/or influent particles are small and subject to detachment through hydraulic shear. A filter aid may be necessary or beneficial even when the incoming floc particles are of near-neutral charge. This is hard to determine without pilot- or full-scale trials.
- The selection of a coagulant and the pH of coagulation should consider the pH required for other process considerations such as iron and manganese oxidation, pH control to limit bromate formation in an intermediate ozonation step, etc.
- The chloride-to-sulfate mass ratio in the finished water can affect the control of lead in the distribution system (Edwards and Triantafyllidou 2007). In some instances, use of sulfate-based coagulants such as ferric sulfate and alum, rather than PACls or ferric chloride, may be beneficial.

KEY PERFORMANCE INDICATORS

Historically, coagulant dose and pH were selected on the basis of achieving low settled water turbidity in conventional jar tests. This approach was effective in the past when filtered-water turbidity regulations were higher than they are today and when NOM removal for the control of DBPs was less critical. In a modern water treatment plant (WTP), coagulation processes affect the performance of many aspects of the overall treatment process. Because of advances in the understanding of coagulation, improvements in instrumentation, and tighter regulations for multiple treatment parameters, additional key performance indicators (KPIs) are available. These are discussed in the following sections.

The best measure of coagulation effectiveness in a granular media filtration plant is the performance of the filters, and the best KPIs are the unit filter run volume (UFRV) and filter effluent turbidity and/or particle counts.

Filter Performance: UFRVs

The UFRV is calculated by dividing the volume of water produced during the filter run by the filter area. The UFRV begins after the filter has ripened and ends either when the effluent turbidity (or effluent particle count) exceeds the user-defined (or regulated) maximum value or when the maximum head loss is achieved. UFRVs are much better able to capture subtle changes in filter performance over the entire filter run than individual measurements of ripening time or head loss rates alone. In the water treatment industry, a benchmark for a conventional treatment plant is to achieve a UFRV of greater than 300 m^3/m^2 (7,500 gal/ft^2).

Filter Effluent Turbidity and Particle Counts

Filter effluent turbidity and particle count concentrations are important KPIs for filter performance, as they indicate the number of particulate contaminants in the treated water. Turbidity is highly regulated as an indicator of the degree of overall pathogen removal. Particle counts, although typically not included in most regulations, can provide information on small coagulant underdoses and early particle breakthrough. More information on these KPIs is provided in Chapters 4 and 9.

Organics Removal

Organics concentration and removal—as measured by TOC, UV absorbance, or UV transmittance—are also important KPIs. Monitoring NOM concentrations in the source water is important for variable-quality source waters. A rapid change in source water NOM concentration can result in an increase in the coagulant demand, and if the coagulant dose is not increased, an underdose can occur and filter performance can decrease. If source water NOM is not being monitored, the first indication of a problem could be early turbidity breakthrough for a granular media filtration plant.

In addition to monitoring source water NOM, postcoagulation or postfilter NOM can be measured. Increases in treated water NOM or decreased percentages of removal may indicate a coagulation dose adjustment is necessary. NOM should always be measured during bench-scale jar tests. In many cases, good filter performance in the full-scale plant will correspond to the point of diminishing returns for NOM removal.

UV absorbance, UV transmittance, and TOC instruments are discussed in more detail in Chapter 9.

Clarified Water Turbidity

As mentioned, settled water turbidity measurements from jar testing or from operating clarifiers have historically been the main KPI for evaluating coagulation performance. Although clarified water turbidity measurements indicate the performance of those units, they do not provide any information on the filterability of the floc particles that are present in the clarifier effluent. It is quite common for coagulant doses that are adequate to achieve typical clarified water turbidity goals of 1–2 ntu to still produce floc particles that are not readily filtered or significantly decrease filter production efficiency. Therefore, clarified water turbidity should be considered as a "secondary" KPI only for granular media filtration plants.

Table 1-3 Zeta potential ranges for effective WTP operations

Utility	Type of WTP	Zeta Potential Target *mV*	Reference
Calgary, Alta.	Conventional (ballasted clarification)	–5 to +5	Kundert 2014
Epcor (Edmonton, Alta.)	Conventional with seasonal direct filtration	–5 or greater	James et al. 2015
Tittesworth, UK	Conventional	–10 to +3	Sharp et al. 2015
Various Great Lakes, Ont.	Conventional	–7 or greater	Waller et al. 2012
Muskoka, Ont.	Direct filtration	–5 or greater	Pernitsky et al. 2010
Halifax, UK	Conventional	–10 to +5	Sharp et al. 2005
Fort Collins, Colo.	Conventional	–5 to +5	Morfesis et al. 2009
Adelaide, Australia	Conventional	–10 to –5	Holmes et al. 2015
Waterford, N.Y.	Conventional	0±5	Riddick 1964
ASTM	Generic standard	–5 to +5	ASTM 1987
Laboratory	Algae settling and dissolved air flotation	–10 to +2	Henderson et al. 2008

Source: Adapted from Pernitsky 2019
WTP—water treatment plant

Floc Charge Measurement

For clarification and granular media processes to work efficiently, source water particles and the precipitated floc particles resulting from coagulant addition must have net surface charges that are close to neutral. Particle charge can be measured in the WTP on the laboratory bench or on-line with zeta potential and streaming current instruments. Both instruments measure the electrostatic charge of particles in solutions and are described in detail in Chapter 9. Zeta potential and streaming current measurements will be referred to generically in this section as charge measurements. Charge measurement testing can help determine whether the coagulant dose has achieved charge neutralization.

Particles do not need to have a zero charge to flocculate and settle effectively or to filter effectively in a WTP; however, they do need to be "near neutral." This near-neutral operational particle charge range is site specific and must be determined empirically on the basis of actual operation at each WTP or on similar waters. However, many published studies are available from the water treatment industry to provide reasonable guidelines. Table 1-3 summarizes zeta potential ranges reported in the literature for effective WTP operation, as well as selected laboratory studies. The exact particle charge value that will enable proper plant performance is site specific and will depend on local source water quality, WTP infrastructure condition, and overall treatment targets.

A tighter control range for particle charges may provide increased operational robustness. In one study, operators at a WTP who were trying to minimize the cost of chemicals for water purification observed and concluded that water treatment control was optimized at a zeta potential range from 5 to +5 mV and that concern for plant operation increased if the zeta potential was outside the limits of –5 to +5 mV (Malvern Instruments 2011). Although plants can often meet regulations while operating at zeta potentials between –15 and –10 mV, close to the edge of the reported operational windows, they are more susceptible to upsets from relatively small changes to the input conditions. Also, the operational zeta potential window for adequate clarification performance is much broader than the operational window for ensuring good filter performance (Malvern Instruments 2011).

CHEMISTRY OF PARTICLES

This section and subsequent sections of this hour provide additional background on the fundamentals of coagulation chemistry that provide a theoretical basis for the guidelines described in the "Coagulant Selection Guidelines" section.

Particles of many types are found in natural waters and can be considered as inorganic, organic, or biological.

- Most inorganic particles in natural water are mineral particles. Examples include silicates, clays, iron oxides, aluminum oxides, and calcites.
- Most organic particles in natural water are the result of degradation of plant and animal materials. Soil humus and animal manure are examples.
- Biological particles include microorganisms such as viruses, bacteria, protozoa, and algae. Microorganisms may also be attached to suspended particulate matter.

Pure inorganic or organic particles seldom exist in natural water. For example, most inorganic particles in water have NOM sorbed to their surfaces. These organics play a significant role in the surface properties of these particles. Furthermore, NOM, regardless of whether it is in dissolved or particulate form, can chemically bind with many inorganic contaminants such as metals. These interactions can alter the surface properties of either.

In addition to natural particles that occur in water, particles are often added or created in water treatment processes. Metal-based coagulants precipitate as metal hydroxides to form "floc" particles. Oxidants/disinfectants such as chlorine may be added to water to oxidize soluble inorganic metals (e.g., iron and manganese) into insoluble particles. The particles added or created during the treatment process can also be removed by settling and filtration processes.

Particles can also be altered through the treatment process. The use of oxidants before filtration has been shown to benefit filter performance at numerous plants. Oxidants can react with functional groups on the inorganic particle surface or with the functional groups of the sorbed NOM, altering the particle charge (Becker et al. 2004).

Regardless of particle type, all particles have a surface charge that influences their behavior in coagulation processes, and for most particles, this surface charge varies with pH. For example, a clay particle will have hydroxyl (OH) groups on its surface, as shown in Figure 1-1, and the charge of the hydroxyl groups will change as the pH of the bulk water changes, as shown in Figure 1-2. Recall that pH is simply an expression of the concentration of hydrogen ions (H^+) present in the water. At low pH values (acidic), the water contains a relatively high number of H^+ ions, and at high pH values (basic), there is a relatively low concentration of H^+ ions. At high pH values, the lack of H^+ ions in solution causes some of the H^+ ions on the particle surface to leave and migrate into the water. This leaves a net negative charge on the particle surface. Most natural particles start out with a negative charge. At higher pH, this negative charge increases in magnitude, and at lower pH, the magnitude of the negative charge decreases.

CHEMISTRY OF NOM

Coagulant dosages are generally controlled by NOM concentration rather than by turbidity, as NOM molecules carry a significant amount of negative charge. In addition to controlling coagulant doses, NOM forms regulated DBPs when it reacts with free chlorine, so NOM removal is regulated in the United States and many other countries. The US Environmental Protection Agency (USEPA) Disinfectants and Disinfection Byproduct Rules (Stages 1 and 2) require that utilities minimize DBP production by achieving specific levels of TOC removal on the basis of source water TOC and alkalinity. Improving the

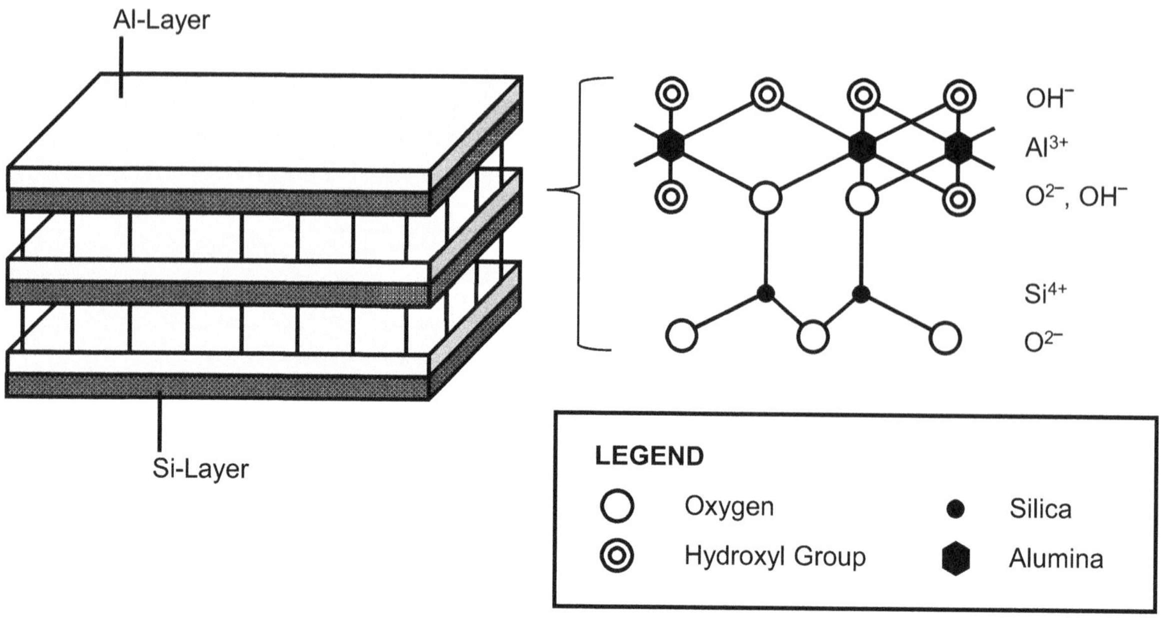

Source: David Pernitsky, adapted from Stumm and Morgan 1996

Figure 1-1 Schematic diagram of inorganic clay particle surface showing OH groups

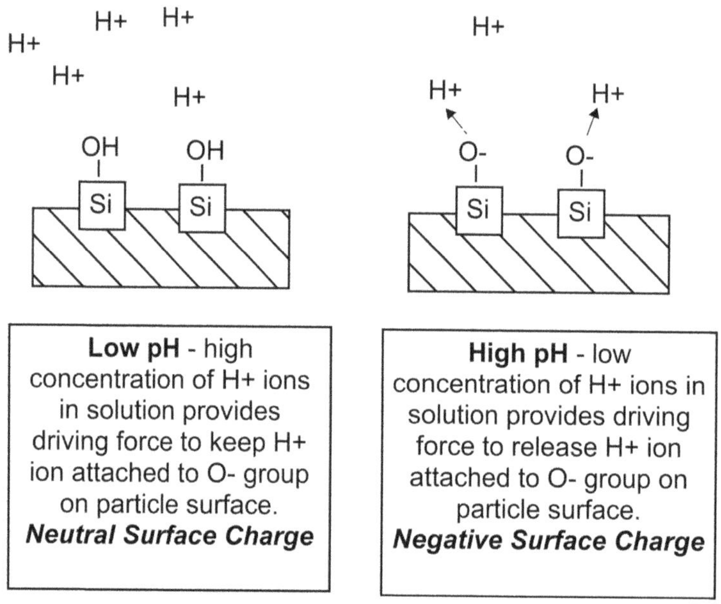

Source: David Pernitsky

Figure 1-2 Conceptual illustration of the relative effect of pH on the surface charge of an inorganic particle

removal of DBP precursors by adding coagulant doses greater than those needed strictly for meeting turbidity regulations is referred to by the USEPA as "enhanced coagulation."

NOM is a complex mixture of natural constituents originating primarily from the degradation of vegetation in the watershed and the growth and decomposition of aquatic organisms such as algae and weeds within the water body. NOM is generally classified into two main components: humic substances and nonhumic substances. More information on these components of NOM is available in *Water Quality and Treatment* (AWWA 2011).

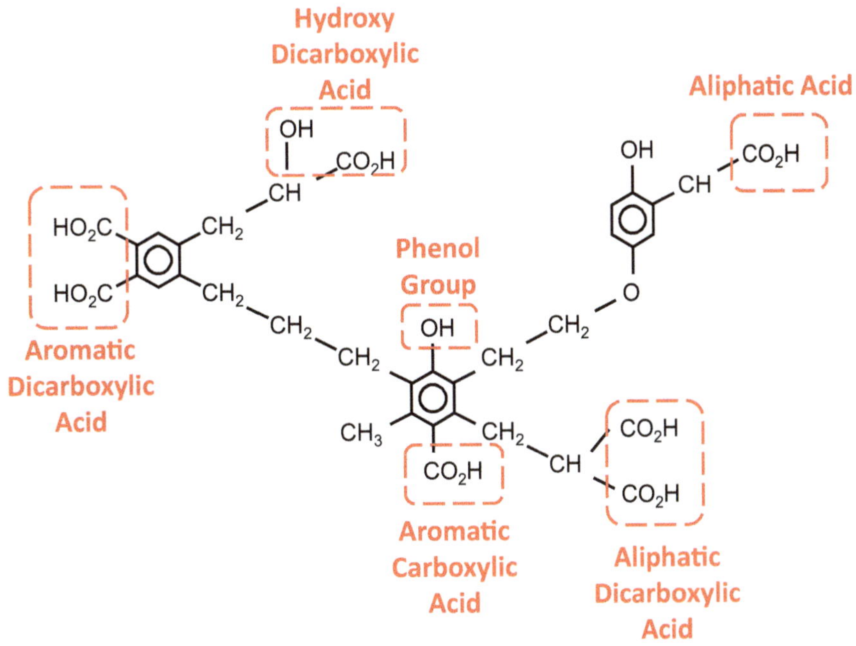

Source: Meric Selbes

Figure 1-3 Schematic diagram of NOM molecule showing ionizable functional groups

Because of NOM's complexity and heterogeneity, its chemical structure cannot be well characterized. The most important properties of NOM are molecular weight, functional groups, and charge behavior. NOM contains weakly acidic functional groups such as carboxylic and phenolic groups, as shown in Figure 1-3. The charges of these functional groups change with pH, as shown in Figure 1-4. In a manner similar to particle surface charges discussed earlier, as the pH increases, the negative charge of the NOM molecule increases. Because of this negative charge, the presence of NOM in water will induce an additional demand for coagulant dose.

As illustrated in Figures 1-2 and 1-4, both inorganic particles and organic molecules carry a negative charge. However, the charge demand of NOM is typically 10–100 times greater than the charge demand of inorganic particles. For this reason, coagulant dosages are generally controlled by NOM concentration rather than by turbidity in many surface waters (Pernitsky and Edzwald 2006, Edzwald and Van Benschoten 1990). Some utilities use preoxidation with chlorine, chlorine dioxide, ozone, or potassium permanganate to react with the source water NOM and thereby eliminate some of the coagulant demand, although DBP formation must be considered.

NOM from different water sources will have different characteristics. The concept of specific ultraviolet absorbance (SUVA) has been developed as an operational indicator of the nature of NOM and the effectiveness of coagulation for removing NOM, dissolved organic carbon (DOC), and DBP precursors (Edzwald and Tobiason 1999, Edzwald and Van Benschoten 1990). SUVA values offer a simple characterization of the nature of the NOM on the basis of measurements of the UV254 and DOC. SUVA is defined as the UV absorbance of a water sample normalized with respect to the DOC concentration. SUVA is normally calculated on samples of source water before the addition of any treatment chemicals. It is expressed in units of m^{-1} of absorbance per mg/L of DOC.

$$\text{SUVA} = [\text{UV254}(\text{cm}^{-1}) \times 100\ (\text{cm/m})]/\text{DOC}\ (\text{mg/L})$$

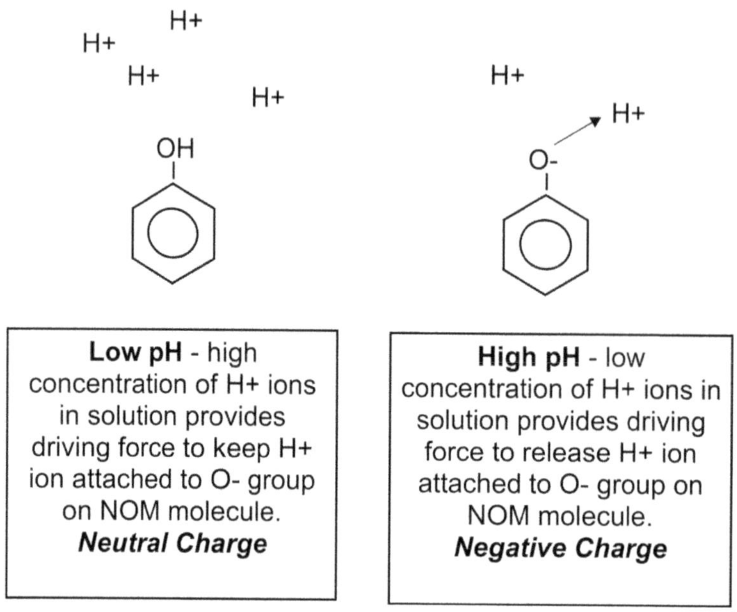

Source: David Pernitsky

Figure 1-4 Conceptual Illustration of the relative effect of pH on the charge of an NOM molecule

Table 1-4 Guidelines on the nature of NOM and expected DOC removals

SUVA	NOM Effect on Coagulation	DOC Removals*
<2	NOM has little influence	<25% for alum
2–4	NOM influences	25–50% for alum
>4	NOM controls	>50% for alum

Source: Adapted from Edzwald and Tobiason 1999

DOC—dissolved organic carbon, NOM—natural organic matter, SUVA—specific ultraviolet absorbance

* Removals are typically higher with iron coagulants.

General guidelines for the interpretation of SUVA values are presented in Table 1-4. A general estimate of the expected TOC removal via optimized, alum-based coagulation has also been developed on the basis of source water SUVA (Letterman and Yiacoumi 2011):

$$\text{Percentage of TOC Removal} = 10.8\ (\text{SUVA})^{1.32}$$

CHEMISTRY OF COAGULANTS

Coagulants provide a source of positive charge to neutralize the negative charges of particles and NOM. Types of water treatment coagulants in general use are metal salts (either aluminum- or iron-based) and organic polymers. Aluminum- and iron-based coagulants react with water to form a variety of dissolved metal-hydroxide species as well as solid-phase metal-hydroxide precipitates. The charge on the dissolved coagulant species, the charge of the floc particles themselves, and the relative amount of floc formed are a function of pH. Therefore, the pH at which coagulation occurs is one of the most important parameters for proper coagulation performance. See Figure 1-5 for how to read a solubility diagram. See Figure 1-6 for locations of different ionic species.

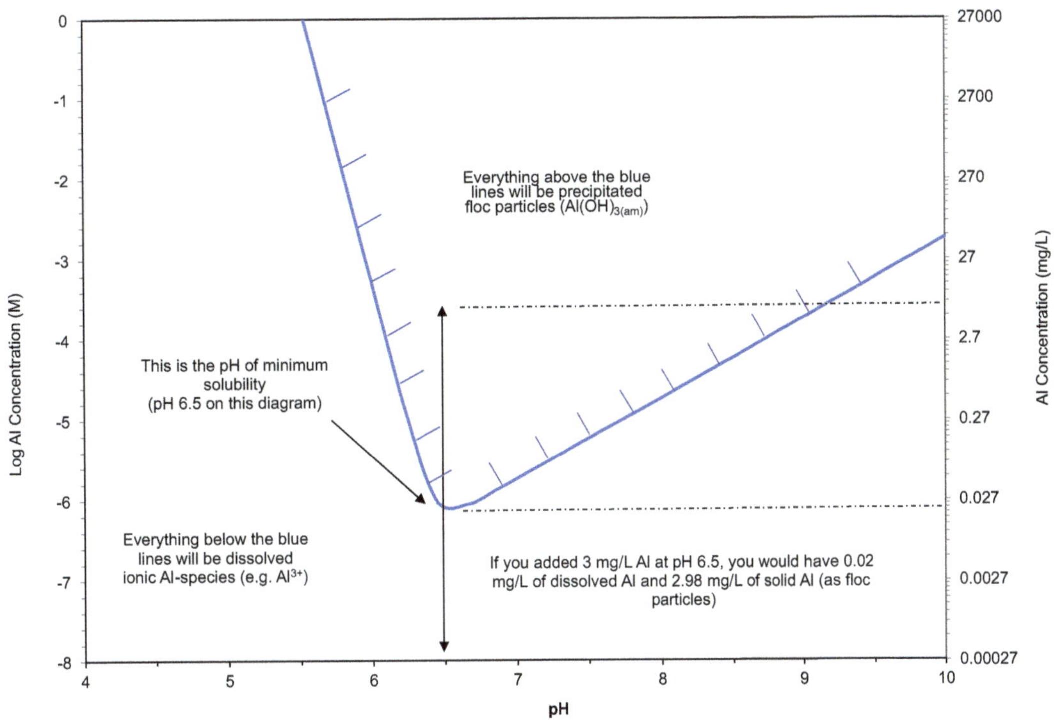

Source: David Pernitsky

Figure 1-5 How to read a solubility diagram

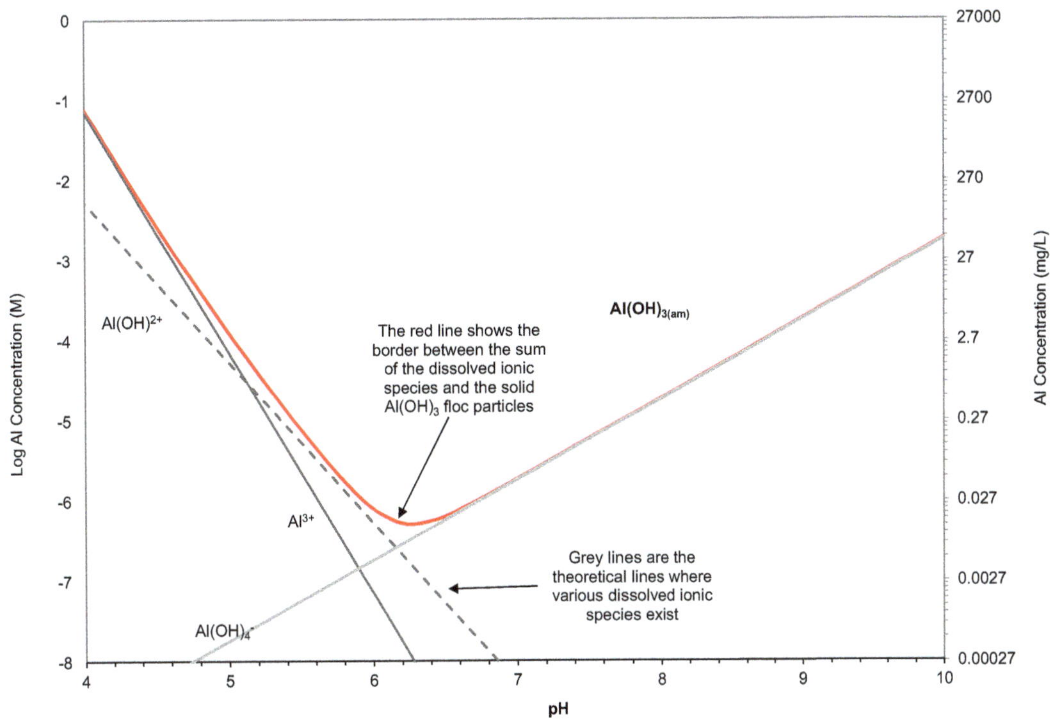

Source: David Pernitsky

PACl—polyaluminum chloride

Figure 1-6 Location of different ionic species

Solubility diagrams show the division between dissolved and solid species for a range of pH and coagulant doses. At any coagulation pH, you can determine the split between how much of the metal (Al or Fe) will exist as a dissolved ion and how much will exist as solid floc particles.

For example, if you added 3.0 mg/L as Al to water and the coagulation pH was 6.5, 0.02 mg/L of the Al would be in the dissolved form and 2.98 mg/L would precipitate as solid floc particles.

ALUMINUM METAL-SALT COAGULANTS

Aluminum-based coagulants include alum (aluminum sulfate) and many polyaluminum coagulants, such as PACl, polyaluminum sulfate (PAS), and ACH. Polyaluminum coagulants differ from alum in that they contain a highly charged Al_{13}^{7+} polymer species that is formed by adding base (OH^-) to a solution of aluminum chloride and/or aluminum sulfate under controlled reaction conditions. The higher the concentration of base added, the higher the concentration of the Al_{13} polymer formed and the less alkalinity consumed by the coagulant after addition. For commercial polyaluminum coagulants, basicity ranges from 20 (low) to 50 (medium) to 75 (high) to 85 percent (very high).

Figure 1-7 shows a solubility diagram for aluminum sulfate (alum) and a 75 percent (high) basicity PACl. At any given pH, at concentrations above the colored lines, precipitated floc particles will form. Below the V-shaped curves, only dissolved-phase species will exist. The pH of minimum solubility for alum is approximately pH 6.0 at 20°C, but

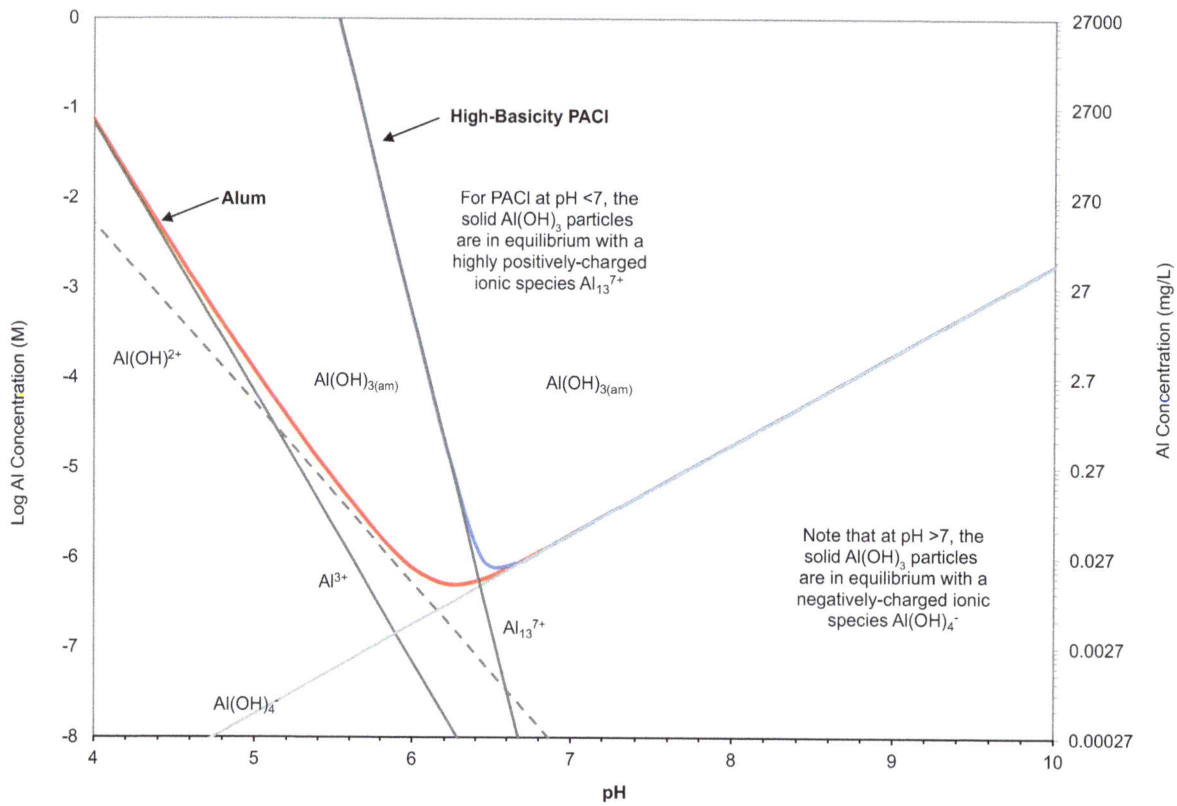

Source: David Pernitsky. Adapted from Pernitsky and Edzwald 2003

PACl—polyaluminum chloride

Figure 1-7 Solubility diagram for alum and high-basicity PACl at 20°C

the solubility curve is wide, indicating that large amounts of floc will form over a broad pH range. For high-basicity PACl, the pH of minimum solubility is significantly higher than for alum. Also, solubility rapidly increases on the acidic side of the pH of minimum solubility; poor floc formation would therefore be expected at pH values lower than the pH of minimum solubility. Although not shown on Figure 1-7, the solubility curves for medium- and low-basicity PACls would fall between the high-basicity PACl and alum, and the solubility curve for ACH would fall slightly to the right of the high-basicity PACl.

The pH of minimum solubility increases as temperature decreases. This is especially important in cold climates because of the wide range in source water temperatures experienced. For example, the pH of minimum solubility for alum increases from 6.2 at 20°C to 6.5 at 5°C. Over that same range, the pH of minimum solubility for high-basicity PACl changes from 6.4 to 6.9. This is shown graphically in Figure 1-8.

For aluminum coagulants, performing coagulation as close to the pH of minimum solubility results in the maximum conversion of chemical to useful floc particles, minimizes dissolved aluminum residuals, and has been shown to result in good overall treatment performance in the field (Pernitsky and Edzwald 2006). More information on aluminum coagulant chemistry can be found in Pernitsky and Edzwald (2003), and the importance of controlling residual aluminum concentrations is discussed in depth in Edzwald (2020).

As shown in the solubility diagram, the amount of dissolved-phase Al species is much lower than the amount of solid-phase floc particles. For metal-salt coagulants, therefore, the parameter that is most important for coagulation performance is the charge on the floc particles themselves rather than the charge of any dissolved species. Figure 1-9 shows the floc surface charge (as measured by zeta potential) versus pH for several coagulants added

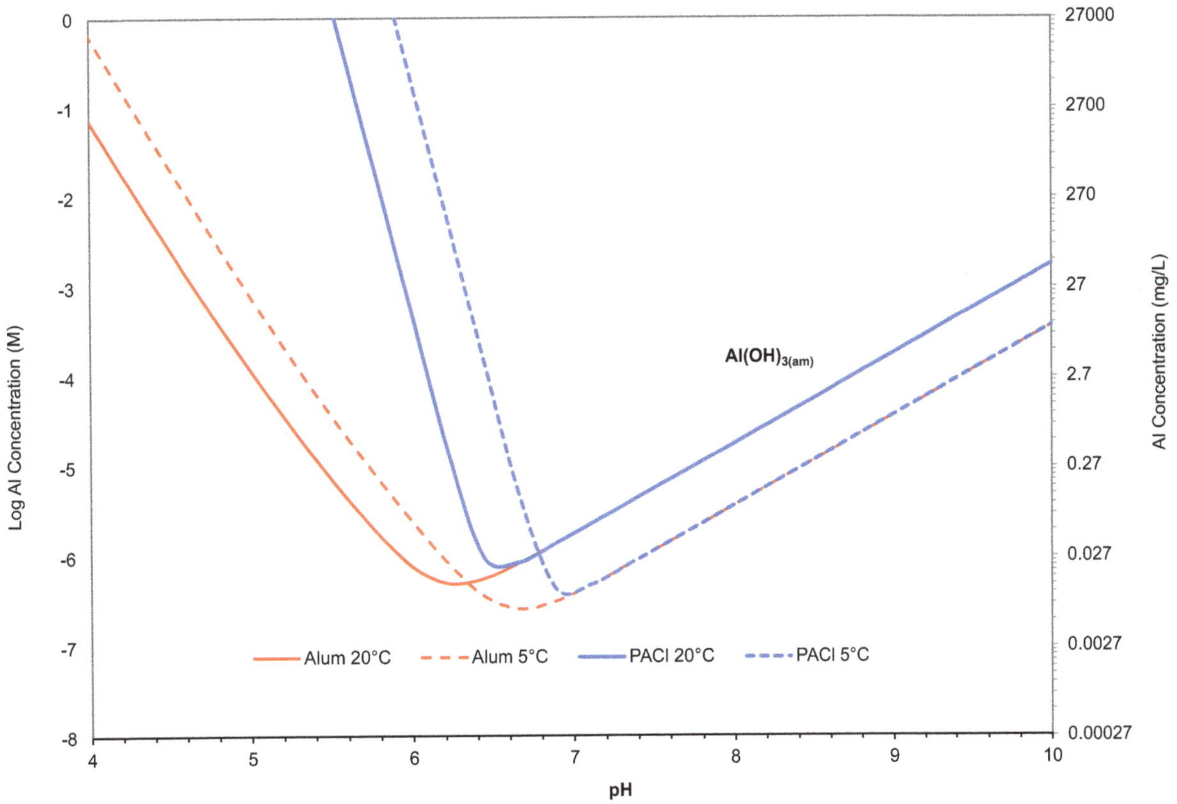

Source: Adapted from Pernitsky and Edzwald 2003

Figure 1-8 Variation in solubility for alum and high-basicity PACl with temperature

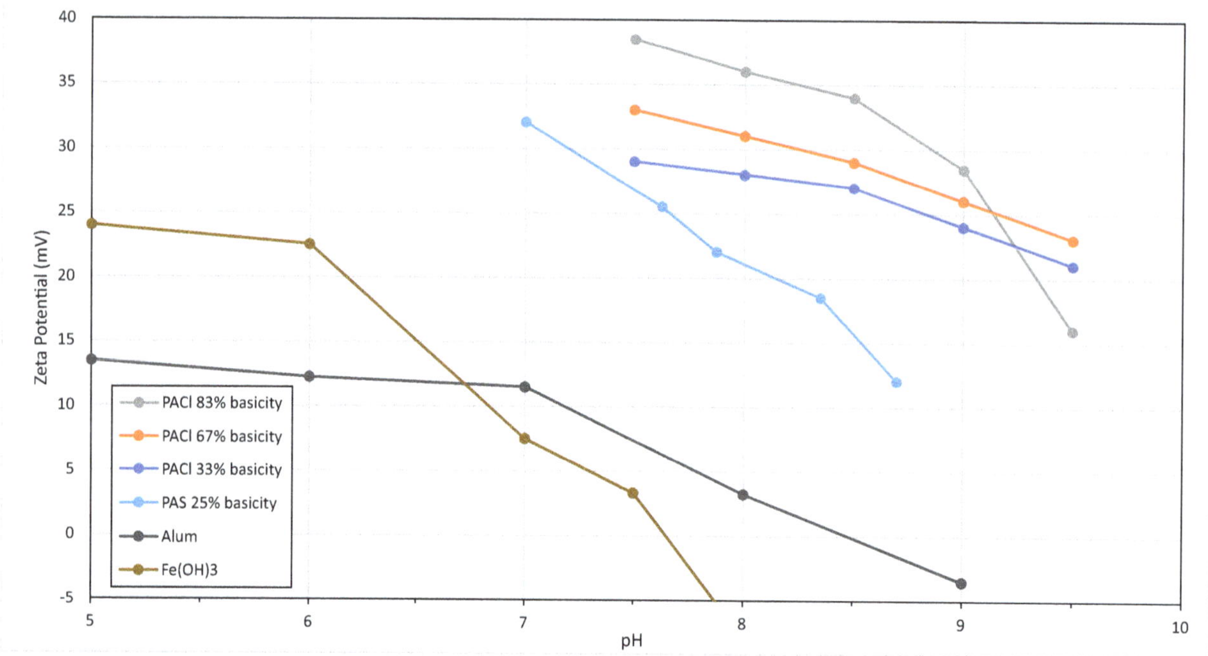

Source: Data for PACls from Solomentseva et al. 1999; data for PAS from Solomentseva et al. 2004; data for alum from Duan et al. 2014; data for $Fe(OH)_3$ from Guan et al. 2008

PACl—polyaluminum chloride

Figure 1-9 Coagulant floc charge versus pH for various coagulants

to water free of particles and NOM. The positive charge of the alum floc particles decreases rapidly as the pH increases greater than 7.0, which explains why alum is more effective at lower pH. The PACls and the PAS show significantly higher positive floc charges than the alum, which is a result of the influence of the Al_{13}^{7+} polymer. These coagulants can be effective even at pH values greater than 7.5, although dissolved Al concentrations can be high.

IRON METAL-SALT COAGULANTS

Iron-based metal-salt coagulants are also used in water treatment, with iron (III) ferric chloride and ferric sulfate being the most popular. Iron (II) ferrous coagulants are not common, as their base Fe^{2+} ion carries less positive charge than the Fe^{3+} base ion for the ferric coagulants. Figure 1-10 shows solubility data for ferric coagulants. Iron coagulants are much less soluble than aluminum-based ones, meaning that for a given pH, a greater percentage of the metal added will be present as particles. Thus, iron-based coagulants can be used over a much greater pH range without worrying about residual dissolved metal concentrations in the finished water, and changes in solubility with water temperature can effectively be ignored. The pH of minimum solubility for ferric iron is near pH 8.8. However, unlike Al-based coagulants, ferric chloride ($FeCl_3$) and ferric sulfate ($Fe_2(SO_4)_3$) are not always effective water treatment coagulants near their pH of minimum solubility because of the weak positive charge of the resulting floc particles with pH greater than 6.5. This can be observed in Figure 1-10. Although iron coagulants have been used successfully at higher pH values for low-NOM waters and in lime-softening applications, more effective performance is generally seen at lower pH, typically 5.5–6.5, at which more positively charged species are present.

Iron coagulants are more acidic than alum or PACls and therefore result in greater pH depression after addition.

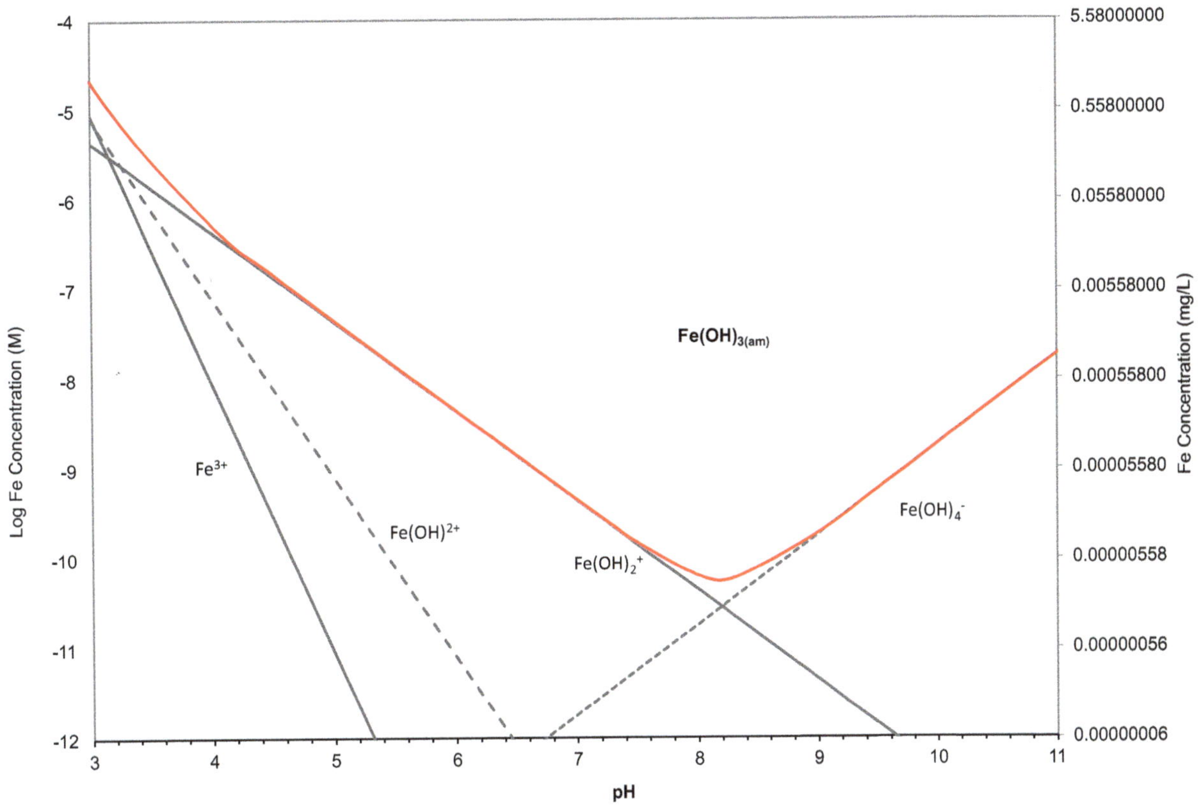

Source: Adapted from Stumm and Morgan 1996

Figure 1-10 Speciation and solubility diagram for ferric coagulants at 20°C

ORGANIC POLYMERS

Many types of organic polymers are commercially available for use in water treatment applications. They are manufactured with various chemical backbones in a wide variety of molecular weights and are available with cationic, anionic, or nonionic charges. In general, cationic polymers of low to medium molecular weight are used as primary coagulants in conjunction with metal-salt coagulants. These polymers are used to neutralize the negative charges on particles and dissolved NOM with a mechanism similar to the metal-based coagulants discussed in the previous section.

Higher-molecular-weight polymers are typically used at much lower doses as flocculant aids or filter aids. These larger polymers are used to bridge metal-salt floc particles to increase their size in sedimentation applications or to increase the size and strength of the floc particles in filter-aid applications. These polymers are usually nonionic or low-charge anionic polymers. Low-charge cationic polymer can also be used as flocculant aids or filter aids, but this is not common.

There are several advantages to the use of polymers compared with metal-salt coagulants. First, the charge on most polymers is relatively unaffected by pH in the range of water treatment applications. This makes cationic polymers good choices for applications in which metal-salt coagulants have low levels of cationic charge (pH greater than 7–7.5 for alum and greater than 6.5 for iron coagulants). Second, they do not consume significant amounts of alkalinity. And last, they produce less sludge. Table 1-5 shows a summary of the most common types of organic polymers and their typical applications.

Table 1-5 Overview of common water treatment organic polymers

Type	Charge	Molecular Weight Range	Uses	Observations
Polyacrylamide	Nonionic or anionic	10^5–10^7	Flocculant aid or filter aid	Powder or emulsion form
Poly(DADMAC) or poly(DMDAAC)	Cationic	10^4–10^6	Primary coagulant or coagulant aid	Liquid form. Range of charge densities available
Quaternized polyamines (Epi/DMA)	Cationic	10^4–10^5	Primary coagulant or coagulant aid	Liquid form. Range of charge densities available

Source: Adapted from James M. Montgomery Consulting Engineers 1985

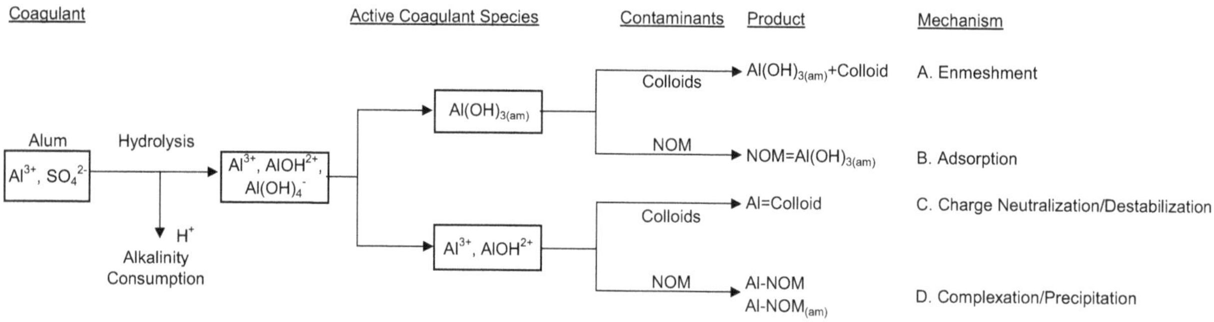

Source: David Pernitsky

$Al(OH)_{3(am)}$ refers to amorphous aluminum hydroxide floc particles.

Figure 1-11 Conceptual view of coagulation reactions for alum

Coagulation Mechanisms

The removal of dissolved NOM and particles in drinking water coagulation is thought to occur via five primary mechanisms:

- enmeshment (particles),
- adsorption (NOM),
- charge neutralization–destabilization (particles),
- precipitation (NOM), and
- interparticle bridging.

These coagulation mechanisms have been studied extensively, and many good references are available (Letterman and Yiacoumi 2011, Hundt and O'Melia 1988, Dempsey et al. 1985, Edwards and Amirtharajah 1985, Amirtharajah and Mills 1982). The first four mechanisms are presented conceptually in Figure 1-11, using alum as the coagulant example.

Research and observations in the field suggest that for most water supplies, the coagulant dose is controlled by NOM and not by particles (Edzwald 2013). The negative charges on the NOM molecules create a coagulant demand that must be satisfied by the coagulant. When a sufficient coagulant dose is added to satisfy this demand, then a solid phase precipitates. The resulting precipitated floc particles will then be available for enmeshment of particles (often referred to as sweep-floc coagulation) and further adsorption of NOM onto the floc surfaces. Because the percentage of Al or Fe that is in the form of floc particles is much greater than the percentage that is present as a dissolved ion (Figures 1-7 and 1-10), direct charge neutralization of particles by dissolved-phase coagulant species is limited for almost all waters.

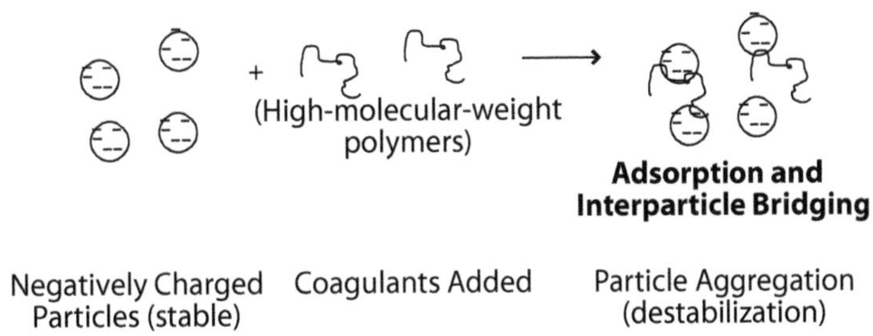

Figure 1-12 Conceptual view of adsorption and interparticle bridging coagulation mechanism

The fifth mechanism, interparticle bridging, can be used to describe the action of high-molecular-weight flocculant aids or filter aids, as shown in Figure 1-12. The selection of cationic, anionic, or nonionic flocculant aid or filter-aid polymers depends on the charge characteristics of the source water particles and floc particles present.

COAGULATION AND FILTRATION APPLICATIONS IN WASTEWATER, REUSE, STORMWATER, AND SEAWATER TREATMENT

Wastewater Treatment and Reuse

Coagulation is often used to treat wastewater in both primary and tertiary treatment. In primary treatment, the main goal is conditioning suspended and dissolved solids for removal by sedimentation. In tertiary treatment, coagulation is used both for chemical precipitation of phosphorus (P) and for the preparation of solids leaving biological treatment for removal by filtration.

Municipal wastewater differs from raw surface water in several important ways. First, the concentrations of both dissolved and suspended solids are much higher. The suspended solids tend to be larger in diameter and contain more organic compounds than in water treatment. More coagulant is necessary not only for these higher concentrations of organic particles but also to react with the significant concentrations of dissolved organic matter and phosphate. The flocs formed from these constituents tend to be more fragile and less settleable than the flocs created in drinking water treatment (Adin and Asano 1998). However, Ghosh et al. (1994) found that general principles of coagulation for drinking water can be applied to wastewater successfully. They completed jar tests using alum to understand the effect of coagulant dosage and pH on coagulation, flocculation, sedimentation, and filtration for tertiary treatment of wastewater.

Cikurel et al. (1996) did further work with alum for tertiary filtration of secondary effluent. They focused on the influence of polymers and noted that in combination with alum, a high-molecular-weight anionic polymer strengthened the flocs, keeping them from detaching in the filter. They also looked at using cationic polymers in place of alum and found that high-charge cationic polymers of medium to high molecular weight (2.5×10^5 to 6×10^6 Da) were just as effective as alum plus anionic polymer and more effective than alum alone at lower concentrations. The higher the molecular weight, the lower the necessary dose of cationic polymer, but also the higher the filter head loss.

Stormwater

Several regulatory jurisdictions have guidance for reducing stormwater runoff volume. However, not all stormwater runoff can be eliminated; thus, some of these entities are trying to improve the quality of the runoff before it enters the receiving bodies. Stormwater runoff can contain an array of constituents, depending on the watershed and precipitation intensity. Stormwater runoff can potentially have high levels of particles, nutrients (phosphorus and nitrogen because of agricultural application), salinity (because of road salt applications), oil and grease, synthetic organic contaminants, metals, and pathogens. Typical stormwater treatment includes a wet or dry pond to settle out the particles with the purpose of minimizing discharge of total suspended solids (TSS) to the receiving water. As indicated previously, if particle charges are not destabilized, removal of particles can take substantial time via settling. For this reason, treatment of stormwater runoff may be needed. Treatment processes can be implemented to enhance the removal of particles and remove them quicker. If the main goal is to remove particles and reduce discharge of TSS, the treatment process is likely to include coagulation, flocculation, and sedimentation. However, additional treatment processes (such as oxidation, disinfection, flotation) may be needed to remove other constituents (nutrients, metals, organics, pathogens, etc.).

As stormwater runoff's composition is different than surface water, some changes are anticipated with the treatment processes. One of the main differences with stormwater treatment compared with surface water is the lack of alkalinity. In general, the alkalinity of rainwater and therefore stormwater runoff is relatively low compared with that of surface water or groundwater. For this reason, the alkalinity of a surface water typically decreases after precipitation. If only stormwater is to be treated, the alkalinity can be in the range of 10–25 mg/L as calcium carbonate ($CaCO_3$). Therefore, controlling pH during coagulation is more challenging compared with surface water treatment. A broad range of particles can also be found in stormwater, with different sizes that depend on the rainfall intensity. This can create differences in settling velocity, which may need to be compensated for via coagulation. Therefore, coagulant dose may be different depending on the intensity. Furthermore, there can be constituents that can affect coagulation chemistry. For example, during treatment of runoff with road salt, the salinity can be higher, which can affect solubility of the coagulants, which will be discussed in the following section. There can be also nutrients like phosphates that can increase the coagulant demand.

While the general principles of runoff water coagulation are similar to drinking water, caution must be taken. Because of fluctuations in rainfall intensity, runoff water quality can change from one rainfall to another. Even within a single rainfall, the water quality may change with time. For example, more contaminants are anticipated at the beginning of a runoff (the "first flush"), and the concentration of contaminants can decrease with continuing runoff. To accommodate these changes, the coagulant dose may need to be adjusted during the rain. Jar testing using coagulant to understand the effect of coagulant dosage and pH on coagulation, flocculation, sedimentation, and filtration is recommended. Jar testing may even be needed multiple times in a single rainfall to accommodate for the changing runoff water quality (beginning runoff versus continuing runoff).

Seawater and Saline Waters

Seawater coagulation is focused on removing suspended solids, algae cells, and NOM, typically as pretreatment for reverse osmosis (RO) seawater desalination processes. The coagulation of seawater and other saline waters differs from the coagulation of fresh waters in several respects. Most importantly, seawater and brackish waters have much a higher amount of total dissolved solids, typically expressed as salinity, ranging between 31 and 45 ppt parts per thousand (Edzwald and Haarhoff 2011). The high salinity results in an increase in ionic strength, which decreases the magnitude of the surface charge on particles, allowing them to come closer to each other before repulsion forces are felt. Typically, in the case of inorganic particles such as clay, sand, aluminum silicates, and calcium carbonate, the attraction forces allow these particles to aggregate and settle without chemical coagulation, as evidenced by the settling of sediments at the mouth of a river that enters the ocean (i.e., estuaries). Increases in salinity explain a much larger proportion of the observed sedimentation of particles at a river delta than what would be predicted on the basis of the decrease in river velocity alone.

The higher salinity of seawater also affects the ion product of water and the buffering capacity. For seawater, buffering capacity is affected by the presence of boron and silica as well as carbonates. Seawater's ability to resist drastic pH changes is largely a result of the shift in the equilibrium constant for water and carbonate system.

Increased salinity also increases the solubility of aluminum and iron metal-salt coagulants, with the magnitude of the increase greater for aluminum salt. Therefore, aluminum coagulants are rarely used in seawater coagulation applications. The higher concentrations of dissolved aluminum present in seawater may be available to form precipitates and scales that can foul piping or RO desalination membranes. Iron coagulants are used in practice on the basis of their much lower solubility. Organic cationic polymers formulated to maintain their charge in high-salinity conditions are also occasionally used, but they are used under conditions of tight process control, as any polymer carryover to downstream membrane processes can cause severe fouling issues.

Unless influenced by the discharge of large solids-bearing rivers, seawater turbidity tends to be lower than found in surface supplies. Turbidity removal is still important, however, as membrane systems typically operate better with turbidities closer to 0.1 ntu to prevent fouling. Algae blooms are common in many seawater sources and can lead to severe fouling of downstream membrane processes. Coagulation pretreatment followed by dissolved air flotation and filtration is common for seawater treatment plants subject to algae blooms. Influent suspended solids concentrations for desalination membrane processes are often measured using the silt density index, a value defined in ASTM D4189 that estimates the fouling potential of the feedwater. Seawater NOM concentrations are influenced by the proximity of freshwater river discharges and algae blooms. Seawater NOM concentrations are typically 2–5 mg/L but can increase significantly during algae blooms (Edzwald and Haarhoff 2011).

The coagulation of nonseawater, high-salinity waters is also affected by the parameters discussed in this section. More details regarding the effects of salinity on coagulation processes can be found elsewhere (Edzwald and Haarhoff 2011).

KEY TAKEAWAYS

Coagulation is used in water treatment processes to neutralize the negative charges on source water particles and to convert source water dissolved NOM to a particulate form that can be removed by clarification and filtration processes. The addition of metal-salt coagulants and organic polymeric coagulants produces floc particles that can enmesh source water particles and provide surfaces for the adsorption of NOM. These floc particles must be near neutral in charge for clarification and granular media filtration processes to work efficiently. Membrane filtration processes are not dependent on floc charge for particle removal.

Coagulant dose is usually controlled by the source water NOM concentration, rather than the source water turbidity. The pH at which coagulation occurs is also critically important. Alum is generally most effective when coagulation occurs at less than pH 7.0, and PACls are effective at pH greater than 6.5, with higher pH values possible for higher basicity products. Iron coagulants are generally most effective at pH 6.5 or lower. Coagulant conditions must be determined on a case-by-case basis for each individual water and treatment plant, and these often change seasonally. Coagulant conditions include the coagulant type, dose, and coagulation pH. Ensuring that the coagulant will be operating in the appropriate pH range involves examining the source water alkalinity, the expected coagulant dose, how much alkalinity the coagulant will consume, and whether any supplemental acid or base is required to get to the target pH. The effectiveness of coagulation should be evaluated considering filter performance, not just from clarifier performance. The best KPI is therefore the UFRV, as this parameter accounts for variations in clarifier turbidity and therefore solids loading to the filters, filter ripening time, filter effluent turbidity, filter head loss rate, filter loading rate, and overall filter run time.

REFERENCES

Adin, A., and Asano, T. 1998. "The Role of Physical-Chemical Treatment in Wastewater Reclamation and Reuse." *Water Science & Technology*. 37(10):79-90.

Amirtharajah, A., and K.M. Mills. 1982. "Rapid-Mix Design for Mechanisms of Alum Coagulation." *Journal AWWA*. 74(4):210-216.

ASTM (ASTM International). 1987 (v11.01). *Annual Book of ASTM Standards*. Philadelphia: ASTM.

ASTM 2023. D4189. Standard Test Method for Silt Density Index (SDI) of Water. West Conshohocken, Pa.: ASTM International.

AWWA. 2011 (6th ed.). *Water Quality and Treatment: A Handbook on Drinking Water* (J.K. Edzwald, editor). New York: McGraw-Hill.

Becker, W.C., C.R. O'Melia., K.-K. Au, and J.S. Young Jr. 2004. *Using Oxidants to Enhance Filter Performance*. Denver: Water Research Foundation and AWWA.

Cikurel, H., M. Rebhun, A. Amirtharajah, and A. Adin. 1996. "Wastewater Effluent Reuse by In-line Flocculation Filtration Process." *Water Science & Technology*. 33(10-11): 203-211.

Cornwell, D.A. 2006. *Water Treatment Residuals Engineering*. Denver: Water Research Foundation and AWWA.

Dempsey, B.A., H. Sheu, T.M. Tanzeer Ahmed, and J. Mentink. 1985. "Polyaluminum Chloride and Alum Coagulation of Clay-Fulvic Acid Suspensions." *Journal AWWA*. 77(3):74-80.

Duan, J., J. Wang, T. Guo, and J. Gregory. 2014. "Zeta Potentials and Sizes of Aluminum Salt Precipitates – Effect of Anions and Organics and Implications for Coagulation Mechanisms." *Journal of Water Process Engineering*. 4:224-232.

Edwards, G.A., and A. Amirtharajah. 1985. "Removing Color Caused by Humic Acids." *Journal AWWA*. 77(3):50-57.

Edwards, M., and S. Triantafyllidou. 2007. "Chloride-to-Sulfate Mass Ratio and Lead Leaching to Water." *Journal AWWA*. 99(7):96-109.

Edzwald, J.K. 2020. "Aluminum in Drinking Water: Occurrence, Effects, and Control." *Journal AWWA*. 112(5):34-41.

Edzwald, J.K. 2013. "Coagulation Mixing Revisited: Theory and Practice." *Journal of Water Supply: Research and Technology–AQUA*. 62(2):67-77.

Edzwald, J.K., and J. Haarhoff. 2011. "Seawater Pretreatment for Reverse Osmosis: Chemistry, Contaminants, and Coagulation." *Water Research*. 45:5428-5440.

Edzwald, J.K., and J.E. Tobiason. 1999. *Enhanced Coagulation: USA Requirements and a Broader View. Removal of Humic Substances From Water*. Trondheim, Norway: IAWQ/IWSA Joint Specialist Group on Particle Separation.

Edzwald, J.K., and J.E. Van Benschoten. 1990 (1st ed.). "Aluminum Coagulation of Natural Organic Matter." In *Chemical Water and Wastewater Treatment* (H.H. Hahn and R. Klute, editors). Berlin: Springer-Verlag.

Ghosh, M., A. Amirtharajah, and A. Adin. 1994. "Particle Destabilization for Tertiary Treatment of Municipal Wastewater by Filtration." *Water Science & Technology*. 30(9): 209-218.

Guan, X.H., J. Wang, and C.C. Chusuei. 2008. "Removal of Arsenic From Water Using Granular Ferric Hydroxide: Macroscopic and Microscopic Studies." *Journal of Hazardous Materials*. 156(1-3):178-185.

Henderson, R.K., S.A. Parsons, and B. Jefferson. 2008. "Successful Removal of Algae Through the Control of Zeta Potential." *Separation Science and Technology*. 43(7): 1653-1666.

Holmes, M., P. Reeve, C. Pestana, C. Chow, G. Newcombe, and J. West. 2015. "Zeta Potential Measurement for Water Treatment Coagulation Control." OzWater 2015 Proc., Adelaide, Australia.

Hundt, T.R., and C.R. O'Melia. 1988. "Aluminum-Fulvic Acid Interactions: Mechanisms and Applications." *Journal AWWA*. 80(4):176-186.

James, M. Montgomery Consulting Engineers. 1985 (1st ed.). *Water Treatment Principles & Design*. New York: John Wiley & Sons.

James, W., S. Craik, S. Heidary-Monfared, G. Solonynko, and M. Pors. 2015. "Using Zeta Potential to Optimize Seasonal Direct Filtration." Western Canada Water Conf., Winnipeg, Man.

Kundert, K. 2014. "Alberta Flood 2013 – City of Calgary Water Treatment System Resiliency." 16th Canadian National Drinking Water Conf., Gatineau, Que.

Letterman, R.D., and S. Yiacoumi. 2011 (6th ed.). "Coagulation and Flocculation." In *Water Quality and Treatment: A Handbook on Drinking Water* (J.K. Edzwald, editor). New York: McGraw-Hill.

Malvern Panalytical. 2011. *The Role of Zeta Potential in Water Treatment Process and Control*. Horsham, PA: Water Online.

Morfesis, A., A.M. Jacobson, R. Frollini, M. Helgeson, J. Billica, and K.R. Gertig. 2009. "Role of Zeta (ζ) Potential in the Optimization of Water Treatment Facility Operations."

Industrial and Engineering Chemistry Research. 48(5):2305-2308.

Nyström, F., K. Nordqvist, I. Herrmann, A. Hedström, and M. Viklander. 2019. "Treatment of Road Runoff by Coagulation/Flocculation and Sedimentation." *Water Science & Technology*. 79(3):518-525. doi:10.2166/wst.2019.079.

Pernitsky, D.J. 2019. "Coagulation 101." 2019 AWWA Annual Conf., Denver.

Pernitsky, D.J., R. Cantwell, E. Murphy, N. Paradis, J. Boutilier, and G. Bache. 2010. "First Aid for Direct Filtration – Using Zeta Potential to Improve Operation." 2010 AWWA Water Quality and Technology Conf., Savannah, Ga.

Pernitsky, D.J., and J.K. Edzwald. 2006. "Selection of Alum and Polyaluminum Coagulants: Principles and Applications." *Journal of Water Supply: Research and Technology–AQUA*. 55(2):121-141.

Pernitsky, D.J., and J.K. Edzwald. 2003. "Solubility of Polyaluminum Coagulants." *Journal of Water Supply: Research and Technology–AQUA*. 52(6):395-406.

Riddick, T.M. 1964. "Role of Zeta Potential in Coagulation Involving Hydrous Oxides." *TAPPI Journal*. 47(1):171A.

Sharp, E.L., R. Rigby, A. Hall, K. Moore, and B. Jefferson. 2015. "Using Online Zeta Potential Measurements for Coagulation Control: A First for the UK Water Industry." IWA Specialist Conf. on Natural Organic Matter in Water, Malmo, Sweden.

Sharp, E.L., J. Banks, J.A. Billica, K.R. Gertig, R. Henderson, S.A. Parsons, D. Wilson, and B. Jefferson. 2005. "Application of Zeta Potential Measurements for Coagulation Control: Pilot-Plant Experiences from UK and US Waters with Elevated Organics." *Water Supply*. 5(5):49-56.

Solomentseva, I.M., S. Barany, and J. Gregory. 2004. "Surface Properties and Aggregation of Basic Aluminum Sulphate Hydrolysis Products: 1. Electrokinetic Potential and Hydration of BAS Hydrolysis Product Particles." *Colloids and Surfaces A: Physicochemical and Engineering Aspects*. 230(1-3):117-129.

Solomentseva, I.M., N.G. Gerasimenko, and S. Barany. 1999. "Surface Properties and Aggregation of Basic Aluminum Chloride Hydrolysis Products." *Colloids and Surfaces A: Physicochemical and Engineering Aspects*. 151(1-2):113-126.

Stumm, W., and J.J. Morgan. 1996 (3rd ed.). *Aquatic Chemistry: Chemical Equilibria and Rates in Natural Waters*. New York: John Wiley and Sons.

Waller, M., R. Cantwell, Q. Muylwyk, and D.J. Pernitsky. 2012. "Rethinking Coagulation Practices on the Great Lakes Using Zeta Potential." 2012 AWWA Water Quality and Technology Conf., Toronto, Ont.

CONVERSION AND CALCULATION TABLES

Table 1-6 Coagulant unit conversion table

Multiply concentration as:	By	To obtain:
Al^{3+}	1.00	Concentration as Al^{3+}
Al_2O_3	0.529	
$Al_2(SO_4)_3$	0.158	
$Al_2(SO_4)_3{\cdot}14H_2O$	0.0909	
$Al_2(SO_4)_3{\cdot}18H_2O$	0.0811	
$AlCl_3$	0.202	
$AlCl_3{\cdot}6H_2O$	0.112	
Fe^{3+}	1.00	
$FeCl_3$	0.345	
$FeCl_3{\cdot}6H_2O$	0.207	
Fe_2O_3	0.700	
$FeSO_4$	0.368	
$Fe_2(SO_4)_3$	0.280	
$Fe_2(SO_4)_3{\cdot}4.5H_2O$	0.233	
$Fe_2(SO_4)_3{\cdot}9H_2O$	0.199	Concentration as Fe^{3+}
Multiply concentration as:	**By**	**To obtain concentration as:**
Al^{3+}	1.00	Al^{3+}
	1.89	Al_2O_3
	6.33	$Al_2(SO_4)_3$
	11.0	$Al_2(SO_4)_3{\cdot}14H_2O$
	12.3	$Al_2(SO_4)_3{\cdot}18H_2O$
	4.94	$AlCl_3$
	8.94	$AlCl_3{\cdot}6H_2O$
Fe^{3+}	1.00	Fe^{3+}
	2.90	$FeCl_3$
	4.83	$FeCl_3{\cdot}6H_2O$
	1.43	Fe_2O_3
	2.71	$FeSO_4$
	3.57	$Fe_2(SO_4)_3$
	4.29	$Fe_2(SO_4)_3{\cdot}4.5H_2O$
	5.02	$Fe_2(SO_4)_3{\cdot}9H_2O$

Table 1-7 Solids production factors

Coagulant Type	Solids Production Factor* *Unitless*
Aluminum-based	4.89 × Aluminum content†
Iron-based	2.89 × Iron content†

* Cornwell 2006
† Content expressed as a percentage in decimal form.

Example:
Alum dose = 15 mg/L as alum ($Al_2(SO_4)_3 \cdot 14H_2O$); Al content = 9.09% or 0.0909
Solids production factor = 4.89 × 0.0909 = 0.445
Solids production = 15 mg/L as alum × 0.455 = 6.675 mg/L of solids

Table 1-8 Alkalinity consumption

Coagulant	Alkalinity Consumed per Coagulant Dose*
Aluminum-based	
0% Basicity	5.56
30% Basicity	3.89
50% Basicity	2.78
70% Basicity	1.67
82% Basicity	1.00
Iron-based	
Ferric type	2.68
Ferrous type	1.79

* Unit expressed as mg/L as $CaCO_3$ consumed per mg/L as metal dose

Example:
Alum dose = 22 mg/L as alum ($Al_2(SO_4)_3 \cdot 14H_2O$); Al content = 9.09% or 0.0909
Al dose = 22 mg/L as alum × 0.0909 = 2 mg/L as Al
Amount of alkalinity consumed (mg/L as $CaCO_3$) = 5.56 × 2 mg/L as Al = 11.1 mg/L as $CaCO_3$

Chapter 2

Rapid-Mixing and Flocculation Processes

INTRODUCTION

Rapid mixing, also known as flash mixing or initial mixing, includes the addition of a coagulant chemical and dispersing the chemical quickly and evenly to the influent water. Rapid mixing is followed by flocculation, under which chemically conditioned particles will come into contact to build floc of appropriate size for removal in downstream particle removal processes.

Proper rapid-mix and flocculation performance is critical to overall plant performance. Removal of turbidity, organic carbon, iron (Fe), and manganese (Mn) depends on the addition of pretreatment chemicals. Proper selection and dosage of these chemicals not only can improve contaminant removal but also can reduce sludge production and reduce operations and maintenance (O&M) costs.

Chapter Summary

- Mixing intensity (agitation) is characterized with G and Camp number (GT).
    - $G = 300 - 1{,}000\,s^{-1}$ for rapid mixing.
    - $G = 10 - 100\,s^{-1}$ for flocculation.
- The degree of blending is characterized by coefficient of variation (COV), with good performance around COV less than 0.05.
- Rapid mixing and flocculation can be accomplished with hydraulic or mechanical mixing.
- Coagulation and flocculation performance can be assessed through changes in ultraviolet absorbance and unit filter run volume.

Rapid-Mixing and Flocculation Goals

Rapid-mixing and flocculation processes have different objectives. In the case of rapid mixing, the objective is near-instantaneous dispersion of a chemical to provide relatively even contact with the particles. Flocculation is the addition of mixing energy to promote contact and aggregation of the chemically conditioned particles into floc that can be removed by clarification and/or filtration.

Optimal floc size is specific to the site and is influenced by the nature of the source water, the coagulation chemicals used, and the physical configuration of the water treatment plant; however, some general guidelines exist. The attributes of the flocculation step are primarily dictated by the requirements of the downstream processes. As an example, dissolved air flotation (DAF) targets smaller floc sizes than conventional sedimentation because of the requirement that the floc attach to air bubbles and float. Direct filtration is a special case in which a filterable floc size (typically referred to as "pin floc") is the target. In this case, large flocs may be strained in the top layer of the filter (a phenomenon called filter blinding), which increases loss of head across the filter, necessitating premature backwash. Even in the case of conventional settling, there is usually a maximum desirable floc size based on floc composition and characteristics of the settling process. This is because as flocs grow, they incorporate more water in their structure, bringing their density closer to that of water and resulting in slower floc settling velocities. Larger flocs are also more susceptible to breakup because of shear. Therefore, it is important that the floc size objectives be understood in configuring and operating the flocculation step within a plant.

Performance Considerations

The best measure of rapid-mix and flocculation effectiveness is the performance of the overall production of filtered water. For granular media filtration plants, overall plant performance is best measured by unit filter run volume (UFRV), and rapid-mix and flocculation parameters should be optimized to improve UFRV. For membrane plants, optimization should be focused on maximizing flux and production between cleanings. In addition to UFRV, there are two major parameters used for evaluating the performance of coagulation rapid-mix and flocculation basins: removal of particles and removal of natural organic matter (NOM) in the clarification and filtration processes.

Velocity Gradient in Mixing and Flocculation

The concept of the hydraulic velocity gradient is important for both rapid mixing and flocculation. The intensity of mixing is generally quantified by the velocity gradient, which is often expressed as G with units of s^{-1} (i.e., 1/s or "per second"). The velocity gradient is calculated using the energy dissipation rate in the fluid. A higher number indicates more intense mixing.

Multiplying G by the hydraulic residence time of a reactor, T, results in the unitless parameter GT. Because this is a multiplication of the mixing intensity by the time allowed for mixing, this gives a rough sense of the overall fluid deformation and makes for ready comparison of reactors.

Hydraulic explanation of energy dissipation. To understand the concept of energy dissipation rate, it is useful to look at a hydraulic flocculator, as seen in Figure 2-1. The water level at the beginning of a hydraulic flocculator is higher than the water level at the end of a hydraulic flocculator because as water travels through the flocculator, it loses energy through fluid shear and, to a lesser extent, wall friction, converting kinetic energy into heat. The resulting drop in water level is a measure of the loss of energy and is referred to as head loss, h_L. This is a difference in potential energy, so multiplying by the acceleration of gravity gives a measure of the energy loss per mass of the fluid (i.e., ft-lb/lb = ft or J/kg).

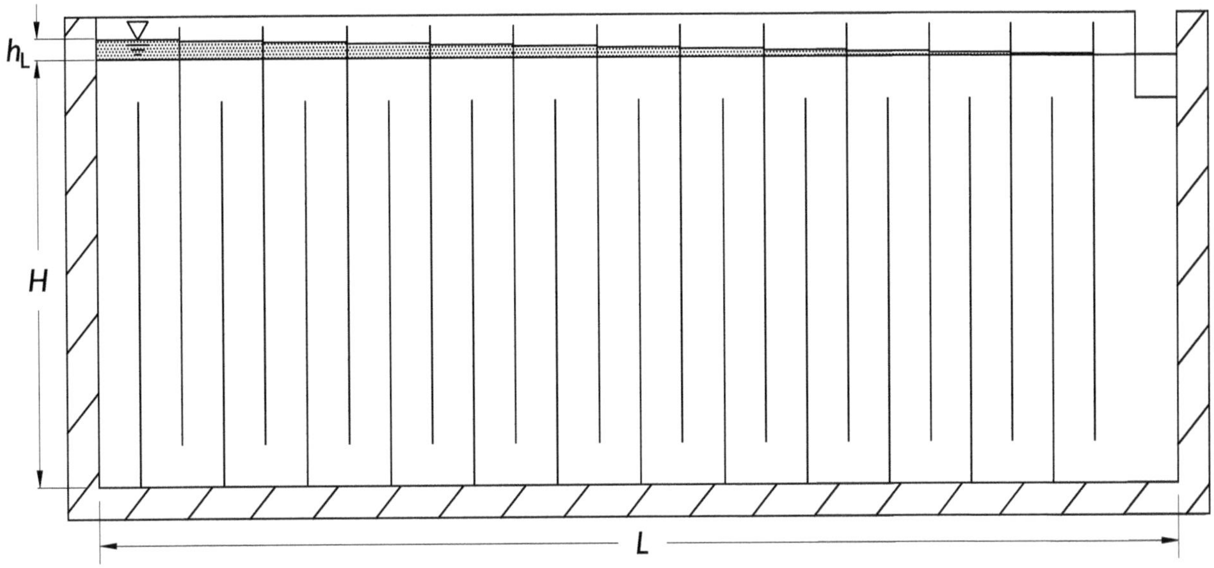

Source: Pennock et al. 2021

Figure 2-1 Channel of a vertically baffled hydraulic flocculator

To convert this to energy dissipation rate, a sense of the time to dissipate this energy to heat is needed. For this, the average time water spends traveling through the flocculator, *T*, is used. Thus, the average energy dissipation rate of the hydraulic flocculator is calculated as

$$\varepsilon_{avg} = \frac{g^{\circ} \cdot h_L}{T}$$

Where: g = acceleration caused by gravity
ε = energy dissipation rate, measured in units of ft-lb/(lb·s) = ft/s (W/kg)

One way to obtain a measure of *G* from the average energy dissipation is to use the Camp–Stein definition of *G* (Camp and Stein 1943):

$$G = \sqrt{\frac{\varepsilon_{avg}}{\nu}}$$

Using this relationship to calculate the mean value of *G* is not strictly accurate because taking the square root of an average will not mathematically give the same result as taking the average of a square root. This should not be a problem when comparing similar flocculator geometries, which would have the same distribution of energy dissipation rate, but caution should be used when comparing average *G* values between reactors with different geometries, such as baffled hydraulic flocculators and mechanically mixed flocculators.

Mechanical mixing: power and velocity gradient. Calculation of *G* for mechanically mixed reactors is based on empirical relationships that have been established over the years. A procedure recommended by D.K. Nix (2009) for calculation of *G* in full-scale basins or mixing chambers is

$$G = C_{TEMP} \cdot P^{0.5}$$

Where: P = the horsepower per 1,000 gal (3,785 L) of volume in the basin or chamber being mixed

C_{TEMP} = a constant based on temperature, which converts $\left(\frac{\text{hp}}{1{,}000\ \text{gal}}\right)^{\frac{1}{2}}$ to s^{-1}

Viscosity of water varies with temperature, so the temperature constant varies. An empirical equation for C_{TEMP} for water temperatures from 39.2 to 70°F (4–21.1°C) is

$$C_{TEMP} = 5.4845\,TEMP\,(°C) + 332.48 \approx 3.0476\ TEMP\ (°F) + 234.99$$

The velocity gradient varies significantly with water temperature (viscosity); therefore, mixer speed may need to be changed seasonally where significant changes in source water temperatures occur throughout the year. When the optimal velocity gradients have been identified by jar testing as described in Chapter 7, the plant mixer(s) and flocculator speed and power can be set accordingly by adjusting the unit's variable-frequency drive (VFD) speed up or down to match the power indicated by the following formula:

$$P(\text{kW}) = \frac{\left(G^2 \cdot \mu \cdot V\right)}{0.7(1{,}000)}$$

or

$$P(\text{hp}) = 1.341 \cdot \frac{\left(G^2 \cdot \mu \cdot V\right)}{0.7}$$

Where:

0.7 = nominal overall loss efficiency with the mixer VFD, motor, and power train

μ = dynamic viscosity in newton-seconds per square meter (e.g., when water temperature = 40°F [4°C], μ = 0.001569; when water temperature = 50°F [10°C], μ = 0.001308; and when water temperature = 60°F [16°C], μ = 0.001109)

V = specific mixing chamber volume in cubic meters

P = power (kilowatts or horsepower) applied to the mixer motor—usually read directly at the motor VFD display or calculated from amperes displayed at the drive ammeter where speed adjustment is by mechanical speed variators rather than VFD speed

To calculate horsepower per 1,000 gal (3,785 L) for a basin with one mixer, divide horsepower of the mixer or flocculator motor by the volume of the basin in gallons and multiply that result by 1,000. For a baffled flocculation basin with three separate chambers, each with a motorized flocculator, the *G* value of each chamber must be calculated separately using the horsepower of the motor driving the flocculator in the chamber and the volume of that chamber.

While the optimal *G* value developed in a jar test provides valuable information and a general indication of proper mixing, the optimal full-scale *G* values may be somewhat different because of scale-up issues (Amirtharajah et al. 1991). The relationship between optimal jar test *G* values and full-scale *G* values varies on the basis of several factors and is often best understood through site-specific experience. Flocculation intensity is discussed in more detail later in this chapter.

Blending

Hydraulic gradient, *G*, is an important measure for both the rapid-mixing and flocculation processes in characterizing agitation—the forcefulness with which fluid is mixed. To characterize blending, the degree to which two fluid streams are evenly combined, another metric—coefficient of variation (COV)—is used and is particularly useful in rapid mix. When the aim of coagulation is precipitation, it is important that coagulant is dispersed evenly throughout the volume (Edzwald 2013). This is because hydrolysis and precipitation reactions are time-dependent and irreversible; therefore, the coagulant's uniform distribution throughout the volume being treated is important so that the whole volume is coagulated to the same degree COV is defined as

$$\mathrm{COV} = \frac{\sigma}{C_{avg}}$$

Where: σ = the standard deviation of the concentration
C_{avg} = the average concentration (MWH 2012)

This is measured by performing enough concentration measurements (at least five) of coagulant (or a tracer such as fluoride) to produce a reliable mean and a standard deviation (Edzwald 2013). A standard criterion for COV is to specify that it should be less than 0.05 to ensure adequate blending (Edzwald 2013).

RAPID MIX

The main consideration for the rapid-mix portion of a water treatment facility is ensuring that the correct amount of chemicals (i.e., coagulant) is dispersed evenly into the water (Clark et al. 1994, Amirtharajah et al. 1991). Generally, this is accomplished by adding the chemical into a hydraulic space where the chemical can quickly be distributed to the entire flow volume. This can occur in a pipe, in a small rapid-mix basin, or along the crest or nappe of a weir. A significant amount of research into rapid-mix design has been done in the industry, and designs have changed over time. In some cases, older rapid-mix basins that were designed for detention times of 1 min or more (as was required under some older criteria) are not as efficient in accomplishing this goal as compared with newer concepts that tend to be based on short detention time and rapid dispersion methods. However, research has also shown that these older designs are completely adequate in many plants, and as such, mixing performance must be evaluated on a case-by-case basis (Edzwald 2013).

Types of Rapid Mixers

The types of rapid-mixing systems generally consist of basin mechanical mixing systems, inline mechanical mixing, jet mixing, and static hydraulic mixing, which are discussed in this section.

Basin mechanical mixing. The basin mixing systems often consist of two stages, with each stage providing 15–30 s of detention time. Each mixing stage generally provides a G of 300–1,000 s^{-1} (GLUMRB 2012, Kawamura 2000). The mixers can be paddle or turbine type. An example of a typical basin mixing system with a pitched-blade impeller is presented in Figure 2-2.

Inline mechanical mixing. Inline mechanical mixers can be thought of as a "mixer in a pipe." Figure 2-3 presents a diagram of a typical inline mechanical mixer system. The inline system has been used in many plants throughout the United States and provides a much smaller footprint than basin mixing systems. However, the detention time of an inline system is less than 1 s. Inline mechanical mixing systems can be configured with different types of impellers and baffles to optimize mixing. The mixing systems are often equipped with taps for chemical injection. Generally, the G is 300–2,000 s^{-1}.

Figure 2-2 Image of a vertical basin mixer used in softening

Source: Hayward Gordon

Figure 2-3 Example of an inline mechanical mixer

Jet injection mixing. Jet injection mixing is a method of rapid mixing that uses a jet of water to create turbulence for mixing chemicals, particularly metal-salt coagulants, into the process stream. This type of system is convenient for use in a closed pipeline where the symmetry of the process flow matches that of the water jet but is also commonly applied in open channels. Images of this type of system are shown in Figures 2-4 through 2-6.

Water for the jet may originate in the plant water supply system or may be from a small loop with a pump. The typical flow rate for the water jet is approximately 2–4 percent of the process flow. With a pumped system, often called pumped diffusion, the pump suction is connected shortly upstream of the mixing zone. Water is then pumped in a loop and sprayed back into the process stream. A schematic of this type of system is shown in Figure 2-6. Nozzle selection and design flow rate for the pump determine the amount of

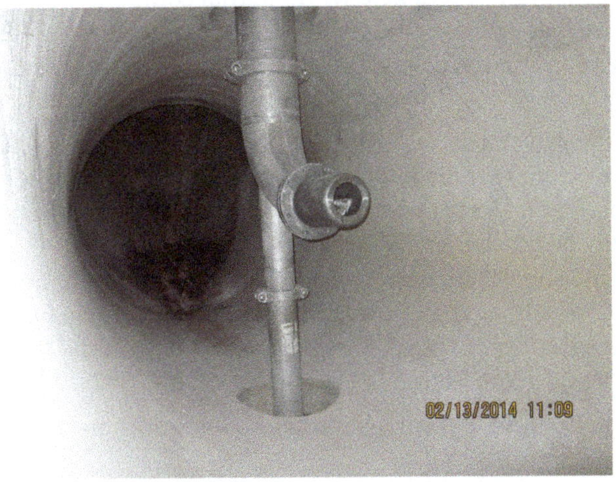

Source: Carollo Engineers

Injection nozzle directed upstream inside the source water piping.

Figure 2-4 Austin's Water Treatment Plant 4

Source: Carollo Engineers

Mixing water injection nozzle connects to the top of the pipe. Chemical injection is shown at the side of the pipe.

Figure 2-5 Colorado Springs Edward Bailey Water Treatment Plant

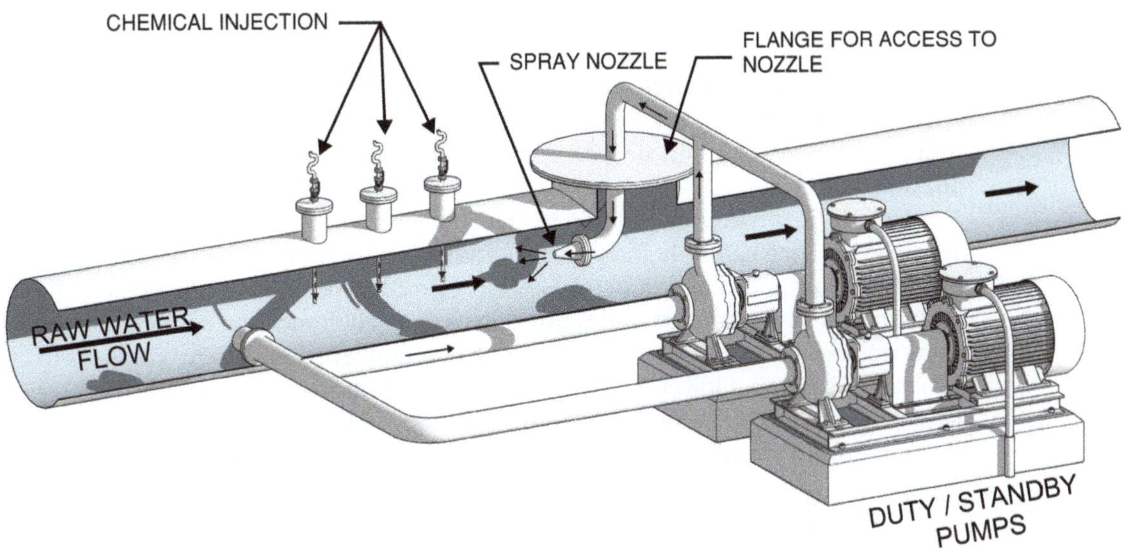

Source: Carollo Engineers

Figure 2-6 Pumped diffusion schematic

turbulence generated (i.e., the *G* value), which can be estimated using the equations presented previously in this chapter; typical *G* values range from 300 to 2,000 s^{-1}.

Many different configurations have been used for this type of system:

- Nozzle facing cocurrent or countercurrent: Computational fluid dynamics modeling typically shows a slight, but not significant, improvement in mixing with the nozzle facing into the flow in a countercurrent fashion. Operationally, when chemicals with high clogging or precipitation potential are used, cocurrent orientation generally reduces the maintenance required.
- Chemical injection into the pumped stream: Significant clogging often results from injecting chemical directly into the pumped stream, as coagulation with concentrated chemical begins immediately.
- A deflector plate to spread out the flow or a nozzle with internal components that produce a cone spray: Both of these options have been used successfully with the most significant challenges coming from small particle passages in the nozzle clogging.

Operationally, the *G* value of the system is adjusted by changing the flow rate through the nozzle, with a higher *G* associated with a higher flow rate. This flow rate may be adjusted by using a valve to control flow if a plant water system is the source of mixing water. If a pump is used, the pump may be placed on a VFD.

Static mixers. Static mixers are motionless mixing devices that allow for the inline, continuous blending of fluids within a pipeline. With no moving parts, static mixers use the energy of the flow and flow pattern disruption to disperse chemicals within the flow stream. Considerations when using a static mixer include

- the configuration of the element used for mixing, as the configuration varies among the suppliers;
- the number of elements needed to achieve the necessary chemical dispersion;
- the head loss across the mixer under the full range of plant flow conditions; and
- *G* value over the full range of treatment flow rates.

The amount of mixing depends on the plant flow rate, and static mixers in parallel may be necessary to handle the full range of flows. Typical *G* values are set by the mixer geometry and range from 300 to 2,000 s^{-1}. In general, static mixers do not provide the operational control that is provided by the previously discussed mixing systems, and some result in suboptimal mixing conditions and excess chemical usage. Figure 2-7 presents a diagram of a wafer-type pipe static mixer. Open-channel static mixers generally have an order-of-magnitude less head loss than pipe static mixers and also have less head loss variation with changes in flow rate, although they also produce lower velocity gradients (Edzwald 2013).

Hydraulic mixing. In addition to static mixers, there are several flow configurations that achieve mixing passively. Plants can take advantage of the mixing that happens in pumps, at bends in pipes and channels, at weirs (see Figure 2-8), and at valves. Mixing can also be accomplished at Venturi meters (see Figure 2-8) or parshall flumes, the open-channel analog of Venturi meters. Weirs and Venturi meters tend to produce higher velocity gradients than pipe and channel bends but lower than pipe static mixers. Compared with pipe static mixers, they also tend to use less hydraulic head (Edzwald 2013). Typical *G* values are lower than those of other mixers and are very dependent on geometry of the passive mixing area.

Chemical Injection

Chemicals are typically injected into the water by inline chemical quills, diffusers, or jet injectors before or during rapid mix. Many different designs have been successfully used in practice for the addition of multiple chemicals. In many cases, coagulants, pH-adjustment

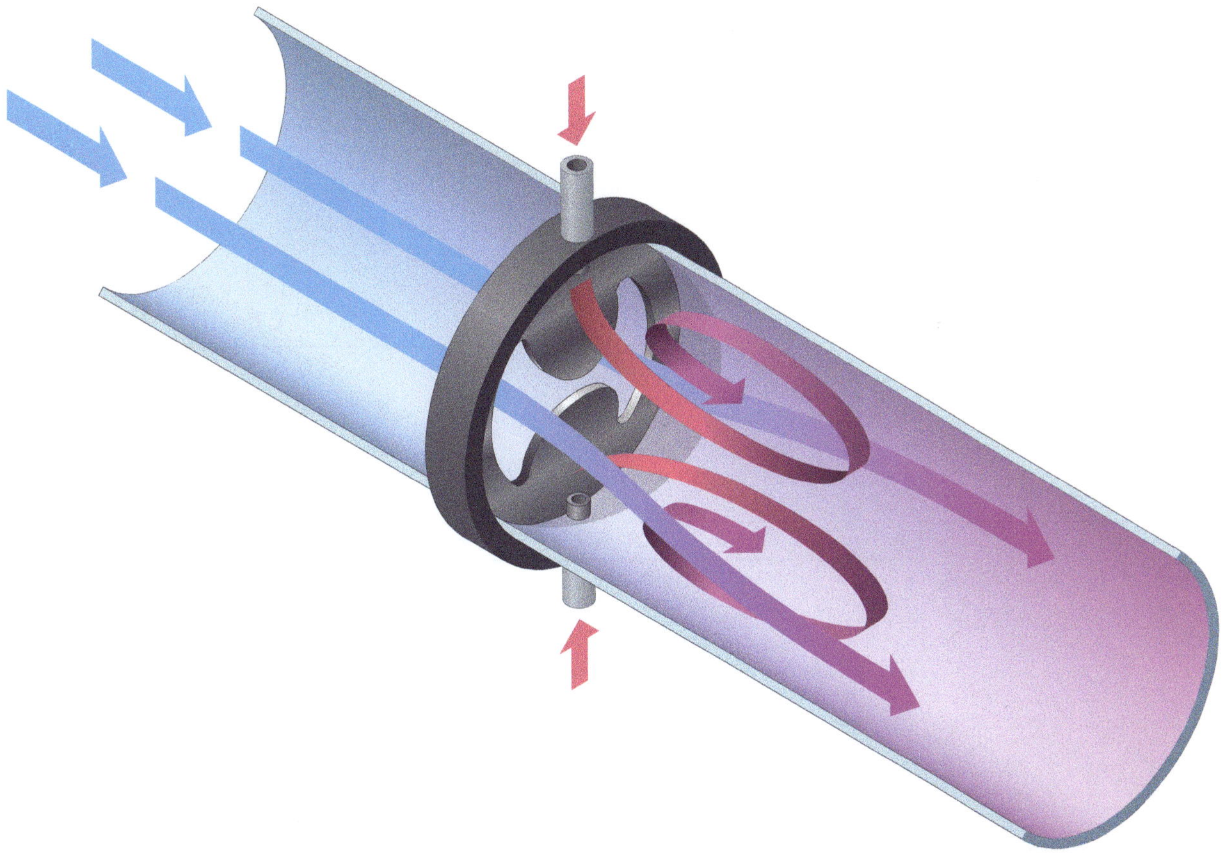

Source: Westfall Manufacturing

Figure 2-7 Mixing of chemicals with static mixers

chemicals, and polymers are added together at a single rapid-mix location. However, in some plants, bench- or pilot-scale testing has determined that staggered chemical addition can result in performance improvements. For example, when an alkaline pH-adjustment chemical is needed to provide alkalinity, the pH-adjustment chemical is often added before the coagulant, so alkalinity is available to react with the coagulant. In some instances, reversing this order of chemical addition may provide some treatment benefit. See the "Coagulant Selection Guidelines" section of Chapter 1 for details on coagulants and their chemistries.

Polymers are sometimes used to enhance the removal of particles (i.e., turbidity) or organic matter within coagulation, flocculation, sedimentation, and/or filtration. When they are used for enhancing a unit operation or treatment process, they are typically referred to as an "aid." For example, if a polymer is used during coagulation, it is typically referred to as a "coagulant aid." Polymers used as coagulant aids tend to be positively charged, while flocculant aids tend to be neutrally or negatively charged.

Oxidants and disinfectants. Another consideration in the optimization of coagulation or rapid-mix system is the use of preoxidants. Preoxidants such as free chlorine, permanganate, ozone, and chlorine dioxide can be used to oxidize metals and organics, immobilize algae, and improve coagulation. The injection location, required reaction time, and rapid-mixing system are important considerations in determining the effectiveness of a preoxidant. For retrofitting an existing plant, full-scale trials with testing at multiple application points may be beneficial.

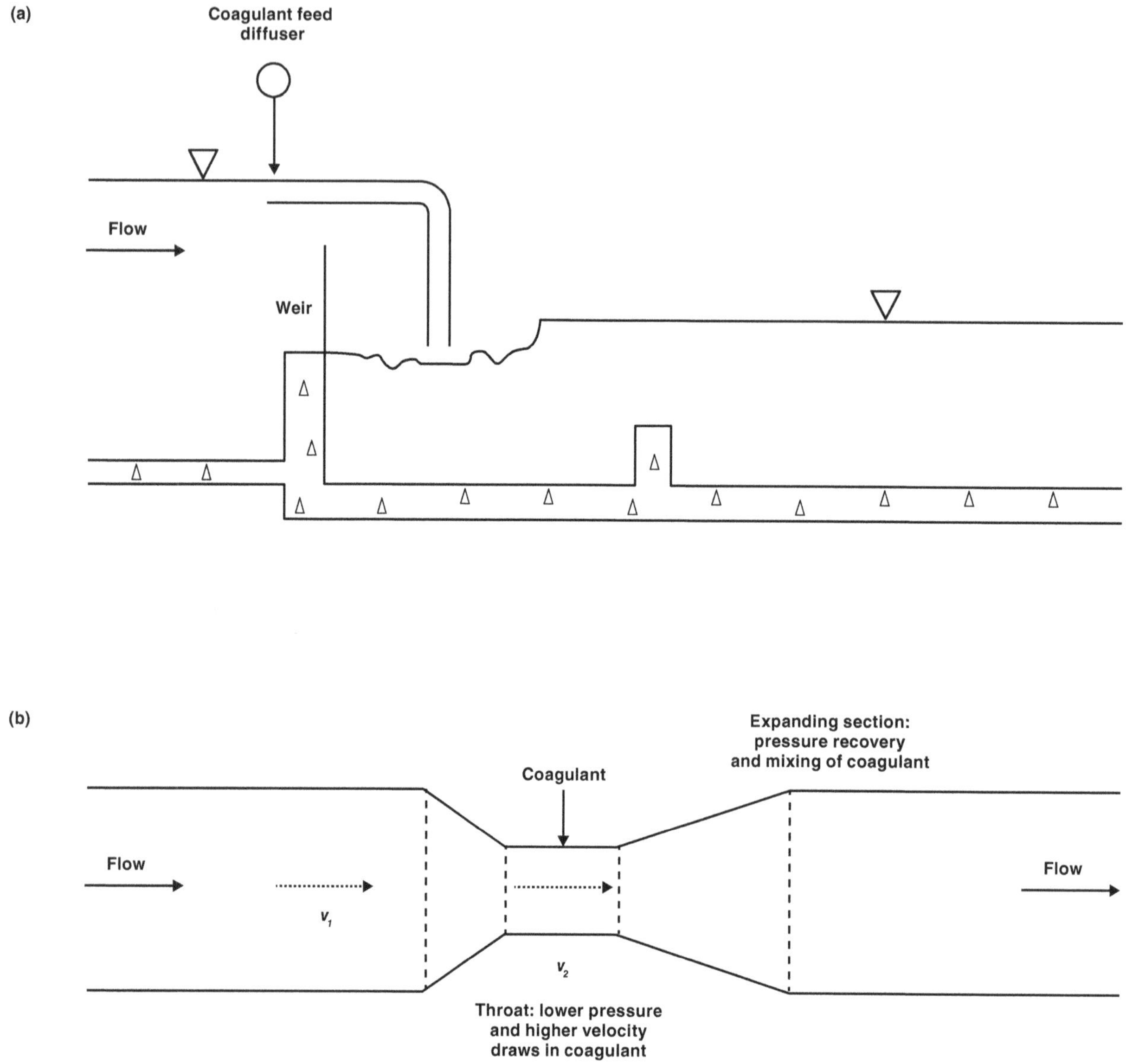

Source: Edzwald 2013

Figure 2-8 Hydraulic mixing over a weir (a) and through a Venturi meter (b) (Edzwald 2013)

Adsorbents. Activated carbon is commonly used to adsorb color-causing organic compounds, taste-and-odor compounds, disinfection byproduct precursors, and synthetic organic chemicals in drinking water treatment. Powdered activated carbon (PAC) is typically used in conjunction with the coagulation process. The rapid-mixing system is an important process for dispersing PAC evenly to the water to maximize the removal of organic compounds.

Carrier water. Carrier water is typically used as follows.

- To dose the chemical: Some chemicals would be impossible to feed without carrier water (e.g., PAC). For such chemicals, carrier water is blended with the chemical to form a slurry, which can be pumped from the chemical storage location to chemical injection location.
- To provide or promote dispersion of the chemical at the dosing point: If the chemical feed rates are low, carrier water can be used to increase the injection velocity to the process. Increased velocity improves the dissipation of the chemical

to the water. For example, diffusers typically require a minimum injection velocity at each orifice, which can be achieved by using carrier water.

The use of carrier water with some chemicals can result in scaling problems if the hardness of the carrier water is high. For example, because sodium hypochlorite is basic, it will increase the pH of the carrier water, and this may be enough to cause precipitation of the calcium carbonate ($CaCO_3$) in the carrier water. Therefore, attention must be paid to the chemical feed system using carrier water. If scaling is a concern, softened or decarbonated dilution water should be used. Some utilities have also had success with tempered water to prevent freezing of carrier water in cold climates.

Use of carrier water can cause premature hydrolysis of coagulants, which can affect the performance of coagulation (Swetland et al. 2013). Therefore, alum- or ferric-based coagulants, polyaluminum chloride coagulants, and some polymers are commonly dosed as neat solutions, although many plants use coagulants with carrier water with no adverse effects.

Operational Considerations

Variation in water quality. Chemical dose is dependent on the source water's characteristics. Turbidity, NOM content of the water (specific ultraviolet [UV] absorbance), alkalinity, temperature, and pH can affect the required coagulant dose. It is important to adjust the coagulant dose as the source water quality changes. Jar testing can be conducted if fluctuations in the source water quality are observed, and jar test results can be used to determine the optimal coagulant dose.

Coagulation pH. The pH at which coagulation occurs is a critical process control parameter, as it affects (1) the surface charge of the particles present in the source water, (2) the charge of source water NOM, (3) the charge of dissolved-phase coagulant species, (4) coagulant solubility, and (5) the surface charge of floc particles (see Table 1-1).

On-line pH sensors can be very valuable for monitoring the pH during rapid mix and continuous monitoring of coagulation. In addition, grab samples can be collected to confirm optimal coagulation conditions are met during rapid mix. In cases with more than one rapid-mix basin, it is important to measure pH at each rapid mix to ensure that equal chemical doses and mixing are occurring in each basin. Significant pH differences can potentially hinder the removal of particulates and organics. If differences are detected, the amount of acid or base dosed for each treatment train may need to be determined individually.

Particle charge measurement. Streaming current and zeta potential monitors measure particle charge and therefore can be used to monitor changes in the overall charge of particulates in the water after coagulant addition. Their application for drinking water treatment has been increasing, as they are most often used after coagulant addition and within the rapid-mix basin. If the streaming current or zeta potential instrument indicates a change in the charge of the floc particles, then the coagulant type, dosage, or both can be adjusted to compensate for the change in charge in the water. The streaming current or zeta potential reading associated with coagulation conditions that result in good treatment plant performance can be used as an operational set point. If the reading changes from this set point, dosages of the treatment chemicals can be adjusted to bring the particle charge back to the previously determined operational set point. Additional information can be found in Chapters 1 and 9.

Rapid-mix intensity (velocity gradient, *G*). Maintaining constant mixer speed does not ensure the same mixing quality as water temperature changes. Viscosity of water increases with decreasing temperature, and thus, it is not uncommon for optimal coagulation conditions to vary seasonally, particularly for coagulation performance to decrease in cold water. If applicable, the mixer speed can be varied to maintain proper mixing (i.e., *G* values) in the rapid-mix basins with seasonal viscosity changes.

FLOCCULATION

In flocculation, chemically conditioned particles come into contact to build a floc. As flocculation progresses, particles come together to form larger floc. Larger flocs are more susceptible to shearing effects unless the flocculation intensity is gradually reduced. This is typically achieved through a staging of flocculation that allows a progression from higher to lower levels of flocculation intensity (mixing energy). The flocculation intensity is controlled at each stage to provide the best balance between the effect of agitation in promoting particle contact and the floc shear that becomes more significant as floc size is increased through successive stages. The term *tapered flocculation* is often used to describe this approach.

Chemical Injection

Polymers used as flocculant aids are usually the only chemical added during flocculation. To ensure adequate dispersion of flocculant aids without shearing the floc particles already formed, they are often added at the openings between floc cells, at weirs separating floc cells, or across the entire floc basin width. The dose of the flocculant aid depends on the polymer structure, water quality, and treatment system characteristics. Bench-scale and full-scale testing are necessary to determine the dose and injection point of the flocculant aid.

Types of Flocculators

Flocculation configurations generally tend to fall into two general categories: mechanical flocculation and hydraulic flocculation. The main advantage of mechanical mixing is greater flexibility for adjusting flocculation speed to match variations in both water quality and flow rate as well as seasonal changes in viscosity (see the "Rapid-Mix Intensity [Velocity Gradient, *G*]" section). The main advantage of hydraulic mixing is both simplicity and economy of operation because energy input is attained through head loss during flow rather than by motor-driven mixing devices. This also means that there is no need for moving parts, which periodically fail.

The most common types of mechanical flocculators are vertical turbine and paddle-wheel flocculators. Walking-beam flocculators or hydraulic flocculators are also present at several older facilities but are not often installed in new facilities. Upflow clarifiers, which include flocculation inside a clarifier, are described in detail in Chapter 3.

Vertical turbine flocculators. Vertical turbine flocculators have a significant O&M benefit when all of the serviceable moving parts are placed above the water surface, as is typical of designs in recent years. This advantage, however, may be offset to an extent in larger plants because several flocculators may be required.

Of the impeller types used in vertical turbine flocculators, the axial-flow, foil-type impellers have the lowest power requirement at a given level of flocculation intensity (Clark et al. 1994). This can simplify the overall mechanics of equipment and reduce its size. The treatment result with this equipment is also good when properly configured with multiple cells that are operated in series and have appropriate impeller diameter, impeller placement at a depth that yields the most effective cell-wide agitation, and appropriate cell geometry. A photograph of this type of flocculator is shown in Figure 2-9.

Paddle-wheel flocculators. Paddle-wheel flocculators can be provided in both horizontally (see Figure 2-10) and vertically mounted configurations. The process performance of this equipment varies widely and tends to be sensitive to flow pattern through the basin, but a high level of performance can be achieved when the flow pattern approximates a plug-flow condition. This can be achieved when horizontally mounted paddle flocculators are installed in a basin with a serpentine baffling pattern that directs flow along the shaft

Source: Jacobs

Figure 2-9 Foil-type axial vertical turbine flocculator

of the paddle. Lower process effectiveness is observed in other configurations such as a horizontally mounted flocculator in which flow is perpendicular to the shaft and in vertically mounted paddle flocculators. This reduction in performance is the result of a greater tendency toward short-circuiting in these basins; the most significant deterioration occurs with flow across a horizontal flocculator.

While the operational, mechanical, and potential process advantages of vertical turbine flocculators have led toward common use of this configuration, a properly configured horizontal flocculator with a plug-flow character may provide slightly higher flocculation efficiency (Clark et al. 1994). This configuration can be easily arranged through the use of serpentine baffling to provide a plug-flow characteristic that yields the greatest control over the floc size distribution. Floc size distribution, particularly maintenance of a uniform floc size, can be important for optimal operation of downstream clarification processes, particularly for high-rate plate settlers. While the submerged bearings associated with paddle flocculators are not as easily maintained as the above-water assemblies of vertical turbine flocculators, bearing assemblies are available that provide a reasonable service life for this application. In the case of the drive units, direct drive systems have become more widely available as a replacement for the older chain-and-sprocket type.

Source: Jacobs

Figure 2-10 Horizontal paddle flocculator

Hydraulic mixing. Hydraulic mixing is typically accomplished through either horizontally baffled (around-the-end) channels or vertically baffled (over-and-under) channels, although several other configurations also exist. The flow through a vertically baffled channel is shown in Figure 2-1. The main advantages of these types of systems are that there are no moving parts and no direct power usage. The main disadvantages are that there can be significant head loss in the baffled basins (as much as 2–3 ft or 0.6–0.9 m), and mixing intensity is a function of plant flow rate, meaning that lower mixing intensity and longer retention time are provided at low flow rates, and the reverse is true at high flow rates. However, experience has shown that good flocculation often occurs when *GT* (velocity gradient multiplied by time of mixing of flocculation) values range from 3×10^4 to 2×10^5, meaning that the low intensities at low flow rates can work acceptably if available detention times provide *GT* values within the indicated range (Kawamura 2000). If flocculators do not have sufficiently high flow velocities or sufficient underdrains, flocs can settle at the bottom of these flocculators, requiring periodic cleaning.

Flocculation can be affected by the detention time, which is affected by basin volume, baffling, and staging of flocculation cells. Baffling and staging are necessary to control short-circuiting and to achieve the most effective use of basin volume.

In situations with nearly constant flow and where precise control of flocculation energy is not required, another way to accomplish mixing without external power input or moving parts is what is called static mixing. This is accomplished by placing specially designed inserts into pipes to accomplish inline mixing. These are more compact, as they are placed within plant piping, but they are harder to directly observe for O&M purposes. In comparison with the static mixers used for rapid mix, these static mixers generally are longer and occupy less cross-sectional area in the pipe because they are required to generate lower values of *G* for longer residence times. Figure 2-11 illustrates a static mixer configured with a wafer-type static mixer upstream for rapid mix. Figure 2-12 illustrates a low head loss static mixer used for flocculation.

Operational Considerations

Flocculation intensity (velocity gradient, *G*). Adjustment of flocculation intensity for best conditions in each stage is an important consideration in both design and operation of flocculation facilities. This may even involve seasonal adjustments to accommodate changes in water quality and the associated changes in floc character. These types of adjustments are especially necessary in locations where a seasonal variation in cold-water coagulant performance is encountered.

General guidelines for full-scale flocculation basins are *G* values that vary between approximately 10 and 70 s^{-1} (Kawamura 2000), although *G* values as high as 100 s^{-1} are used in DAF plants. General guidelines for *G* values may be used for setting initial levels of flocculation, followed by minor adjustments until a best condition for operation is achieved.

Visual observation. Visual observation of floc and clarified water can be an important part of flocculation process control, particularly during times of changing source water quality. While the desired physical appearance of a floc can vary greatly depending on the source water, the coagulant used, and the downstream removal processes, one measure of good flocculation conditions is often the observation of "seams" or "marbling effect" during flocculation. This effect generally consists of defined clusters of floc with "clear" water evident between the clusters. In mechanical flocculators, there is a lag between changes made in rapid mix and resulting changes in observations of floc. For hydraulic flocculators, this lag tends to be quite small.

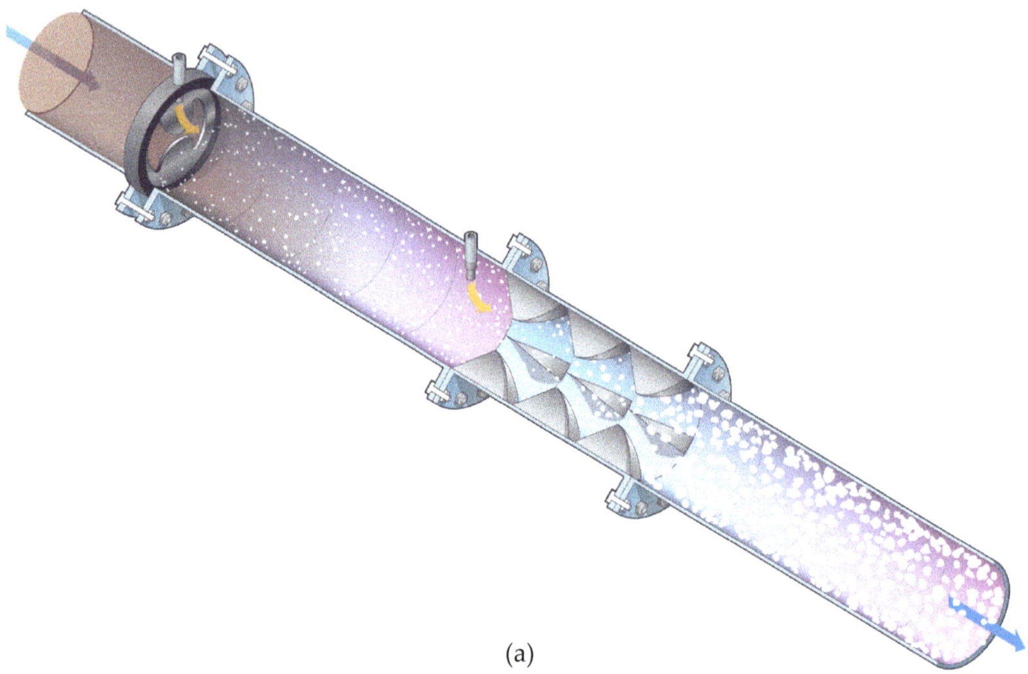

(a)

Source: Westfall Manufacturing

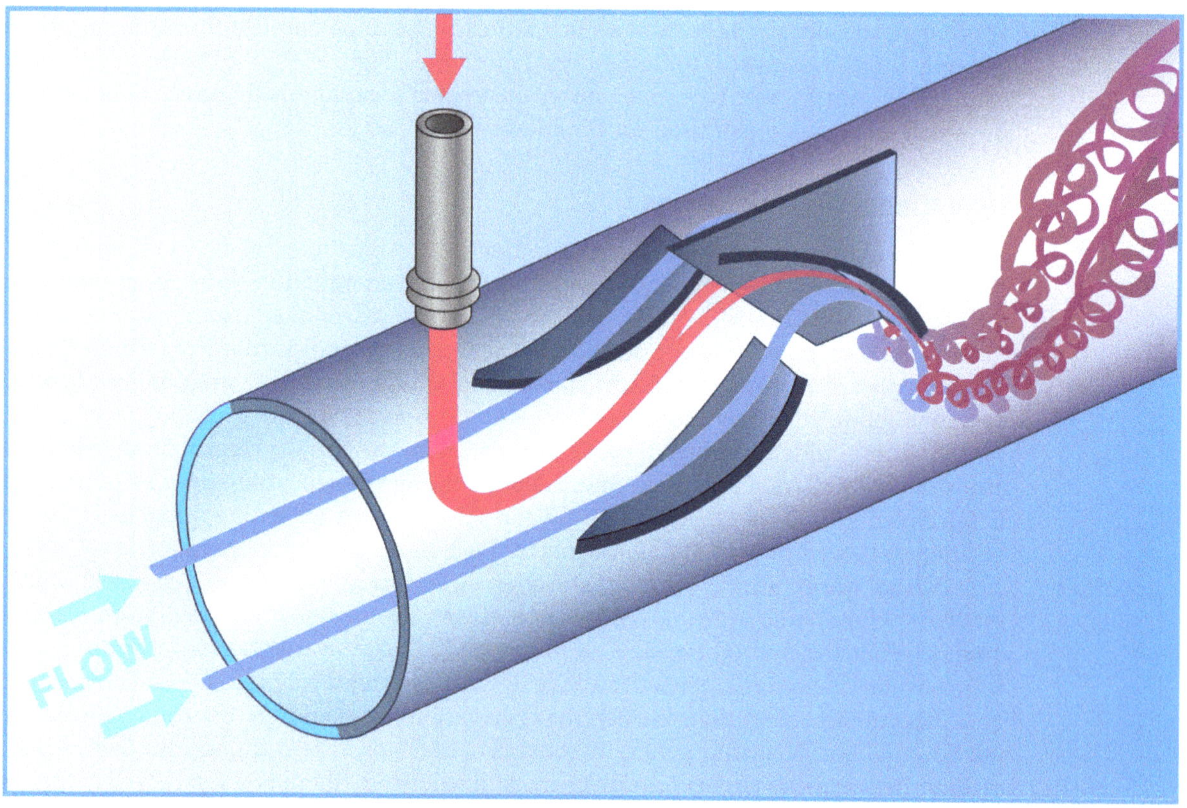

(b)

Source: Westfall Manufacturing

Figure 2-11 Diagram of static mixer for flocculation preceded by a wafer-type static mixer for rapid mixing (a) and diagram of low head loss static mixer used for flocculation (b)

TROUBLESHOOTING

Rapid Mix

Rapid mixing is difficult to troubleshoot, as the effects of inadequate mixing are seen in the downstream clarification and filtration steps. Common problems in full-scale rapid-mix processes include

- inadequate dispersion of chemicals and
- inaccurate dosing of chemicals.

Poor dispersion of coagulation chemicals results in a portion of the water not being adequately coagulated. This untreated portion of the flow may then settle and filter poorly. As discussed previously, this may be caused by poor injector design, inadequate mixing energy, or hydraulic short-circuiting. Similarly, discrepancies between the intended chemical dose and the actual dose delivered may be the result of poor chemical feed pump calibration, inaccurate flowmeter readings, or unequal hydraulic splitting between process trains. Problems of these types can be indicated by high COV values.

Initial investigations should include verification of all water and chemical flow rates, chemical dosing calculations used in supervisory control and data acquisition displays, stock chemical concentrations and dilutions, and on-line process instrument calibrations. These should be checked over a range of chemical doses and process flow rates.

An effective way to determine whether rapid mixing is not adequate or whether chemical dosing is inaccurate is to compare the removal of UV-absorbing substances in the full-scale rapid-mix process with the removal seen in jar tests. The reactions between coagulant chemicals and NOM are completed within seconds, and UV absorbance measurements are easy, accurate, and repeatable. If there are significant variations between the UV removal in the full-scale plant and the jar tests, more detailed investigations are warranted. For large basins, several samples at different locations in the basin may be needed.

For troubleshooting rapid-mix problems related to mixing energy, comparing UFRVs under two mixing conditions may be useful. For static mixers, this can be accomplished by varying the flow rate. For rapid-mix basins, it may be possible to add mixing energy on a temporary basis by installing a temporary propeller mixer into the rapid-mix basin or adding a temporary submersible recirculation pump that discharges back into the rapid-mix basin and increases turbulence and mixing. If UFRVs do not change significantly between the two different mixing conditions, efforts to optimize mixing may not be worthwhile.

Flocculation

Flocculation is a physical process that affects the performance of clarification and filtration processes, so troubleshooting efforts should be focused on measurable effects on these processes. Although operators who have extensive experience at a plant may be able to relate visual observations of floc form to performance, quantifiable measures such as clarifier turbidity or UFRV are preferred. UFRVs are a better measure of overall plant performance than clarified water turbidity, so flocculator speed optimization and troubleshooting should be based on maximizing UFRV.

REFERENCES

Amirtharajah, A., M.M. Clark, R.R. Trussell, and AWWA Research Foundation. 1991. "Mixing in Coagulation and Flocculation." Water Research Project Report No. 316. Denver: Water Research Foundation.

Camp, T.R., and P.C. Stein. 1943. "Velocity Gradients and Internal Work in Fluid Motion." *Journal of Boston Society of Civil Engineering*. 30(4):219-237.

Clark, M.M., R.M. Srivastava, J.S. Land, L.J. McCollum, D. Bailey, J.D. Christie, and G. Stolarik. 1994. "Selection and Design of Mixing Processes for Coagulation." Water Research Project Report No. 613. Denver: Water Research Foundation.

Edzwald, J.K. 2013. "Coagulant Mixing Revisited: Theory and Practice." *Journal of Water Supply: Research and Technology–AQUA*. 62(2):67-77. https://doi.org/10.2166/aqua.2013.142

GLUMRB (Great Lakes–Upper Mississippi River Board). 2012. *Recommended Standards for Water Works*. Albany, N.Y.: Health Research, Health Education Services Division.

Kawamura, S. 2000 (2nd ed.). *Integrated Design and Operation of Water Treatment Facilities*. New York: John Wiley & Sons.

Nix, D.K. 2009. Personal communication.

Pennock, W.H., L.W. Lion, and M.L. Weber-Shirk. 2021. "Design Algorithm for Vertically-Baffled Flocculators." *Environmental Engineering Science*. 38(7):592-606. https://doi.org/10.1089/ees.2020.0288

Swetland, K.A., M.L. Weber-Shirk, and L.W. Lion. 2013. "Influence of Polymeric Aluminum Oxyhydroxide Precipitate-Aggregation on Flocculation Performance." *Environmental Engineering Science*. 30(9):536-545. https://doi.org/10.1089/ees.2012.0199

Chapter 3

Clarification Processes

Chapter Summary

- Both hydraulic and solids loading affect clarifier performance.
- Many different clarifiers geometries have been used, and specific treatment goals influence optimum geometry.
- Optimization requires monitoring of solids in basins, which can be facilitated with instrumentation.

Clarification is a critical step in most water treatment plants (WTPs), typically following coagulation and flocculation and preceding filtration. However, in some applications, such as wastewater and stormwater treatment, clarification processes are the primary treatment processes.

Clarification generally refers to the physical removal of suspended particles from the water column. While coagulation and flocculation are generally chemically dependent treatment processes, clarification itself is generally a physical process, although often highly reliant on chemical preconditioning during the flocculation stage (see Chapter 2).

The following sections highlight the general concepts for achieving an efficient clarification process while characterizing unique optimization techniques for specific clarification technologies.

GENERAL CONCEPTS OF CLARIFICATION

Separation of particles from water by gravity is one of the most widely used and oldest processes in water treatment. Historically, clarification has used large quiescent basins where particles settle as the water moves horizontally through the basin. This theory was first developed by Alan Hazen in 1904, and according to his concept, particles would settle to the bottom and be removed more quickly in shallow basins. This concept is limited by larger plant footprints and hydraulics; however, all clarifiers operate under the fundamental fluid mechanics principles initially developed by Hazen—that is, that

Table 3-1 Types of sedimentation

Settling Type	Description	Typical WTP Locations
Discrete settling (type 1)	Particles settle as independent units with little interaction with neighboring particles.	Grit chambers, presedimentation basins for sand removal, settling of sand particles during backwash
Flocculant settling (type II)	Flocculant settling occurs when particles start interacting with one another.	Most conventional settling basins, metal-salt coagulation processes
Hindered or zone (type III)	Strong interparticle forces result in the formation of a matrix of particles that settle together.	Thickeners (sludge disposal), bottom of some sedimentation basins
Compressive settling (type IV)	Not true settling; water drains slowly from a mat of particles.	Dewatering operations

WTP—water treatment plant

particles are acted upon by drag and buoyant forces, which affect the ability of the particles to be removed from the water column.

High-rate clarification instruments, such as tube and plate settlers, were developed to improve solids removal with a smaller basin geometry (Hansen and Culp 1967) but still rely on sedimentation as the only particle removal process. Further advances in particle separation include various configurations of solids-contact clarifiers, sludge blanket clarifiers, and ballasted flocculation.

Sedimentation occurs in regimes based primarily on the concentration and flocculation tendency of particles under four types, as shown in Table 3-1.

While most sedimentation basins in use in WTPs experience type II settling, and newer technologies capitalize on the increased particle removal efficiency of this type of settling, many design and operational characteristics are based on equations developed around type I settling, as equations that accurately describe type II settling have proved to be impractical.

The simplifying assumptions of Hazen's concept are that water flows horizontally across the quiescent zone under plug-flow conditions, each particle in the water is discrete, and no flocculation occurs during settling. These simplifications are applicable in most situations and are suitable for design and benchmarking, but they do not capture all of the complexities of a real sedimentation basin. In a deep sedimentation basin, sedimentation velocity has the possibility of increasing with depth. This means that jar testing is challenging to extend to true sedimentation basin performance. Samples are extracted after only a short travel distance in the jar test procedure, so the opportunity for additional flocculation is much less likely in the test jar as compared with the full-scale basin. This may be offset, however, by nonhorizontal flow in a full-scale basin, whereas if the jar test jar is manipulated carefully, vertical currents can be minimized.

Some particles, such as algae or lightweight particles in cold water, have physical characteristics that lead them to float rather than settle. Particle separation using dissolved air flotation (DAF) has been developed to float these particles instead.

HYDRAULIC AND SOLIDS LOADING

Properly designed and operated clarification basins provide effective removal of particulate solids and low effluent turbidity. The following paragraphs define and discuss typical metrics used to describe clarification basin parameters and their effects on clarification process performance. These parameters are summarized in Table 3-2.

The detention time of a sedimentation basin is the average time it takes for water to move from the inlet to the outlet. The detention time must be long enough to allow solids

Table 3-2 Hydraulic and solids loading parameters

Clarification Parameter	Typical Values
Hydraulic loading rate	0.3–0.5 gpm/ft^2 (0.7–1.2 m/h)
Weir overflow rate	10,000–20,000 gpd/ft (125–250 m^3/d/m)
Solids loading rate	25–30 lb/d/ft^2 (100–150 kg/m^2/d)

to settle in the basin, or they will be carried over into downstream processes. The basis for calculating detention time is the total volume of the basin (typically in gallons or cubic meters) divided by the flow rate (typically in gallons per minute [gpm] or cubic meters per second [m^3/s]). This provides the theoretical hydraulic detention time. However, various geometric, hydraulic, and operational factors can affect the actual detention time. These factors can lead to "short-circuiting" within the basins, allowing fractions of the flow, and the associated suspended solids, to move through the basin in a shorter time than the theoretical detention time. Short-circuiting can affect overall process performance and is discussed in detail in later sections.

Hydraulic loading rate, called the surface overflow rate or surface loading rate, is the primary design parameter for sizing sedimentation basins. It is the rate of flow of sedimentation basin effluent per unit of sedimentation basin surface area. Units are generally gallons per minute per square foot (gpm/ft^2) or cubic meters per hour per square meter (m/h). The hydraulic loading rate is essentially the upflow rate that settling solids need to overcome to settle before the water is carried out of the basin. Typical overflow rates are dependent on the type of solids being settled, water temperature, and configuration of the basin but generally range from 0.3 to 0.5 gpm/ft^2 (0.7 to 1.2 m/h). Typical loading rate range is affected primarily by two factors: the density of the water and the density of the particle. Higher loading rates are generally acceptable for warm waters with heavy suspended solids loads. Conversely, colder waters that are high in color or organics frequently require lower loading rates. The temperature effects are due to an increase in the density of the water, which will increase the buoyant force and will result in a lower settling velocity. To overcome the potential adverse effects of water temperature, the density of the particle can be modified. Coagulants and coagulant aid chemicals are available that can result in denser flocs that settle faster.

Weir loading rates are typically calculated as gallons per day per linear foot of weir (cubic meters per day per linear meter of weir) or gallons per minute per linear foot of weir (liters per minute per linear meter) and represent the rate of flow over the clarification basin effluent weirs. Typical values are 10,000–20,000 gal per day (gpd)/ft (125–250 m^3/d/m). Weir performance can be monitored by observing the flow and solids carryover of the weir. If the weir loading rate is too high, there may be water surging across the weir. Inlet conditions may also contribute to surging or resonance within the basin, which are visually evident through highly variable flow over the weirs, even at constant influent flow rates, or noticeable ripples across the basin water surface. Surging or resonance waves can be alleviated by placing baffles at the inlet structure or at key locations within the basin to prevent the formation of resonating waves. Finally, uneven weirs can also cause uneven distribution of overflow rates, leading to additional solids carryover and poor settling basin performance.

The solids loading rate of a basin is an important design criterion in water clarifier design. It is calculated as the total influent solids, in pounds per day or kilograms per hour, divided by the basin area, in square feet or square meters. (The number of solids is calculated as the concentration of solids times the flow, times the density of water.) Typical design values are 25–30 lb/d/ft^2 (100–150 kg/d/m^2).

Sedimentation basin performance relies on uniform plug flow. Nonuniform flow can occur from hydraulic inconsistencies within the sedimentation basin inlet geometry, basin configuration, or outlet geometry; from wind acting on exposed basins; and from differences in density between water in the basin inlet and water already in the basin (Hudson 1972). Density differences can be caused by temperature, solids loading, or salinity. Nonuniform flow through the clarification process can lead to short-circuiting in the basins, which ultimately leads to deteriorated clarified water quality. Short-circuiting can be moderated with the inclusion of baffled influent walls and longitudinal baffles in the main body of the basin. Longitudinal baffles can be retrofitted by installing wooden planks in existing basins, taking care to avoid generating serpentine flow which increases turbulence. Baffles installed across the end of the basin will prevent surface currents from disturbing particles. Additional considerations to reduce short-circuiting are discussed later in this chapter.

Lastly, increases in plant flow decrease the settling time in an individual basin. If possible, controlling flow through the basins by adjusting the plant's production or bringing another basin on-line can improve performance in periods of high flow. Large swings in flow can disrupt the layer of sludge settled on the basin floor (sludge blanket), leading to solids carryover and poor basin performance. Hydraulic controls, such as baffles and orifice walls, can be used (and installed as a retrofit) to limit disruptions caused by changes in plant flow.

TYPES OF CLARIFICATION PROCESSES

The clarification process generally consists of water-holding basins designed with the specific intention of removing flocculated particles. Clarification processes are based on a physical means for removal of these solids, through either settling or flotation mechanisms. At its simplest, clarification, in the form of conventional sedimentation basins, is achieved through the use of very large basins, which provide several hours of detention time to achieve efficient settling of the flocculated particles within the basin. However, because of the area required for these large basins, design engineers and water treatment equipment suppliers have developed a variety of technologies that decrease the time required for sedimentation and thus decrease the size of the process basin needed.

The specific parameters for optimized operations can vary significantly for each clarification process. As such, the following sections provide details and insights for optimized operation of several popular types of clarification processes.

Conventional Sedimentation Basins

Conventional sedimentation basins rely solely on gravity to achieve low settled water turbidity. They have long residence times, with the intent that large floc in the water will settle as flow passes along the length of the basin. Conventional sedimentation basins rely on a smooth, unmixed flow to achieve good performance, as shown in Figure 3-1.

Components and function. As shown in Figure 3-1, conventional sedimentation basins have very few components:

- an influent baffle (or other method) to evenly distribute incoming flow,
- a long, narrow basin to promote quiescent flow,
- an effluent weir to remove clarified water from the top of the water column, and
- a sludge-removal mechanism to remove settled solids.

Typically, the sludge blanket will be thickest near the entry to the basin, with the quantity of solids decreasing along the flowpath, so uniform flow entry to the basin, above the basin floor, is important to avoid disturbing the sludge blanket and resuspension of the settled solids.

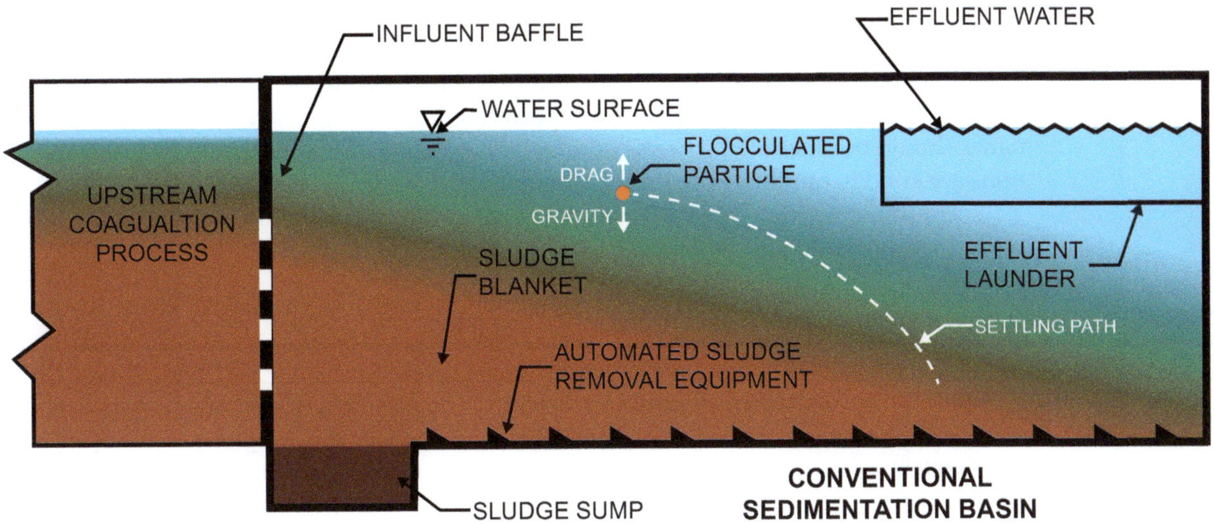

Source: Carollo Engineers

Figure 3-1 Diagram of typical conventional sedimentation basin

Many plants make use of "settling basins" or "contact basins" for pretreatment and removal of heavier particles. These basins typically have a much lower residence time than a conventional sedimentation basin, and while solids are removed, a significant fraction continues on to be removed in the filters, similar to the direct filtration plant. These basins were likely designed with the intention of removing turbidity spikes in otherwise good-quality water.

Coagulation goals. As conventional sedimentation relies on gravity settling of floc particles, upstream coagulation will ideally produce floc large enough to settle in the time frame achievable within the basin. Goals for coagulation include making floc of sufficient size to settle and compact in the bottom of the basin, without overdosing and using unnecessary chemicals. Upstream coagulation often uses a metal salt as a primary coagulant to neutralize charge. Coagulation may also include a coagulant aid or a flocculation aid polymer to provide additional charge neutralization or help bridge across smaller floc to create larger floc.

High-molecular-weight anionic polymers are often best suited for conventional sedimentation basins because these polymers, used in conjunction with metal salt–based coagulants, provide polymer bridging. By providing polymer bridges, the smaller floc particles are able to stick together, rapidly forming larger flocs that settle more quickly in the conventional basins. Nonionic and cationic polymers are also sometimes successful as the primary coagulant in these systems, depending on the specific source water quality. See Chapters 1 and 2 for discussion on dosing polymer. In all cases, jar testing, as described in Chapter 7, can be highly effective in determining the type and dose for both primary coagulant and polymer addition.

General maintenance. There is little mechanical maintenance associated with a conventional sedimentation basin. The most significant operational–maintenance item is sludge removal. To maintain the long process stream residence in these basins, settled solids need to be removed periodically. Frequent solids removal mitigates the solids becoming anoxic and releasing dissolved metals or releasing gas bubbles that can form when the sludge sits for a long time. These gas bubbles can lead to the resuspension of sludge or taste-and-odor issues. Solids are removed either through automated sludge collection or through manual washdown.

In plants that require manual washdown, the basins must be drained, and accumulated solids are then washed to the drain. This process is often labor intensive and results in large quantities of sludge being conveyed to residuals- and/or solids-handling processes several times a year. If sludge removal is performed manually, basin cleaning and maintenance should be scheduled when water demand is below average and source water conditions are good to minimize the load on the remaining units in service.

Troubleshooting. Table 3-3 provides general recommendations for troubleshooting typical sedimentation basin issues.

Table 3-3 Conventional sedimentation basin troubleshooting guide

Problem	Possible Cause	Inspection Items	Remedies
What is the issue	*What might be causing it*	*Items to check to confirm the problem*	*Actions to do to fix the problem*
• Algae growth	• Sunlight • Nutrients in source water • Algae in source water	• Watch for formation of byproducts, such as 2-methylisoborneol, geosmin, or toxic algal metabolites	• Provide a chemical oxidant or algicide in or upstream of the basin • Manually remove algae using a fine-screen net • Drain and clean basin to prevent algae regrowth • Shade covers to prevent algae growth
• Sludge carryover	• Solids buildup in the basin • Anoxic bubbles • Density currents • Loading rate too high	• Look for bubbles in sedimentation • Look for upwelling in sedimentation • Check sludge depth in basin	• Reduce loading rate • Drain basin and remove solids • Improve hydraulics to minimize density differences • Remove solids more frequently
• High settled water turbidity	• Insufficient floc formation • Loading rate too high • Density currents • Poor influent hydraulics	• Confirm chemical feed systems are properly functioning • Assess floc formation through jar testing • Look for upwelling in sedimentation • Visually assess floc formation and settling in the basin	• Adjust rapid-mix and coagulation scheme to achieve better floc formation • Improve hydraulics to minimize density differences • Reduce loading rate
• Increased turbidity during automated sludge removal	• Resuspension of solids blanket by removal mechanism	• N/A	• Slow the mechanism down if possible
• Increases in metal concentrations in the effluent	• Particulate metals are dissolving because of anoxic conditions.	• Perform total and dissolved metals concentration tests at the influent and effluent of the basin	• Avoid anoxic conditions in the sludge through more frequent sludge removal or upstream addition of oxidant
• Solids collection mechanism will not operate.	• Solids are too thick.	• Measure solids percentage with Imhoff cone or graduated cylinder	• Operate mechanism more frequently
• Low percentage of solids removal	• Limited to no sludge blanket on the bottom of the basin; cleaner water is being pulled in during sludge removal	• Use a sludge judge to determine sludge blanket thickness	• Optimize operation to balance blanket depth for compaction, without interfering with removal mechanisms

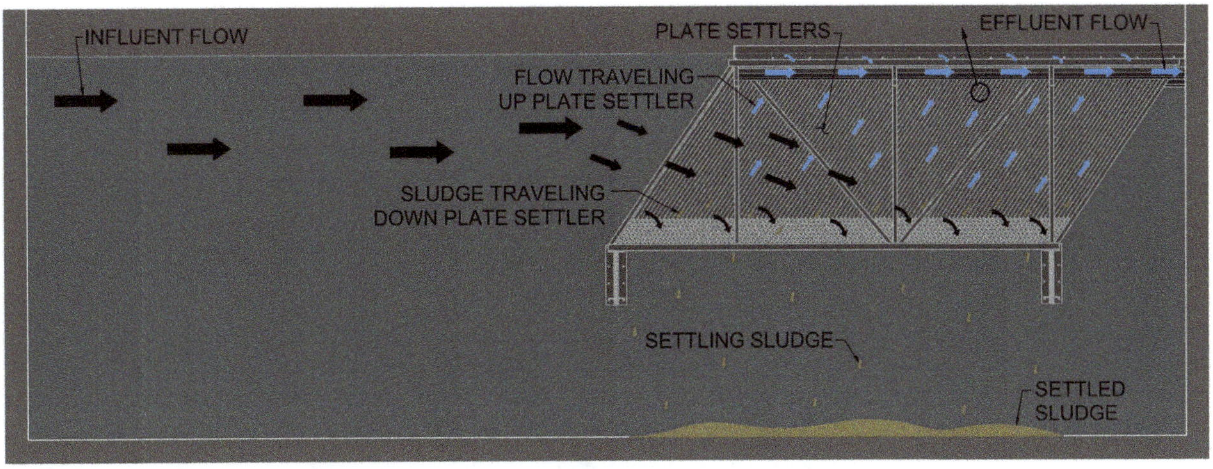

Source: Meurer Research

Figure 3-2 Diagram of typical tube or plate settler basin

Tube and Plate Settlers

Plate and tube settlers reduce the amount basin surface area required for clarification. Sedimentation basins with plate and/or tube settlers generally require less than one-fourth of the surface area of a conventional sedimentation basin. The reduction in surface area is achieved by installation of plates or tubes in a basin, which effectively reduces the settling distance and allows for double counting of the actual clarifier area. See Figure 3-2.

Components and functions. Inclined tube and plate settlers were developed in the 1960s as a practical approach to the problem of providing short distances for sedimentation while also providing for narrow sedimentation basins that had lateral hydraulic stability (Hansen and Culp 1967). They resolved the question of how to apply Hazen's sedimentation theory and increase the settling surface area to achieve settling in a short time. The tube settler concept involved placing a bundle of inclined tubes in a settling basin, positioned so that flocculated water passed upward through the tubes. The vertical distance inside the tube from the top to the bottom of the inclined tube (the distance a particle must settle to reach the "bottom" of the "basin") can be several inches (centimeters) instead of 10–16 ft (3–5 m) from the water surface to the bottom of a conventional settling basin. A bundle of tubes provides many "false floors" onto which floc particles can settle, agglomerate, and slough down to the bottom end of the tube to be discharged and settle to the bottom of the settling basin. Plate settlers operate with similar principle to that of tube settlers; however, these use large rectangular plates, instead of tubes, set at angles typically between 50 and 60 degrees to create the "false floors" to encourage particle settling described previously. Figure 3-3 illustrates the concept of tube and plate settler physics.

Total basin area, plate pack plus approach and maintenance access area, hydraulic loading rates for tube and plate settlers typically range from 1 to 3 gpm/ft^2 (2.4 to 7.3 m/h), depending on the settling characteristics of the floc that is formed in pretreatment. However, hydraulic loading rates for tube and plate settlers are typically measured on the basis of the effective or projected surface area of all the plates or tubs in the system. These effective loading rates also incorporate an efficiency factor but generally are less than 0.5 gpm/ft^2 (1.2 m/h).

Coagulation goals. As tube and plate settler technologies rely on gravity settling of floc particles, upstream coagulation is similar to conventional sedimentation, and plate settlers are ideally suited for retrofit of conventional sedimentation basins. Coagulation will ideally produce floc large enough to settle within the effective surface area of plate and/or tube system. Upstream coagulation chemicals are similar to conventional sedimentation often using a metal salt as a primary coagulant to neutralize charge and at times may also include a coag-

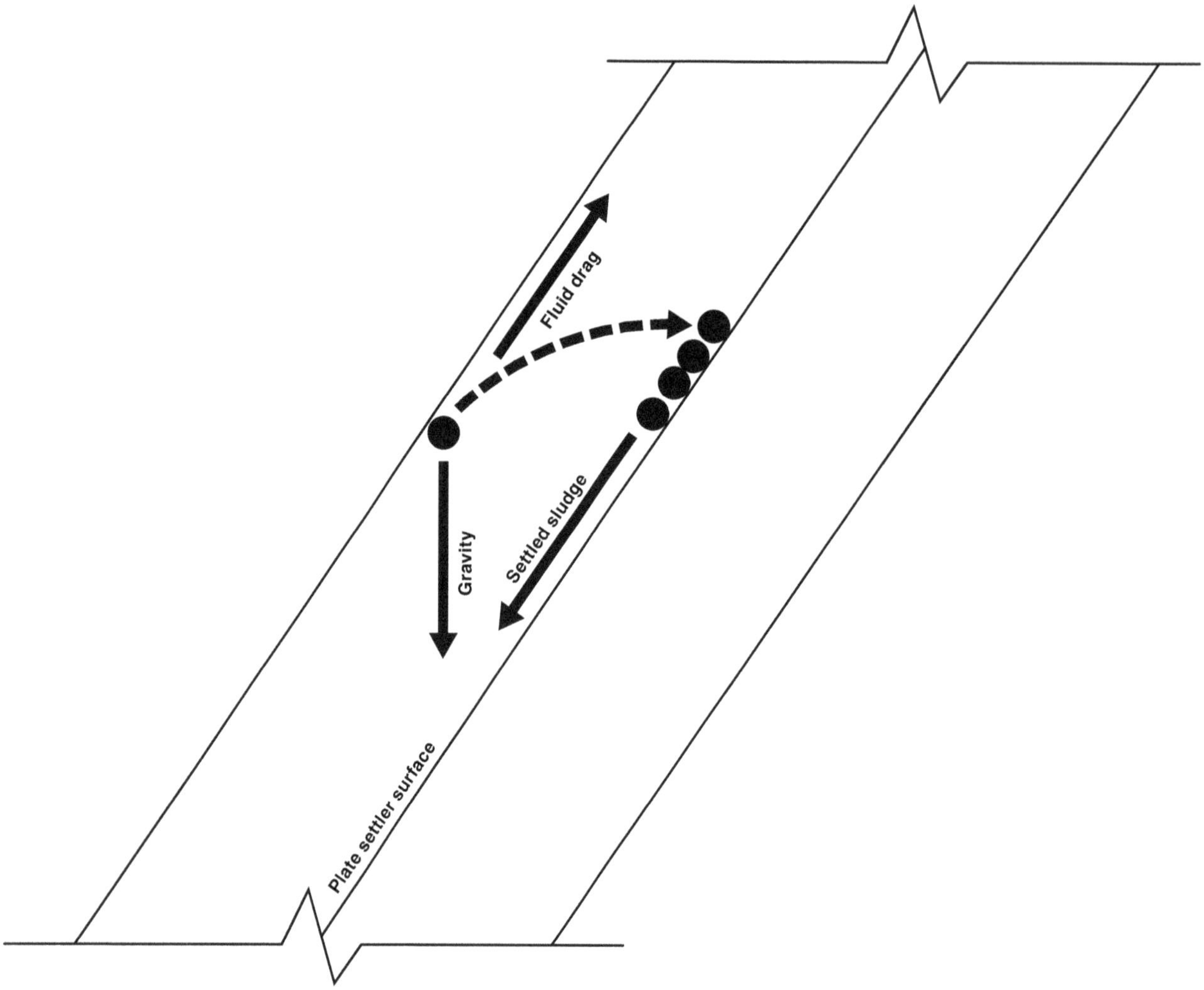

Source: Meurer Research

Figure 3-3 Plate and tube settler theory

ulant aid or a flocculation aid polymer to provide additional charge neutralization or help to create larger floc. In all cases, jar testing, as described in Chapter 7, can be highly effective in determining the type and dose for both primary coagulant and polymer addition.

General maintenance. Plate and tube settlers themselves contain no moving parts. Maintenance is typically associated with removing solids that have accumulated on top of and within the plates and tubes. This can be accomplished by draining the basin and spraying the plates and tubes with water. Tube settlers are often fabricated of a low-density plastic, and low-pressure water should be used in these applications to avoid damaging the tubes. Plate settlers made of stainless steel are more robust and can accommodate higher-pressure washdown. Tubes and plates should be washed frequently to avoid excessive solids buildup. There have been incidences in which significant solids have built up across tube settlers to such an extent that they have collapsed.

Tube settler life is generally 5–10 years and will require replacement. Plate settlers have a life span of at least 20 years. Provisions should be made for full access to the top of tube and plate settlers to allow for maintenance, cleaning, and removal when needed.

Troubleshooting. Settling mechanisms and general operation for systems with plates or tubes are similar for conventional sedimentation basins. See the "Troubleshooting" section under the "Conventional Sedimentation Basin" section, Table 3-2, for general troubleshooting that can be applied to tube and plate settlers as well.

Solids-Contact and Sludge Blanket Clarifiers

Early approaches to finding more economical alternatives to large sedimentation basins with long detention times led to the use of flocculator–clarifiers and solids-contact clarifiers. These high-rate processes combined a central mixing zone, flocculation, and an outer sedimentation zone in one cylindrical basin. In these clarifiers, a baffle wall separates the mixing and flocculating zones from the sedimentation zone. Chemical reactions and floc formation are promoted by recirculating water within the mixing zone and, in some cases, by mixing incoming water with sludge drawn from the bottom of the basin. The units can achieve high solids in the mixing zone and may be operated with a sludge blanket in the settling zone. The desired result of the chemical addition along with the mixing and flocculation zones is to create a floc that settles at a rate higher than the upflow of water within the sedimentation zone.

Components and function. Figure 3-4 provides a general diagram of a flocculator–clarifier. Figure 3-5 provides a general diagram of a blanket clarifier.

Typical components of a flocculating clarifier include a mixer or flocculator; a flocculation chamber, which is often an open-bottom, vertical wall cylinder or cone sized to provide appropriate flocculation time; collection launders; and sludge withdrawal system. Often, the flocculation system is a single-pass design in which settled sludge is not reintroduced into the influent water. The sludge withdrawal system may include mechanical scrapers to move settled sludge to a concentrating hopper. The concentrating hopper accumulates sludge, which is periodically discharged through a pipe and sludge withdrawal control valve.

Solids-contact clarifiers have many similar components. A mixer is used to combine influent water, settled sludge, and treatment chemicals before entry into the flocculation

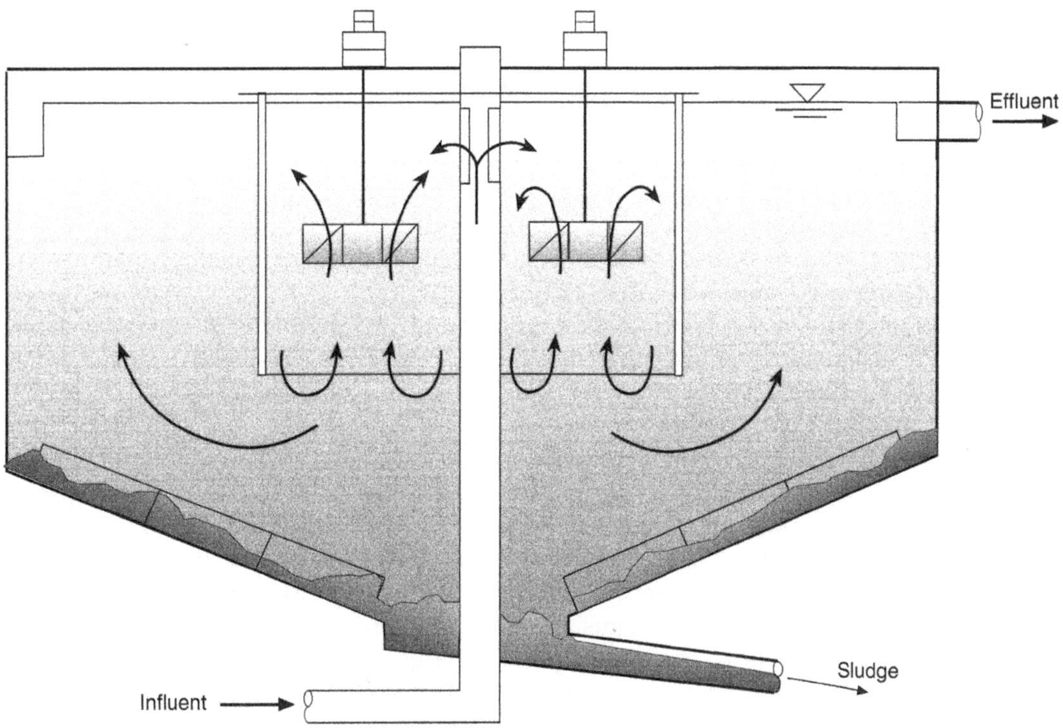

Source: WesTech Engineering

Figure 3-4 Diagram of a flocculator–clarifier

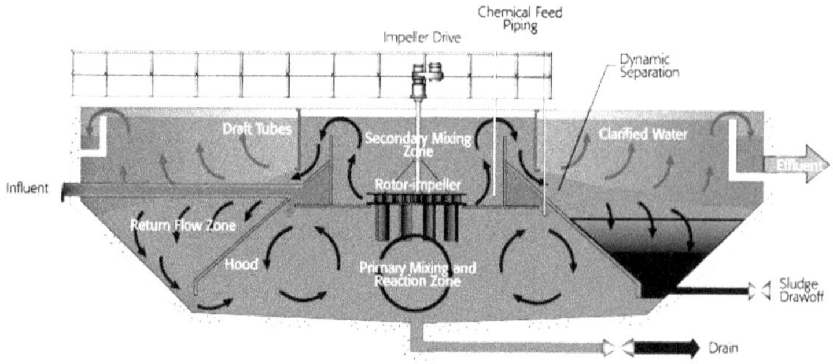

Source: Degremont Technologies

Figure 3-5 Diagram of a blanket clarifier

zone. A sloped dividing wall between the flocculation and sedimentation zones allows tapered flocculation within the flocculation zone while providing a decreasing upflow rate in the sedimentation zone.

In sludge blanket clarifiers, floc that settles in the sedimentation zone is allowed to accumulate in the bottom of the clarifier. The level of this accumulation is controlled by periodic removal of excess floc from the clarifier through the sludge withdrawal system. The accumulated floc is termed the "sludge blanket." As water from the flocculation zone enters the bottom of the sedimentation zone, it is forced to pass through the sludge blanket, which results in further contact between the floc leaving the flocculation zone and the floc in the sludge blanket, improving agglomeration of the floc and its separation from the water. The upward flow of water through the sludge blanket in the sedimentation zone causes the sludge blanket to expand. Some sludge blanket and solids-contact clarifiers are designed so the cross section of the sedimentation zone increases as water moves upward toward the weirs or launders. This causes the rise rate to decrease and is helpful in preventing the blanket from washing out of the sedimentation zone when the rate of flow through the clarifier is increased. Many clarifier basins are equipped with sample taps, which draw off at various elevations within the basin. This allows the operator to determine the location of the sludge blanket and the solids concentration within the blanket. Tools such as a sludge judge may also be used to evaluate the sludge blanket.

Another variation of the blanket clarifier process is pulsed flow, in which inflow to the clarifier basin takes place in periodic pulses, which are intended to impart an agitation to the floc blanket and enhance floc removal.

Blanket clarifier process units have been operated at overflow rates of about 1.8 gpm/ft^2 (4.4 m/h) for softening and about 1.0 gpm/ft^2 (2.4 m/h) for chemical coagulation applications. These overflow rates can be approximately doubled if tube settlers or plate settlers are added to the sedimentation zone. Further information on tube and plate settlers is provided in the previous sections.

Solids-contact and sludge blanket clarifiers have many operating variables such as recirculation rate, sludge blowdown, and coagulation chemistry. To make it easier to understand the operation, consider the three main connections of the tank (influent, effluent, and sludge removal) and how they will interact. What enters the basin (source water including contaminants and chemicals) must leave the basin in the only two outlet points available, namely the effluent or sludge-removal system.

Treatment chemicals should be reviewed first to verify they are being applied at the right dose to form an appropriate floc. This can be verified through jar testing. Without proper chemistry, successful operation will be difficult to achieve. Recirculation rate is the next parameter to consider once proper chemistry is in place. The rate should be sufficient to contact source water, treatment chemicals, and previously formed solids, which result in a floc concentration and size that settles faster than the upflow rate in the clarifier. Lastly, sludge blowdown or removal from the basin should be adjusted to extract solids at the rate they are being produced. The resulting operation will show consistent solids concentration in the mixer recirculation area, a stable sludge blanket level, and a consistent sludge-removal concentration. All of these can be observed through normal operating practices and monitoring at the recirculation column, sludge bed indicators, and sludge-removal discharge. If the influent water quality changes, adjustments in the system may become necessary—starting with review of chemical dose rate and followed by recirculation rate and, finally, sludge removal—to maintain stable operation. Systems operating on variable-quality supplies, such as river water systems, will require more frequent monitoring and adjustments.

Coagulation goals. Blanket clarifiers and reactor clarifiers are intended to operate on a continuous basis, 24 h/day, and if used at coagulation plants, they are most effective if they can be used in a base-loaded mode. When this process equipment is operated at a constant rate of flow, the blanket level within the sedimentation zone should not rise or drop substantially. If the rate of flow decreases, the rise rate decreases, causing the blanket level to drop. If the flow increases, the higher rise rate will cause the top of the blanket to move closer to the effluent weirs or launders. Daily increases and decreases in flow in a blanket clarifier or worse, on-and-off operation of such process equipment can cause serious problems, especially when floc in the blanket is only slightly denser than water. If a blanket clarifier is operated on an on-and-off basis, the mixing device and sludge scraper, if present, should remain in operation even when water is not flowing through the basin. This operating method aids in the restarting and stabilization of the process. Like conventional clarifiers, the performance of blanket clarifiers and reactor clarifiers also can be impaired by rapid changes in water temperature.

General maintenance. Maintenance of the clarifier mechanism is needed to keep the system performing well. As with other mechanical systems, some parts need more frequent attention than others do. There is variation of scheduled maintenance on the basis of the equipment manufacturer, so when possible, consult the maintenance manual provided with the equipment. The following is intended to provide some examples of common maintenance items:

- Grease bearings and chains on a regular interval, weekly to monthly. In addition to providing lubrication of wear surfaces, grease prevents intrusion of water or chemicals into the bearing internals.
- Check gearboxes for proper oil fill level. Replace oil per manufacturer's recommended interval.
- Torque sensors, commonly on scraper drive systems, should be checked for proper operation. The torque sensor is used to detect excessive torque and often to stop the mechanism to prevent catastrophic damage. Remove any chemical scale buildup. For electrically operated systems, ensure wire contacts are free of corrosion to maintain a functioning circuit.
- Observe sludge blowdown valve discharge to ensure operation and free discharge of sludge from the system. If the discharge system is automatically controlled, confirm the timers are opening and closing the valve as programmed.
- At least annually, drain the basin and inspect submerged components for wear and rust. While the unit is drained,
  - check hardware for tightness;
  - clean and touch up rusting areas per paint manufacturer's instructions;
  - remove scale buildup;
  - clean launder collection devices (orifices or weirs) and check for level;

- check bearings for signs of wear; and
- operate the unit and observe for wear points, balanced rotation, and appropriate noise level.

Troubleshooting. Table 3-4 provides general recommendations for troubleshooting solids-contact and sludge blanket clarifiers.

Contact Adsorption Clarifiers

Contact adsorption clarifiers (CACs) have been used as a standalone clarification process and in some pre-engineered (package) plants with a multimedia filter placed in series after the CAC unit. Hydraulic loading rates for CACs can be in the range of 5–15 gpm/ft^2 (12–36 m/h). The higher rates aid in-bed flocculation, and lower rates may actually hinder performance.

Table 3-4 Solids-contact and sludge blanket clarifier troubleshooting guide

Problem	Possible Cause	Inspection Items	Remedies
What is the issue?	*What might be causing it?*	*Items to check to confirm the problem*	*Actions to fix the problem*
• Mixer recirculation solids concentration too low	• Solids concentration in source water and/or basin decreasing • Plugged recirculation column • Mixer rotational speed too low	• Recirculation column solids <5% and cloudy (indicates plugging)	• Decrease sludge-removal frequency or duration • Increase mixer rotational speed to unplug column
• Mixer recirculation solids concentration too high	• Solids concentration in source water and/or basin increasing • Mixer rotational speed too high	• High concentration of solids (>20%) in the recirculation column (indicates solids inventory buildup)	• Increase sludge-removal frequency or duration • Decrease mixer rotational speed
• Solids carryover	• Solids buildup in the basin • Insufficient floc formation	• Check sludge blanket depth in basin • Confirm chemical feed systems are properly functioning and adjusted to water conditions	• Adjust chemical doses to achieve better floc formation • Increase mixer rotational speed • Remove solids more frequently or for longer duration
• High settled water turbidity	• Insufficient floc formation	• Assess floc formation through jar testing • Assess loading rate	• Adjust rapid-mix and coagulation scheme to achieve better floc formation • Reduce loading rate
• Sludge blanket too low	• Too many solids being removed from basin	• Assess solids concentration at the sludge-removal point	• Decrease sludge-removal frequency or duration
• Sludge blanket too high	• Solids inventory building up within the unit	• Assess solids concentration at the sludge-removal point	• Increase sludge-removal frequency or duration
• Solids collection mechanism will not operate.	• Solids are too thick. • Plugged or nonfunctioning sludge-removal system • Bearing or gearbox component failure	• Torque reading on scraper drive • Solids concentration at sludge-removal point • Listen to bearings and gearbox for sounds of grinding or wear	• Operate sludge-removal mechanism more frequently or for greater duration • Unplug or repair sludge-removal system • Repair bearing or gearbox components

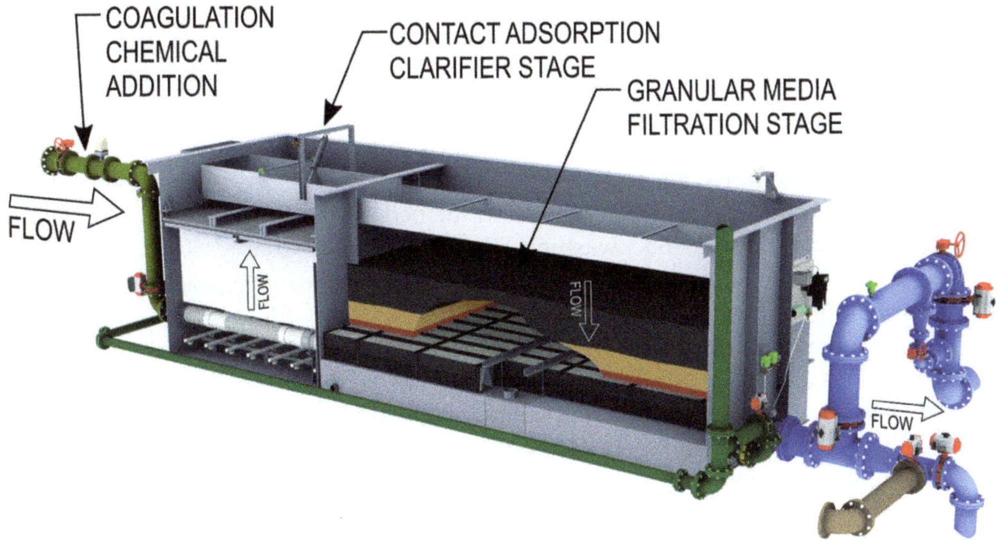

Source: WesTech Engineering

Figure 3-6 Contact adsorption clarifier and multimedia gravity filter in series

Components and function. The CAC generally consists of a bed, 3–4 ft (750–1,200 mm) deep, of coarse media, typically 1.5–3.0 mm, through which coagulated water passes. Before entering the CAC, coagulant is added to the water to begin floc formation. The flow of coagulated water through the coarse media bed aids in flocculation and removal of floc and prepares the water for filtration in a dual- or mixed-media bed. These clarifiers have been produced for operation in both pressure filtration and gravity filtration modes. The coarse media may be buoyant and held in place with a restraining screen system or nonbuoyant and supported by an underdrain system. Buoyant media systems operate in an upward flow direction, while nonbuoyant systems may be operating with either an upward or downward flow direction. Coarse media is typically vendor specific and often proprietary. Media can be plastic pellets or balls or gravel.

When used in smaller pre-engineered package plants, a CAC process unit is placed ahead of a multimedia filter, and the two processes are operated in series. This is shown in Figure 3-6, with water flowing from left to right. Coagulation chemical is added to the influent piping before water enters the upflow CAC and passes downward through the granular media filtration stage.

Coagulation goals. Attaining correct coagulation chemistry is as important, if not more so, for CAC units as it is for other clarification processes. Because of the higher hydraulic loading rate and relatively shallow basins, detention time within the CAC may be as low as 5–15 min. For plants treating cold water or with slow reaction kinetics, the coagulant should be added as far upstream as possible to maximize reaction time. Plants with extremely low influent turbidity or rapidly shifting influent water conditions may benefit form using a streaming current or zeta potential monitor to determine demand and adjust coagulant dose.

CACs have relatively few operating variables, with chemistry, hydraulic loading rate, and terminal head loss set point being the primary adjustable conditions. Chemistry (coagulant and possible polymer dose) can be jar tested to determine whether doses are appropriate. Particular care with respect to detention/reaction times should be used while jar testing for CAC.

The coarse media bed of a CAC performs similarly to a filter by capturing flocculated solids within the media. Unlike many other clarification processes, a large settling floc is not needed or desired because a too-large floc size will clog the coarse media bed and consume storage space. Small floc particles are readily captured within the bed, and due to the higher hydraulic loading rates, a polymer is commonly added to build floc strength to resist shearing forces within the media. Because the coarse media bed has a finite storage capacity, CACs are generally limited to low-turbidity sources with low coagulant demands. To remove the captured floc, the coarse media bed is often backwashed with a combined water and air wash to expand the bed, dislodge solids, and remove them from the CAC (refer to the "Backwashing Filters" section of Chapter 4 for more information). In some instances, the influent water is used for the cleaning process to minimize use of finished filtered water.

General maintenance. The CAC systems have relatively few components; however, each plays an important part in the operation of the system and must remain functioning at all times. Chemical feed systems should be inspected for proper dose rates and removal of sludge from feed tanks. Air wash blowers used for cleaning should be maintained per manufacturer's instructions. If pneumatic valves are used, air supply lines should be checked for moisture accumulation and cleaned out as needed. Any rusting or paint chips should be repaired as soon as possible to prevent excess corrosion. Head loss monitoring and turbidity instruments should be calibrated on a regular basis. The media should be removed on a regular basis for a full interior inspection. Water and air distributions and media should be inspected to ensure they are in operating condition.

Most clarification processes (except for conventional settling basins requiring manual sludge removal) are operated continuously, and the settled sludge or floated floc is removed while the clarifier remains in operation. However, as noted in the previous section, a CAC performs a filtration function, so a CAC requires periodic flow interruption to clean out accumulated floc. When head loss builds up faster in a CAC unit than in the granular media filter following the CAC, the flow of clarified water into the filter must be interrupted while the CAC is flushed to clean out accumulated floc if the CAC and the filter are piped in series. This can result in shutting off the filter when the clarifier is cleaned. For a small system operating a single CAC unit and filter, if the filter is not backwashed each time the CAC unit is taken out of service and cleaned, operators should be very careful when restarting the partially clogged filter after resumption of treatment. If possible, the system controls should be configured to initiate a filter-to-waste step at the restart of the treatment train. For treatment systems that combine a CAC and media filter in one treatment unit, the filter maintenance and troubleshooting should be performed as noted in Chapter 4 of this manual.

Troubleshooting. Table 3-5 provides a general troubleshooting guide for typical issues with the CAC process.

Ballasted Clarification

Ballasted clarification (typically called ballasted flocculation) is a process that applies a relatively high concentration of small-diameter sand or other weighting agents to the flocculation process to accelerate clarification. In this way, ballasted clarification systems can achieve settling in footprints that are approximately 1⁄10 the size of conventional flocculation and sedimentation basins and 1⁄5 the size of flocculation–sedimentation basins with plate settlers. Although this section focuses on sand ballasted clarification, alternate ballasting materials and recovery systems may be used. However, the basic tenets of these alternative ballast systems are similar to those described here. For example, ballasted systems using magnetite as the ballast material are available from some manufacturers, using a magnetic drum to separate the solids from the magnetic ballast.

Components and function. The ballasted clarification process consists of two or three coagulation and flocculation basins followed by a sedimentation basin. The coagulation

Table 3-5 Contact adsorption clarifier troubleshooting guide

Problem	Possible Cause	Inspection Items	Remedies
What is the issue?	*What might be causing it?*	*Items to check to confirm the problem*	*Actions to fix the problem*
• Turbidity reduction across contact adsorption clarifiers is low.	• Insufficient chemical addition	• Confirm chemical feed systems are properly functioning and adjusted to water conditions.	• Adjust chemical doses to achieve better floc formation.
• Turbidity breaks through before terminal head loss.	• Weak floc, insufficient polymer addition	• Confirm polymer feed system is properly functioning.	• Adjust polymer doses to achieve stronger floc.
• Terminal head loss is reached quickly with low influent turbidity.	• Overfeed of coagulant or polymer • Fouled media or retention system	• Confirm chemical feed systems dose rates align with jar test results. • Inspect media and retention system.	• Adjust chemical feed dose. • Clean or replace media or retention system.
• Cleaning operation is insufficient to return starting head loss to normal.	• Maldistribution of air or water	• Observe water surface during cleaning for boils or dead spots.	• Inspect and repair water or air distribution system components.

and flocculation basins are equipped with vertical mixers, and the sedimentation basin features a circular sludge-removal scraper. Ballasted clarification generally follows these treatment steps.

1. A coagulant (typically a metal salt) is added upstream of the flocculation basins to condition and destabilize incoming particles. pH adjustment may also be valuable in this step to improve coagulation, as described in Chapter 1.
2. Flow enters the first chamber of the flocculation basin and is mixed for 1–2 min to produce a small floc. The typical *G* value for flocculation is 150–300 s^{-1} (Young and Edwards 2003).
3. Additional mixing is provided in the second chamber of the basin for an additional 1–2 min. Sand is added, and contact between the sand and floc begins. Sand concentrations used in ballasted clarification are typically in the range of 3 or 4 g/L and up to 6 g/L.
4. Polymer is applied so that floc attachment to the sand can take place. Optimal location for polymer addition can vary, and it is common to have the flexibility of several points of polymer addition.
5. In the third chamber, further mixing for 3–6 min is applied to build the floc size around the sand ballast.
6. The high-density floc suspension that results is conveyed to the final chamber of the basin, the settling tank, where solids liquid separation takes place. Tube or plate settlers are typically used in the settling chamber to enhance the settling process. Loading rates typically range from 20 (49) to 45 gpm/ft^2 (110 m/h) or greater on the basis of the overall surface area of the plates or tubes at the top of the basin (not the projected surface area or individual surface areas of the plates or tubes).

A diagram of a ballasted clarification process train is shown in Figure 3-7.

A critical aspect of the process is collection of the sand or floc mass at the bottom of the settling unit for recycle to separation devices that will recover sand from this stream while discharging less-dense floc that is wasted from the system. In this manner, sand loss from the system is minimized. Hydrocyclones have been applied for this purpose in the

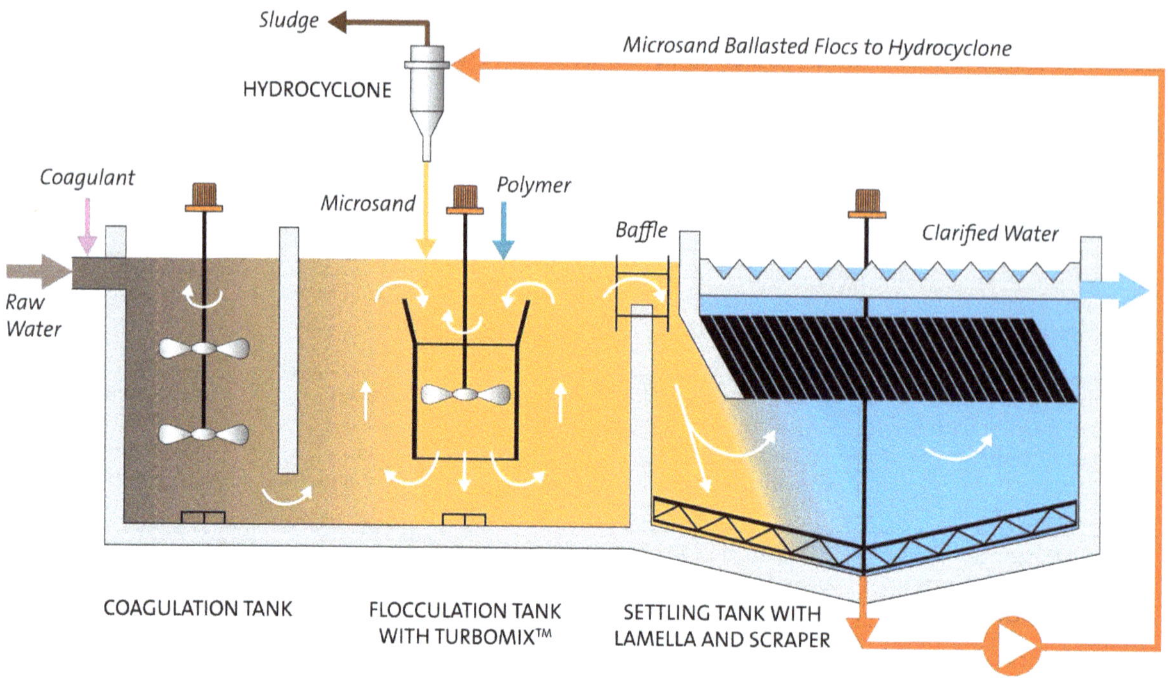

Source: Veolia Water Technologies

Figure 3-7 Diagram of the Actiflo Turbo ballasted clarification system, which combines the second and third mixing chambers into a single flocculation tank

commercial version of this process. Sand recovery using this approach can be relatively effective, requiring limited addition of about 1 mg/L or less to make up for overall losses across the system. Sand addition is typically done as a batch process, with sand being added once or twice per week as needed.

Short detention time in the mixing units combined with high overflow rates produces a clarification sequence that has a very small footprint. The commercial process has also been subject to recent innovations that are intended to improve hydrodynamic features, improve overall process dynamics, and further reduce process footprint.

Crystalline silica, typically found in the sand used in the ballasted clarification process, is a US Occupational Safety and Health Administration–regulated carcinogen because of inhalation hazard. While operator exposure to the dry sand in the ballasted clarification is typically infrequent, during sand replenishment operations only, care should be taken to reduce dust inhalation exposure whenever possible.

Coagulation goals. Optimized flocculation for ballasted clarification is significantly different than that for conventional sedimentation or plate settlers. The initial floc, before polymer and sand addition, should be quite small and compact. Large and fluffy initial floc will still bond with the sand but is often too delicate for the subsequent high-rate processes and often leads to significant turbidity carryover in the ballasted clarification effluent. Polymer is not typically added at the initial chemical conditioning step to avoid too much floc development before the sand is added.

After sand and polymer are added, larger floc will form, with the sand providing the central "seed" for the floc to form around. Again, in this stage, it is important to maintain more compact (although now larger) flocculated particles, as fluffy floc will increase risk of turbidity carryover in the sedimentation step.

Application of proper coagulant dose and pH is critical for this process, as is the case for all clarification processes. Accurate particle destabilization through charge neutralization is necessary to promote particle, sand, and bridging polymer interactions. Selection of an appropriate polymer (Van Cappellan et al. 2008), point or points of polymer application within the sequence, and polymer dose are also essential. Setting the polymer dose requires understanding of certain basic features of operation. A dose that is too small will not adequately attach floc to the sand, while excess polymer can carry over to downstream filters and potentially have an adverse effect on filter run times. Selection of sand size can also be important (Van Cappellan et al. 2008), and care should be exercised to avoid sand of excess size, which lacks surface area for bonding, or sand that is too small and does not provide enough ballast for rapid settling (leading to significant sand carryover). Typical sand grain sizes range from 60 to 120 μm. A well-formed floc ideally suited for settling will include a tight mesh of sand, suspended solids, and bridging polymer.

General maintenance. Typical inspection of a core ballasted clarification system components should include the following areas:

- Inspect the hydrocyclones for proper function and discharge. The discharge from the hydrocyclones should be conical and regular in shape, indicating no obstructions. No significant vibration should be observed at the hydrocyclones. These liners require periodic replacement.
- Observe the floc formation in the process tanks, if possible. Initial floc in the first chamber should be small and well formed. Subsequent floc after sand and polymer addition should be larger but still tightly formed.
- Inspect sand recycle pumps for proper function. Observe evidence of excessive heat or vibration. Correct any leakage for seals or pipe fittings.

It is recommended that influent and effluent turbidity be carefully monitored alongside other water quality parameters, including chemical doses and pH, such that future changes in source water quality can be interpreted through the lens of prior experience.

Because of its compact size, changes to the chemical dose in ballasted clarification can be made relatively frequently. Typical detention times in ballasted clarification systems are 30 min to 1 h. As such, any changes made to the chemical conditioning upstream and within ballasted clarification should be tested for 1–2 h before further optimizing changes are made. Because of this relatively long duration, it is highly recommended that only one process variable be changed at a time while all others remain constant.

Troubleshooting. Table 3-6 provides a general troubleshooting guide for typical issues with the ballasted clarification process.

Dissolved Air Flotation

As opposed to other clarification processes, in which the solids are removed from the bottom of the clarifier, in DAF they are removed from the top. As a DAF clarifier is operated, floc is carried to the top of the water by the air bubbles, where the floc accumulates. A layer of solids called *float* forms on the water surface and gradually grows thicker as the operation continues. The blanket of sludge is supported from beneath by tiny air bubbles.

The DAF process is usually considered when the source water is characterized by low consistent turbidity, such as from a reservoir, and the need to coagulate the water to reduce organics and/or color with hydroxide flocs with inorganic coagulants. In many locations, the source water would be a candidate for direct filtration if not for the required coagulant dose to achieve organics and/or color reduction goals that would overload filters if not clarified. The DAF process is the best available coagulation-removal process for algae reduction because it takes advantage of algae's natural buoyancy to remove it. Gregory et al. (1999) reported that removal of the algae *Microcystis* was 98 percent, as compared with 76 percent through sedimentation. Preoxidation should be avoided with DAF facilities to reduce the likelihood of taste-and-odor compound or cyanotoxin release

Table 3-6 Ballasted clarification process troubleshooting guide

Problem	Possible Cause	Inspection Items	Remedies
What is the issue?	*What might be causing it?*	*Items to check to confirm the problem*	*Actions to fix the problem*
• Elevated ballasted flocculation effluent turbidity	• Incorrect coagulant or polymer dose • Incorrect coagulation pH	• Check chemical feed equipment and doses • Observe shape and character of floc formation • Check sand concentration	• Adjust chemical doses • Change location of polymer addition to increase or decrease time and/or separation between coagulant and polymer addition • Perform jar testing to identify optimal chemical doses • Add more sand to the system if sand concentration is low
• Excessive sand loss	• Clogged hydrocyclone • Incorrect polymer dose • Insufficient sand size	• Check hydrocyclones and confirm they are flowing freely • Check chemical feed equipment and chemical doses	• Clear obstruction from hydrocyclone tip • Refer to previous troubleshooting step for optimizing chemical doses • Consider replacing sand with larger sand size
• Sand pump pressure loss	• Pump inlet clogged	• Check hydrocyclone discharge to confirm pump is flowing • Check pump flowmeter (if available) to identify any loss in flow • Confirm pump inlet is free from obstructions	• Flush sand pump influent piping with water • For major clogs, tank must be drained, and influent piping disassembled

from algae cells. Additionally, research has shown that DAF can be as or more effective than conventional sedimentation for the log removal of giardia- and *Cryptosporidium*-sized particles (Plummer et al. 1995).

The presence of heavier particles like silt from runoff in rivers tends to reduce the efficacy of DAF, as high concentrations of turbidity-causing mineral particles are difficult to float when coagulated. DAF can also be used in the treatment of spent filter backwash water that contains high-turbidity, low-density floc from the filters that is readily floated in the DAF process. There are several pilot tests and full-scale installations of DAF for this application in North America.

Components and function. Figure 3-8 shows a schematic of the DAF system and its individual components. The DAF process starts by feeding coagulant to the source water by either a hydraulic or mechanical rapid-mix system designed to achieve rapid dispersion of the coagulant. This is followed by flocculation in one to three stages, with two-stage flocculation being the most common. Total flocculation hydraulic retention times generally vary between 5 and 20 min of design flow, with tapered velocity gradients (G) from 50–100 s^{-1} in the first flocculator to 20–50 s^{-1} in the last flocculator. Flocculation is most typically achieved with either hydraulic baffles (passive) or vertical mixing (mechanical) for DAF systems. The typical flocculation velocity gradients are higher than conventional sedimentation but with less detention time. This is to allow the creation of a smaller pin floc that will readily attach to the microbubbles that are introduced in the DAF clarification area.

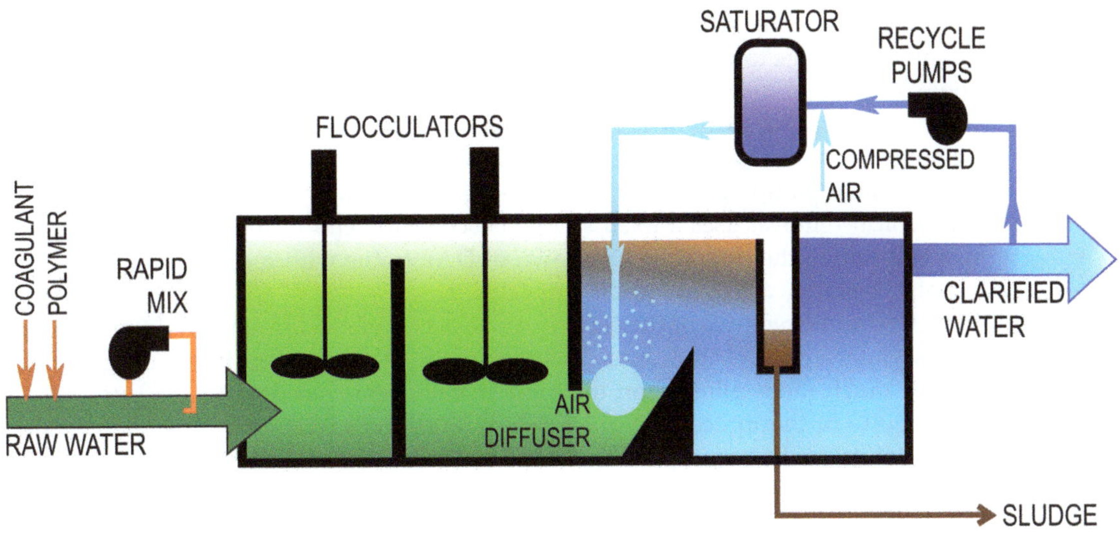

Source: Carollo Engineers

Figure 3-8 Schematic of a typical DAF clarification system

Polymers are often required in the DAF process, most typically as either a coagulant or flocculant that is added in the first stage of flocculation. Having multiple injection locations for the flocculation polymer is important for process optimization of DAF. Dose and performance can vary significantly between the addition of a flocculant polymer at the beginning of the first stage of flocculation and the end of the first stage of flocculation. The polymer is often needed because particles flocculated with metal salt-based coagulants tend to break apart when the air is added to the process. Polymer addition of the right type and dose provides a more robust floc that does not break apart in the turbulence of microbubble injection. Lower-molecular-weight polymers are most frequently used in a DAF process because higher-molecular-weight polymers tend to form larger and/or denser flocs that are also not ideal for flotation.

After a floc is formed, the flocculated water stream enters a "reaction zone," where water that has been saturated with air is added via an array of nozzles. In the DAF process, a clarified water sidestream of 5–15 percent of the source water flow is pumped into a saturator (pressure vessel), where it is mixed with high-pressure air (60–100 psi or 415–690 kPa) and then stored in the bottom of the saturator tank. From the bottom of the saturator tank, water supersaturated with air can be returned to the DAF clarifier reaction zone, where the pressure is dropped to near atmospheric through specially designed orifices or nozzles in a distribution header that spans the width of the tank to generate microbubbles in the size range of about 10–100 μm to float flocs (Zabel 1985). The microbubbles combine with the floc and float them to the surface in the separation zone (also commonly referred to as the "flotation zone") downstream of the contact zone. While conventional DAF systems were originally designed at separation zone loading rates near 4 gpm/ft^2 (10 m/h), depending on the source water to be treated and the design of the DAF, the separation zone loading rates of current high-rate DAF systems can typically vary from 4 (10) to 12 gpm/ft^2 (29 m/h), with rates up to 16–20 gpm/ft^2 (39–49 m/h) under certain conditions. In spent filter backwash applications, typical loading rates start in the conventional DAF range of 4–6 gpm/ft^2 because of the much higher solids loading in equalized spent filter backwash water.

As the system operates, the accumulated floc or float becomes so thick that it must be removed from the water surface before it begins to break apart and sink back down into the clarified water. The float is removed periodically either by a mechanical system or by hydraulic means. While more complex, mechanical sludge-removal systems usually result in obtaining sludge of 2–4 percent dry weight basis. This high solids content reduces the quantity of sludge to be processed in the residuals-handling portion of the plant; however, some plant water may have to be added back to the thicker sludge attained with mechanical sludge removal to allow it to flow from the collection trough into pumping units and be pumped to subsequent processes. Additionally, float of this thickness requires positive displacement pumping, such as hose or progressing cavity pumps. Additionally, because mechanical removal uses a fixed sludge beach at the end of the float zone, which has a fixed elevation, the hydraulic turndown of DAF trains with mechanical float removal is usually limited to a range of approximately 2:1.

Hydraulic removal of the sludge can be accomplished in several different ways, but all use the water level surface to discharge the float over a weir into a trough. This float will be much more dilute than the mechanically removed float (0.1–0.2 percent solids) and require further processing or treatment if residuals treatment is handled on site. The dilute solids can be pumped using centrifugal pumps. If the float is being discharged to a sanitary sewer, the dilute nature of the float is not a major concern.

There have been DAF plants installed that incorporate both a mechanical and hydraulic removal system to allow the operator flexibility. These are rare and increase the system complexity.

Recycle rate. If flow through the process is changed, the recycle rate of flow may need to change as well to maintain a similar percentage of recycle. If changes to the recycle rate are made, care must be taken to ensure that flow through the nozzles and/or dispersion header is within the nozzle operating range. Often, multiple recycle headers are constructed to provide flexibility in adjusting recycle rates to ensure optimal operation of the nozzles and subsequent formation of small bubbles. Bubbles that are too large have a greater tendency to disturb floated floc and prevent floc from rising to the top of the water. A typical design would be two dispersion headers that are designed as one dispersion header to achieve 40 percent of the maximum recycle flow, a second dispersion header to achieve 60 percent of the maximum recycle flow, and the two together to achieve 100 percent of the design recycle flow. Depending on the number and arrangement of the recycle pumps, they can be constant speed or variable speed. Most DAF systems today use variable-speed drives to allow for flexibility in recycle flows while minimizing energy usage. The recycle pumps have the largest energy usage in a DAF system.

Saturator pressure. Another operating parameter that affects process performance is saturator pressure. A sufficient volume of dissolved air should be provided to form air bubbles capable of lifting the suspended solids in the flocculated water to ensure successful operation of the clarifier. This is attained by a combination of the percentage of recycle flow through the saturator, the saturator transfer efficiency, and the air pressure maintained in the saturator. The saturator transfer efficiency is typically defined by the type of saturator (packed versus unpacked) and the method of the air injection into the pressurized water. The higher the saturator pressure, the more air that can typically be dissolved into solution. The warmer the water, the less air that can be dissolved into solution. So, for systems that have a significant water temperature swing, small adjustments in saturator pressure may be needed to obtain similar performance year-round.

Higher concentrations of suspended solids—such as algae or turbidity in the source water or solids in spent filter backwash—may need a greater volume of dissolved air for effective separation and flotation of flocculated solids. In these cases, the amount of air required is typically defined by the air–solids ratio.

Coagulation goals. Several key operational parameters affect DAF performance, including

- flocculation velocity gradients,
- polymer application,

- pretreatment coagulation chemistry,
- recycle rate, and
- float removal (type and frequency).

Correct management of pretreatment chemistry is a must for effective operation of a DAF clarifier, and chemical dosages can be determined by flotation jar tests, as described in Chapter 7. Because particles are floated, it is generally asserted that formation of smaller flocs is preferential to the larger particles typically formed in conventional clarification. Optimal coagulation conditions for a DAF system are those of coagulant dose and pH that produce flocs with a charge near zero and with relatively high hydrophobicity (Edzwald and Haarhoff 2011). In a pilot plant evaluation of both direct filtration and DAF followed by filtration (Black and Veatch 1994), the same pretreatment chemistry that was working well for direct filtration was used when the DAF clarifier was brought on-line for testing, and very effective treatment was attained in the DAF unit.

General maintenance. Maintenance tasks on a DAF system are primarily focused on the mechanical systems, including the recycle pumps, vertical flocculators, air compressors that maintain pressure in the saturator, and the mechanical sludge-removal system. Normal routine maintenance on this equipment is required to keep the DAF system operational. Loss of the recycle pumps or air compressors will mean failure of the treatment system. Redundancy and spare parts inventory for these systems should be considered critical. The mechanical sludge-removal system's maintenance items will depend on the type of system used, which can vary by vendor.

Intermittent maintenance items will include the following.

- Inspection, draining, and cleaning of the saturator to verify that there is no buildup of precipitates and slimes. This can be more significant for packed saturators, as the packing provides more surface area for growth and deposits, which can affect transfer efficiency of air bubbles into the water.
- Inspection and cleaning of the nozzles that deliver supersaturated water into the contact zone for buildup of solids that could affect the microbubble formation. This requires either a draining of the tank or removal of the dispersion header.
- Draining and cleaning of the flocculation and float zones to remove any settled materials that may be present. Depending on the outlet configuration, several inches of solids could be found in the bottom of the float zone.
- Adjustments to the mechanical sludge-removal system. Over time, because of wear on elastomers, expansion and contraction of materials, or other issues, mechanical sludge-removal systems tend to need adjustments to maintain the efficiency of the float removal. This is because of the tight tolerances needed to remove the thin layer of float without affecting the process performance.

Float removal method and frequency. As time passes, the float next to the clarifier wall may begin to adhere to the wall. If this happens, spraying water on the clarifier wall for a short time before the float is removed and in the early stage of removal can help alleviate this condition.

Troubleshooting. Table 3-7 provides a general troubleshooting guide for typical issues with the DAF process.

SLUDGE COLLECTION AND REMOVAL

Either sludge accumulated in clarifiers can be manually cleaned, or a mechanical collection system can be installed. If the basin is to be manually cleaned, basin bottoms typically slope toward the drain at a slope of 1 ft over 12 ft. Mechanical collection systems are subject

Table 3-7 DAF process troubleshooting guide

Problem	Possible Cause	Inspection Items	Remedies
What is the issue?	*What might be causing it?*	*Items to check to confirm the problem*	*Actions to fix the problem*
• DAF effluent turbidity is rising	• Source water changes, suboptimal flocculation, ineffective flotation because of lack of microbubble formation, float layer too thick, hydraulic loading rate too high, recycle rate too low	• Verify any changes in source water conditions • Inspect flocculators and confirm speed or *G* value • Observe float zone for "milky white" color at the front of the float zone after scraping. Observe mechanical scraping mechanism to record frequency of scraping and if the system is effectively removing the float	• Adjust coagulation chemistry • Increase flocculant polymer • Change over to standby saturator if microbubbles are not being formed effectively • Increase scraping and float removal frequency • Adjust mechanical scraper levels
• Lack of whitewater microbubbles in the contact or float zones	• Problem within the recycle–saturator–dispersion nozzle system, oxygen deficit in the source water	• Observe pressure in recycle system; is it adequate? • Is the air pressure in the compressor above the recycle pressure set point? • Is the compressor cycling excessively? • Is the whitewater issue only with a specific dispersion header on-line?	• Switch to standby unit(s) • Switch to second dispersion header and remove the first for inspection and/or cleaning of nozzles
• Float is hard to remove	• Float too thick	• Determine frequency and duration of float removal • Inspect elevation of mechanical removal scraper compared with water surface level of a cleared basin	• Increase frequency and/or duration of the float removal • Adjust the mechanical scraping mechanism lower to remove more volume of float

DAF—dissolved air flotation
Further information on DAF can be found in *Dissolved Air Flotation for Water Clarification*, by James K. Edzwald.

to harsh operating conditions (wear caused by grit, corrosion, clogging) that can result in the need for frequent maintenance. There are several types of sludge collection systems that can be installed in sedimentation basins. However, most mechanical systems fall into the following two main categories:

- paddle-type sludge removal that scrapes solids from the basin floor or
- suction-type sludge removal using tubes or perforated pipe to vacuum solids from the basin floor.

Operation of the sludge-removal systems is generally performed using a timer that establishes the frequency of operation. If the operation is too frequent, only water will be removed, and if the operation is not frequent enough, too much sludge could build up, resulting in potential floc carryover into the settled water, inability to remove sludge, or damage to the sludge collection equipment. The timer frequency is generally established by obtaining frequent samples of the sludge and measuring total suspended solids (TSS). Doing this on a regular basis helps establish a baseline for the plant, which can be used with less frequent sampling as data and understanding progress. A general goal is to have a minimum solids concentration of 0.2–0.3 percent when operating the sludge collectors. As such, timing may need to be adjusted to a longer duration to achieve this minimum solids concentration during periods of low source water turbidity. Conversely, during periods of high source water turbidity, the frequency of operation must be increased to ensure that solids do not become too thick.

Regardless of the removal technology, it is critical that sludge levels remain well below the effluent outlets to avoid short-circuiting; deteriorated clarified water quality; and within the operating parameters of the removal equipment, to prevent damage to the equipment. In addition, poor management of settled solids can lead to carryover of solids to downstream liquid treatment processes, pumping of excessive water to downstream solids-handling processes, anoxic sludge, solids too thick to pump, or loss of basin volume leading to short-circuiting.

Paddle-Type Sludge Collectors

Paddle-type sludge collectors include chain-and-flight type revolving sludge collectors, oscillating sludge scrapers, and circular sludge collection mechanisms.

A typical chain-and-flight system is shown in Figure 3-9. This system uses a series of paddles (flights) to push the settled sludge to a sump, where it is then conveyed either by pumping or by gravity for subsequent handling. In many instances, this sump will have a sludge collector as well. Movement of the flights is achieved using sprockets and chains. Over the past 10 years, there have been significant advancements in the materials of construction of the chain and sprockets using polymeric materials, and the reliability has increased significantly.

Oscillating sludge scrapers work by using a series of paddles on the basin floor that work similarly to chain-and-flight collectors. However, instead of a single paddle traversing the entire length of the basin, oscillating sludge scrapers work by moving sludge incrementally from one paddle to another. In this way, the sludge is moved incrementally across the basin floor from paddle to paddle and ultimately to a sludge sump, where it can then be conveyed to downstream solids-handling facilities.

Suction-Type Sludge Collectors

A typical suction type of collector is shown in Figure 3-10. This system uses perforated pipes on the bottom of the basin that move along the bottom of the basin using a cable-and-pulley system. The perforated pipes are connected to a header that conveys the settled sludge to the residuals-handling system.

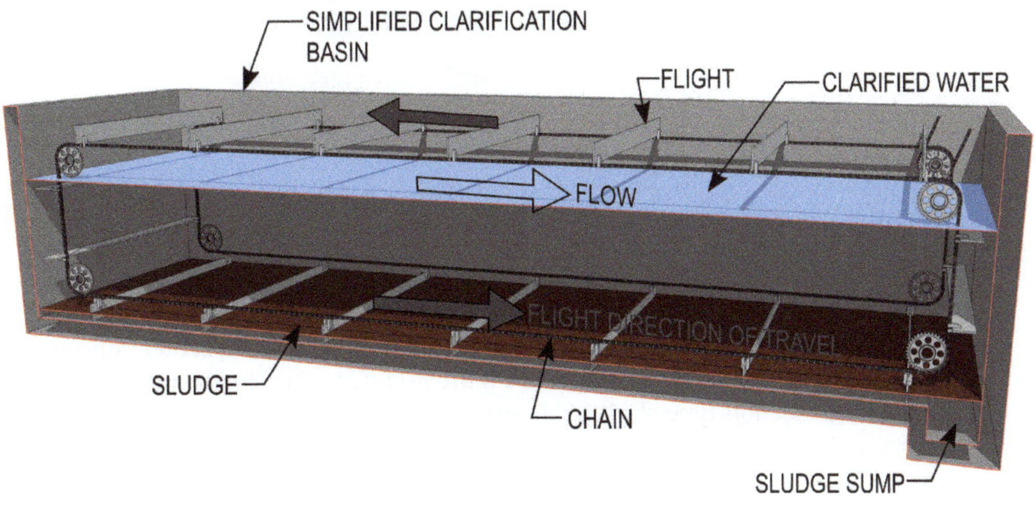

Source: Carollo Engineers

Figure 3-9 Diagram of typical chain-and-flight system

Source: Meurer Research

Figure 3-10 Diagram of suction-type sludge collector system

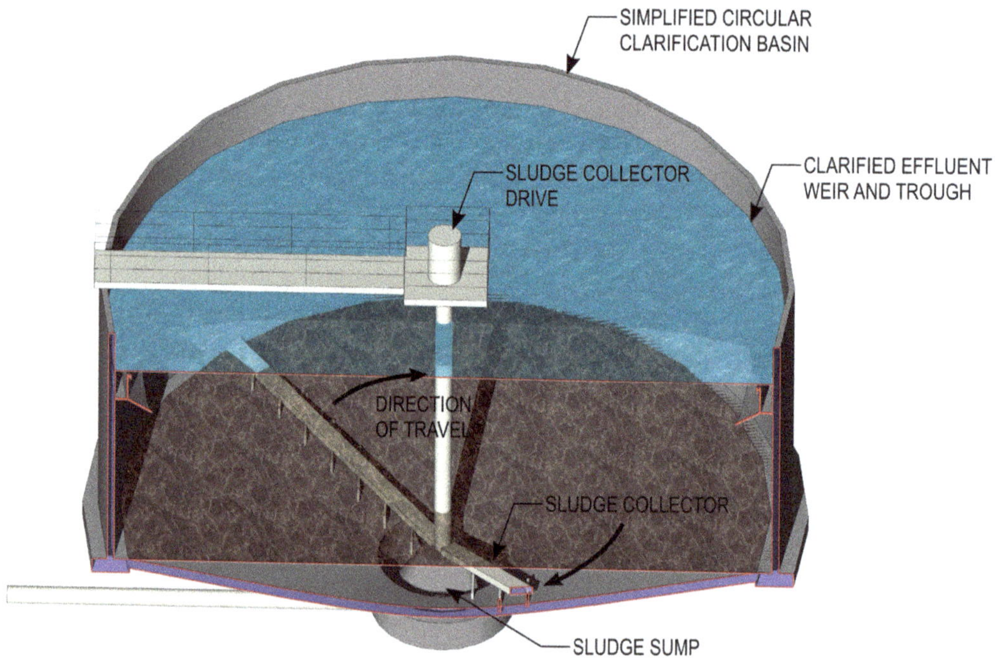

Source: Carollo Engineers

Figure 3-11 Circular sludge collection system

Circular Clarifiers

Circular clarifiers and square or rectangular basins in some instances use a circular type of sludge collection system that moves the settled solids to a center collection sump. A typical circular collector is shown in Figure 3-11. Circular sludge clarifiers can use either paddle- or suction-type sludge removal; however, instead of operating linearly across a rectangular basin, a circular sludge collection system works by fixing rake or suction arms to a rotating central column and scraping or vacuuming the sludge off the circular cross section of floor.

OPTIMIZATION OF CLARIFICATION PROCESSES

Optimized clarification is critical for a well-performing treatment plant. Optimized clarification effectively removes turbidity, reducing solids load on downstream processes like filters, and minimizes chemical, energy, and water use. While the results of the flocculation and sedimentation processes are seen in the clarification process effluent, often the cause of poor performance is found upstream in the coagulation and flocculation stages. However, several key factors, unique to the clarification stage, can be optimized to create better treated effluent and more efficient plant operations.

Optimization Goals

Optimized clarification balances chemical, energy, and water use and waste with low clarified water effluent turbidity and low operator intervention. However, the exact balance point across each of those factors is often unique to each plant's specific requirements and operational goals.

Many clarification success factors are typical across most treatment plants and clarification technologies. This includes reducing effluent turbidity, effectively removing solids from the process stream, avoiding excessive liquid residuals, and accommodating changes in influent water quality without significant operator attention.

Other success factors are often unique to the optimization goals of the treatment plant or operating agency. For example, at a typical conventional water treatment facility, the primary goal of the clarification process may be to provide adequate treatment to reduce filter influent turbidity and maximize filter unit filter run volume (UFRV). Low settled water turbidity is often not enough to achieve this goal and must be coupled with optimized chemical conditioning of the water so as to not clog the filters with chemical polymers in lieu of solids. As mentioned in Chapter 1, one of the best measures of granular media filtration plant performance is the UFRV, which will reflect optimization of the coagulation and clarification processes.

To achieve success in the clarification process, plant operations staff should consider their specific optimization goals for the overall plant and tailor the operational goals for each individual process to meet those unique criteria. There are many different goals for optimization of clarification, but a few references are available to use as targets. The AWWA Partnership for Safe Water uses a settled water turbidity target of 1 ntu; the US Environmental Protection Agency (USEPA) Area-Wide Optimization Program uses a tiered goal of 1 ntu 95 percent of the time when the annual average source water turbidity is less than 10 ntu and a goal of 2 ntu 95 percent of the time when annual average source water turbidity is greater than 10 ntu (USEPA 2004).

Short-Circuiting in Basins

The ideal flow pattern for water in flocculation and clarification basins is plug flow, in which the detention time for all portions of water is the same from entry to exit of the basin. In actual practice, attaining perfect plug flow is not practical, but designers attempt to approximate this flow regime when designing basin inlet and outlet configurations and basin baffling. When plug flow is not attained, short-circuiting is said to occur. In this flow condition, some water passes through the basin in a time shorter than the theoretical detention time (basin volume divided by rate of flow), and other water remains in "dead zones" or zones of little circulation and flow for times that can be considerably longer than the theoretical detention time. Serious short-circuiting problems can impair process basin performance. Such problems can be continuous due to basin design, or they can be intermittent as a result of environmental and water quality conditions. Strategies for dealing with continuous and intermittent short-circuiting are different.

Intermittent short-circuiting problems can be caused by sustained winds and by water quality changes. Just as a sustained high wind blowing for hours in one direction can cause turnover in a large lake, it also can do that in large, open settling basins, especially if the wind is blowing in a direction parallel to the long dimension in the direction of flow of the basin. As with lakes, the longer the distance the wind can blow over the water, the greater the effect on flow patterns in the water. Wind-induced circulation in a sedimentation basin can upset the flow regime and induce short-circuiting.

Short-circuiting in sedimentation basins can be caused also by density differences in water. For river sources, sudden high-turbidity episodes can result in a higher load of suspended solids and slightly greater density of water entering the basin. When this happens, the denser water can go to the bottom of the basin, displace water of lower density, and flow more rapidly to the basin outlet. Rapid changes in water temperature can be induced in lakes by sustained strong winds blowing onshore or offshore and causing turnover. If warmer water suddenly enters a basin, its lower density induces it to pass over the colder, denser water and reach the outlet sooner than normal. A change to colder, denser water would have the same effect as a density current caused by high concentrations of suspended solids, with the colder, denser water entering the basin, sinking to the bottom, and flowing to the outlet in a shorter time than normal.

Intermittent short-circuiting problems are difficult to deal with, but one possible remedy is to decrease the rate of flow in the plant, allowing more time for sedimentation to occur.

If short-circuiting problems impair process performance on a continuing basis, an investigation of short-circuiting may be undertaken. A preliminary approach to this for sedimentation basins is to view the basin from a high point, looking down on the water. When water bodies are viewed from above, changes in color or turbidity are detected much more readily than when water is viewed from ground (water) level. Such high-level viewing might involve observing a basin from the roof of the highest building at the treatment plant or possibly an overview from an aircraft or drone.

With the advances in drone technology, it is now possible to conduct multiple visual inspections for short-circuiting. Software for operation of drones allows for the establishment of waypoints and flight paths that can be repeated. The repetition of the same flight path helps to ensure that a library of inspections and basin performance can be created. Improvements in camera technology also allow for drone observations to include thermal imaging, multispectral imaging, and global positioning system (GPS)–linked high-definition imaging that is survey quality for geographic information system integration.

There are also water-based drones that are essentially a radio-controlled boat with depth sensors and a GPS system. This allows the unit to measure water depth at various points in the settling basin. Those values can be input into survey software packages to create a model of the current water depth. If the basin floor (bottom) is known, the difference between the "constructed" basin depth and the drone collected "water" depth would represent the depth of settled material. Applying the differences across the model of the entire basin at all collected points, the overall volume of settled material to be removed can be calculated. This will also generate a visual representation of where there is more flocculant carryover or short-circuiting if the settled material depths are not uniform over the area of the settling basin. In previous versions of this manual, Hudson (1981) recommended comparing two photographs for comparison of floc patterns. The pattern of the floc clouds can be an indicator of where unequal flow distribution is occurring. Drones and digital media, however, can produce hundreds of images and video to be used for floc pattern determination.

When visual observation has indicated a potential short-circuiting problem, the extent of the problem can be documented by tracer studies. Kawamura (2000) recommends using chemical tracers at concentrations that are safe from a public health perspective and that do not interact with any constituent in the water. Previous tracer studies have used compounds such as lithium and fluoride. However, fluoride can complex with aluminum (Al) when alum is used in high dosages, and lime softening has the potential to remove fluoride. If a tracer study reveals serious short-circuiting, the visual observations of floc clouds and tracer results may be good guidelines for making changes to inlet or outlet configurations. Tracer test procedures may need to be reviewed by drinking water regulatory agencies before such tests are performed. If treatment plant staff are unable to visually identify where short-circuiting is occurring, another option for addressing such problems is to obtain the services of a firm that can provide computational fluid dynamics modeling and advise on how to remedy the short-circuiting. If basins are covered, visual observation of floc patterns would not be possible, and conducting tracer tests could be more difficult because of restricted access for sampling. A tracer study conducted with a compound for which the varying concentrations can be tested could have the resulting concentrations plotted to produce a model of hydraulics. This type of tracer study would not require visual observation.

INFLUENCE OF PRETREATMENT CHEMICALS

The effect of coagulant chemical selection on solids production can be calculated using the following equation. To calculate total daily solids production (in dry pounds of solids per day), the following equation can be used:

$$\begin{gathered}\text{Solids Production}\ (\text{dry pounds/day}) = \\ 8.34 \times Q \times (F \times \text{Coagulant} + \text{TSS} + \text{Polymer} + \text{TOC})\end{gathered} \qquad \text{(Eq 3-1)}$$

Where:

8.34	=	A conversion factor from flow in million gallons per day and doses in milligrams per liter to pounds per day
Q	=	Plant flow in million gallons per day
F	=	A conversion factor for the amount of solids generated per milligram per liter of coagulant added: This value generally ranges from 0.26 to 1, depending on the coagulant. Kawamura (2000) suggests using • 0.26 for aluminum sulfate as dry alum, • 0.54 for ferric sulfate ($Fe_2(SO_4)_3$), • 0.66 for ferric chloride ($FeCl_3$), and • 0.0489 × aluminum concentration for polyaluminum chlorides. Some studies have suggested higher values. Side-by-side jar tests with varying doses of coagulant on the same sample can be used to determine the actual coagulant conversion factor for a particular coagulant and source water.
Coagulant	=	Primary coagulant dose, in milligrams per liter: If additional coagulants are used, they should be multiplied by their unique coagulant factor and added as well.
TSS	=	TSS removed, in milligrams per liter: From plant TSS data or approximated by multiplying turbidity by a conversion factor, which typically ranges between 1 and 2.5.
Polymer	=	Polymer dose in milligrams per liter
TOC	=	Average total organic carbon removed, in milligrams per liter

As shown in the equation, the coagulant selection can have a significant effect on the mass of solids production and the associated volume of solids produced. For example, the amount of solids contributed per milligram per liter of aluminum sulfate is approximately half that as from the same milligrams-per-liter basis of ferric sulfate.

Additionally, use of large amounts of coagulants and polymers can have a significant effect on the quality of sludge produced, especially if low TSS are present. Chemical sludges, such as those produced through enhanced coagulation processes and in which coagulant doses are often quite high, tend to be less dense with a lower viscosity than typical solids produced from source waters with higher concentrations of suspended solids. These viscous chemical sludges can be difficult to remove, depending on the sludge-removal mechanism in the sedimentation process. Longer-duration, lower-flow rate sludge blowdown activities can be effective in improving chemical sludge removal.

INSTRUMENTATION AND MONITORING

To help ensure continuous stable operations throughout the clarification process, there are various types of instrumentation and monitoring that can be used. Like most aspects of the treatment process, each stage or process is dependent on the performance of the previous stage. Clarification is no different. A stable, efficient, and well-operating clarification stage will be dependent on optimized coagulation and flocculation. Because of this relationship, some of the most important instrumentation for successful clarification falls under proper use of streaming current monitors, pH probes, flow indicators, coagulant dose control through supervisory control and data acquisition systems, raw turbidimeters, and raw organics analyzers (TOC or ultraviolet absorbance at 254 nm [UV254]).

There are additional types of instrumentation and monitoring that can be used to help ensure optimized performance through the clarification process. This includes influent and effluent turbidity measurement, influent and effluent pH measurement, settled sludge concentration, and sludge depth, to name a few. From a process monitoring standpoint, there are sensors measuring status of various mechanical devices, valve positions, and gate levels that can lend a great deal of control to the effectiveness of the clarification process. New technologies are being developed at a greater rate than ever before. This, coupled with computing power, means that extremely complex modeling and sensor algorithms are now possible or will soon be available to be implemented and used for optimization.

Table 3-8 provides information related to instrumentation and monitoring.

Table 3-8 Instrumentation and monitoring

Instrumentation and Parameters	Location	Units and Range	How It Is Useful
Streaming current or zeta potential	Coagulation—rapid mix	Near-neutral particle charge	Helps to optimize coagulant dose for influent clarification conditions (see Chapter 1)
Source water turbidimeters	Precoagulation	Nephelometric turbidity unit; range specific to source water conditions	Optimize coagulant dose for best-performing floc formation to improve influent clarification conditions
Source water TOC	Precoagulation	Milligrams per liter; range specific to source water conditions	TOC removal may be regulated depending on concentration. Higher TOC concentrations often lead to higher required coagulant doses (see Chapter 2).

(continued)

Table 3-8 Instrumentation and monitoring (*Continued*)

Instrumentation and Parameters	Location	Units and Range	How It Is Useful
Clarified water turbidimeters	Clarified or settled effluent	Nephelometric turbidity unit; range typically 1–5 ntu depending on performance (ideally <2 ntu)	Indicates whether settled–clarified turbidities meet regulatory requirements and can be first indicators of mechanical failures of DAF, rakes, or various mechanical components in the settled–clarification process
Particle counters	Clarified or settled effluent	Size range and total counts	Helpful in determining size range of removed particles. Provides additional information to be used with turbidity results
pH	Coagulation and clarified or settled effluent	0–14	Primarily used to determine optimized coagulation conditions but can be used in the settled–clarification process to determine pH stability and monitor for pH shift
Clarified TOC	Clarified or settled effluent	Milligrams per liter; range specific to source water conditions and treatment goals	Indicates effectiveness of clarification process for TOC removal, which can be critical for reducing disinfection byproducts
Clarified UV254	Clarified or settled effluent	Reciprocal centimeters; range typically 0.01–1.0 cm^{-1}	Often used as a surrogate for clarified TOC; indicates effectiveness of clarification process for TOC removal, which can be critical for reducing disinfection byproducts
Sludge depth	Settling basin	Feet or meters; range specific to basin configuration	Measures rate of sludge accumulation and indicates frequency of sludge removal
Sludge concentration	Sludge removal effluent–discharge	Milligrams per liter of TSS or %, typically 2,000–20,000 mg/L TSS (0.2–2%)	Confirms effectiveness of coagulation and informs frequency of sludge removal to minimize excess liquid residuals
Level indicators	Clarification–sedimentation basins	Elevation or depth in feet and inches	Water levels affect settling velocities and floc shear and can indicate failures of sediment removal systems.
Flowmeters	Tends to be calculations derived from flows or hydraulic levels before and after settling	Gallons per minute or cubic feet per second	Because of the nature of settling, determination of settling of flow is measured as the difference between influent flows of upstream processes and influent flows of downstream processes. There typically is not a flowmeter for in-process measurement of settling flow rates.

DAF—dissolved air flotation, TOC—total organic carbon, TSS—total suspended solids, UV254—ultraviolet absorption at 254 nm

REFERENCES

Black and Veatch. 1994. "Preliminary Engineering Report—Phase II: Pilot Plant Testing for Filtration of Table Rock and North Saluda Reservoirs." Greenville, S.C.: Greenville Water System.

Clark, M.M., R.M. Srivastava, J.S. Land, L.J. McCollum, D. Bailey, J.D. Christie, and G. Stolarik. 1994. *Selection and Design of Mixing Processes for Coagulation*. Denver, Colo.: AWWA and AWWA Research Foundation.

Gregory, R., T. Zabel, and J. Edzwald. 1999 (5th ed.). "Sedimentation and Flotation." In *Water Quality and Treatment* (R.D. Letterman, editor). New York: McGraw-Hill.

Hansen, S.P., and G.L. Culp. 1967. "Applying Shallow Depth Sedimentation Theory." *Journal AWWA*. 59(9):1134-1148.

Hudson, H.E., Jr. 1981. *Water Clarification Processes: Practical Design and Evaluation*. New York: Van Nostrand Reinhold.

Hudson, H.E., Jr. 1972. "Density Considerations in Sedimentation." *Journal AWWA*. 64(6):382-386.

Kawamura, S. 2000 (2nd ed.). *Integrated Design and Operation of Water Treatment Facilities*. New York: John Wiley & Sons.

Plummer, J.D., J.K. Edzwald, and M.B. Kelley. 1995. "Removing Cryptosporidium by Dissolved Air Flotation." *Journal AWWA*. 87(9):85-95.

USEPA (United States Environmental Protection Agency). 2004. *Handbook: Optimizing Water Treatment Plant Performance Using the Composite Correction Program*. Washington D.C.: Office of Water, USEPA.

Van Cappellan, J., J.K. Clement, S. Kempeneers, G. Budd, W. Bossaerts, M. Huysmans, and G. Milton. 2008. "Long Term Comparison Testing with Sand Ballasted Flocculation (SBF) and High-Rate Dissolved Air Flotation (DAF)." *Water Practice & Technology*. 3(4):wpt2008085.

Young, J.C., and F.G. Edwards. 2003. "Factors Affecting Ballasted Flocculation Reactions." *Water Environment Research*. 75(3):263-272.

Zabel, T. 1985. "The Advantages of Dissolved-Air Flotation for Water Treatment." *Journal AWWA*. 77(5):42-46.

Chapter 4

Granular Media Filtration

INTRODUCTION

Filtration is a widely used solid–liquid separation process in water treatment. Most surface water treatment plants rely on granular media filtration for the removal of microbial, organic, and inorganic particulate matter. Some groundwater treatment plants use filtration to remove particles generated by softening or oxidative processes. The mechanisms for filtration of solids (particles) are complex and are influenced by several factors, including filter media type (sand, anthracite, granular activated carbon [GAC]), depth, effective size, and uniformity coefficient. Fundamentally, media filters remove particles from the influent water and store these particles until they are backwashed and washed out of the system. Granular media filters primarily remove coagulated particles through sedimentation, interception, and diffusion. This is sometimes referred to as depth

Chapter Summary

- Proper pretreatment to prepare particles is critical for optimal removal through granular media filtration.
- Particles either settle on top of filter media or stick to the media; there are many different mechanisms by which this occurs.
- Proper filter operations and management vary by site and are critical for filter performance.
- Filter backwash procedures greatly affect filter performance and must be occasionally reevaluated.
- Biological filters (or biofilters) must first be operated for optimal turbidity and hydraulic and production goals. Biodegradation of contaminants is a secondary goal.

filtration, in contrast to other types of filtration in which straining is the primary removal mechanism. GAC filters have similar removal characteristics for particles and can also remove dissolved species through adsorption. Slow sand and biofilters provide biodegradation of dissolved species in addition to depth filtration. Membrane filters provide a mechanical sieving of particles larger than the pore size of the membrane. Cake filtration (e.g., diatomaceous earth filtration) provides deposition of particles on a precoated filter media, where the deposited particles become part of the filter media.

The range of sizes of uncoagulated particles that can be removed varies. Figure 4-1 summarizes the general uncoagulated particle size range of different types of granular media filters and pore size for other filtering materials (e.g., membranes). Coagulation and its resulting particle sizes and charges are the most important step in water treatment using granular media filters because size exclusion is not the primary mechanism for removal.

Granular media filters are capable of removing particles much smaller than the pore sizes shown in Figure 4-1 with effective coagulation. Figure 4-2 presents sizes of typical particles in water, some of which are known to be removed by filters.

Polioviruses are about 25 nm in size (Young 1961), yet research by Hobeck et al. (1962) demonstrated that poliovirus removal by alum coagulation and filtration could range from 90 to 99 percent, whereas without coagulation, removal was erratic, ranging from 1

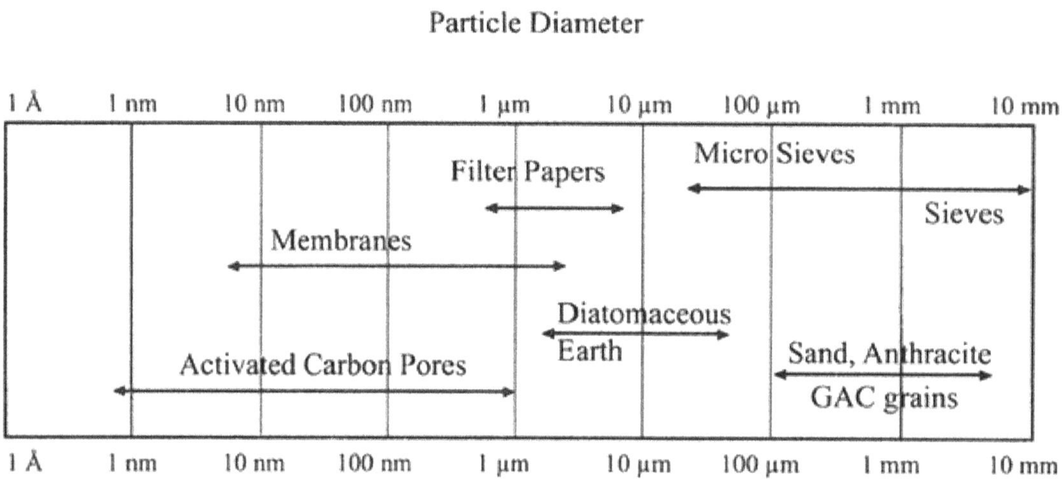

Source: Kevin Castro

GAC—granular activated carbon

Figure 4-1 Uncoagulated particle size range of various filter types

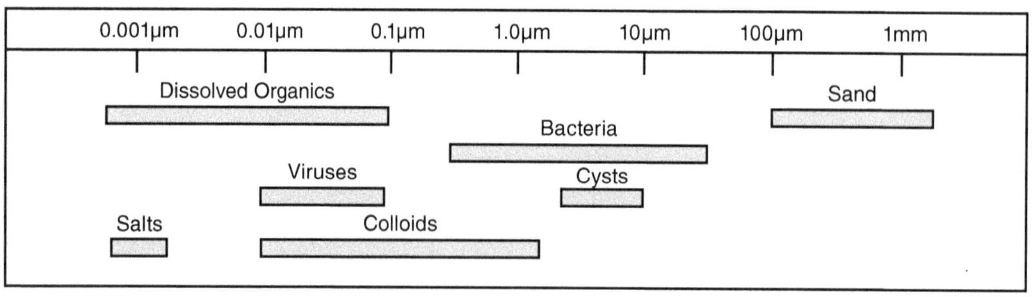

Figure 4-2 Particle types and sizes in natural waters

to 50 percent. With conventional treatment (coagulation, sedimentation, and filtration), removal of poliovirus exceeded 99.7 percent. Research on asbestos fiber removal at the Duluth, Minn., water filtration plant (Logsdon 1979) showed that removal of submicron amosite particles by conventional treatment could be as high as 99.99 percent. Likewise, removal of particles larger than 1 μm by coagulation, flocculation, and rapid-rate filtration could exceed 99.9 percent for giardia cysts (Logsdon et al. 1981) and for *Cryptosporidium* oocysts (Patania et al. 1995). The role of coagulation on the performance of granular media filtration facilities is so crucial that the Surface Water Treatment Rule does not allow any removal credit for giardia cysts at a rapid-rate filtration plant that does not practice coagulation before filtration.

EFFECTS OF COAGULANT CONTROL ON FILTRATION

The primary goal of treatment processes ahead of filtration is to provide charge neutralization, oxidation of some soluble constituents, flocculation, and removal of most larger particles. The type of filtration used and the quality of the source water influence the extent of the pretreatment required for effective filtration of surface waters containing high levels of solids or groundwaters with high concentrations of metals or organics.

For granular media rapid-rate filters, the conventional pretreatment approach will generally consist of coagulation, flocculation, and sedimentation or flotation. In direct filtration applications, the filtration process may only follow oxidation, coagulation, and flocculation. An oxidation–precipitation process is used when dissolved organic or inorganic chemicals are to be removed. For lower-turbidity waters or waters containing algae, dissolved air flotation may be used as the clarification process. Some types of sedimentation are generally used for high-turbidity waters.

Inadequate pretreatment may allow untreated constituents, excessive flocs, or oxidant to be carried over to the filters, which may adversely affect filtration with short filter run length and poor filter effluent quality. A common pretreatment performance goal for many granular media filtration plants is to achieve a filter influent turbidity of less than 1 ntu (95 percent of the time). However, higher influent turbidities can still result in excellent filtered water quality and robust and predictable filter runs.

Filter aid chemicals can be used to enhance filter influent particle conditioning, to strengthen floc and prevent detachment from media because of hydraulic shear, and to improve filter effluent quality. Filter aid chemicals are typically either a polymer or a coagulant and are generally applied to a filter influent in a common channel or pipe with adequate mixing. When high-molecular-weight polymers are used as a filter aid, operators must be careful to apply an appropriate dosage. This is, to some extent, a balancing act because if no filter aid or an insufficient dosage is used, weak floc can cause filters to be susceptible to turbidity breakthrough during filter runs. Using excessive dosages of polymer can result in higher rates of head loss accumulation in filter beds, with possible long-term problems related to backwashing and filter media condition. Either situation can cause filter runs to end prematurely. Chlorine and other oxidants can improve filter performance, in addition to their use as a disinfectant to prevent or control biofouling of filter components. Ozone, in particular, is becoming more popular as an aid to biofiltration. Chlorine pretreatment should also consider the formation of disinfection byproducts (DBPs).

The right type and dose of filter aid can be best identified by conducting pilot column tests on the pretreated water. Such tests can enable the facility to optimize the pretreatment process, which in turn can optimize filtration performance. Pilot tests can be conducted during various seasons to identify any seasonal pretreatment operational requirements and filter aid chemical doses.

Filter effluent quality objectives, filtration rate, media type, and the effective size of media can also influence the extent of pretreatment that is necessary. The effectiveness of pretreatment is monitored by filter performance indicators, such as filtrate quality and quantity objectives (e.g., effluent turbidity), filter run length (time between filter backwashes), and filter ripening characteristics. The unit filter run volume (UFRV, the volume of water filtered through a unit surface area during the run) is used to monitor filter performance and to compare water production from different filter configurations.

The pretreatment for granular media rapid-rate filters as well as appropriate filter loading rates and media bed designs should provide reasonable filter run lengths (more than 24 h under peak flows). A longer filter run length results in more water production per filter cycle. The minimum filter run length may vary among facilities, depending on backwash queuing of filters and backwashing resources. A suggested minimum UFRV goal of 5,000 gal/ft^2 (200 m^3/m^2) is recommended. UFRVs of up to 16,000 gal/ft^2 (650 m^3/m^2) have been documented in biofiltration facilities (Brown et al. 2016). Both the filter run length and the UFRV should be used to manage filter operations.

PARTICLE REMOVAL IN RAPID FILTRATION

The mechanisms of particle removal through a rapid-rate granular media filter are dependent on both particle size to be removed as well as the filtration media size. Particles larger than the void spaces in media are removed through physical straining. Particles smaller than the pore size of media are removed by sedimentation, interception, and diffusion. Figure 4-3 depicts mechanisms of particle capture on a spherical media grain.

Straining

The straining mechanism, characterized by sieving of the removed particles between the media filtration particles, is generally counterproductive to the typical benefits of using granular media filtration. Straining of particles by the media grains cuts off flowpaths through the remaining depth of the media. As a result, the filters will experience a "blinding," a rapid accumulation of head loss. The effect of filter blinding is that particles are removed only at the top of the media bed, and the remaining volume of the media bed is underused. It is typically more beneficial to the overall water treatment facility if larger particles can be removed before granular media filtration via other treatment processes (e.g., sedimentation or flotation).

Interception

Interception occurs as particles are transported to the media surface and the fluid streamlines constrict through the interstitial spacing between media particles. Interception increases as the ratio of particle size to media size increases.

Diffusion

Diffusion is the most dominant mechanism for removal of particles smaller than 1 μm. Charge neutralization of particles in the coagulation step assists in attachment of particles to the media. Thus, chemical addition plays a prominent role in the success of media filtration. Random particle motion, also referred to as Brownian motion, will cause a smaller particle to deviate toward the media surface so that diffusion can take place.

Inertia and Sedimentation

The ratio of settling velocity to superficial velocity has been shown to define the removal efficiency. Particles with mass and velocity great enough can impact directly with the filter

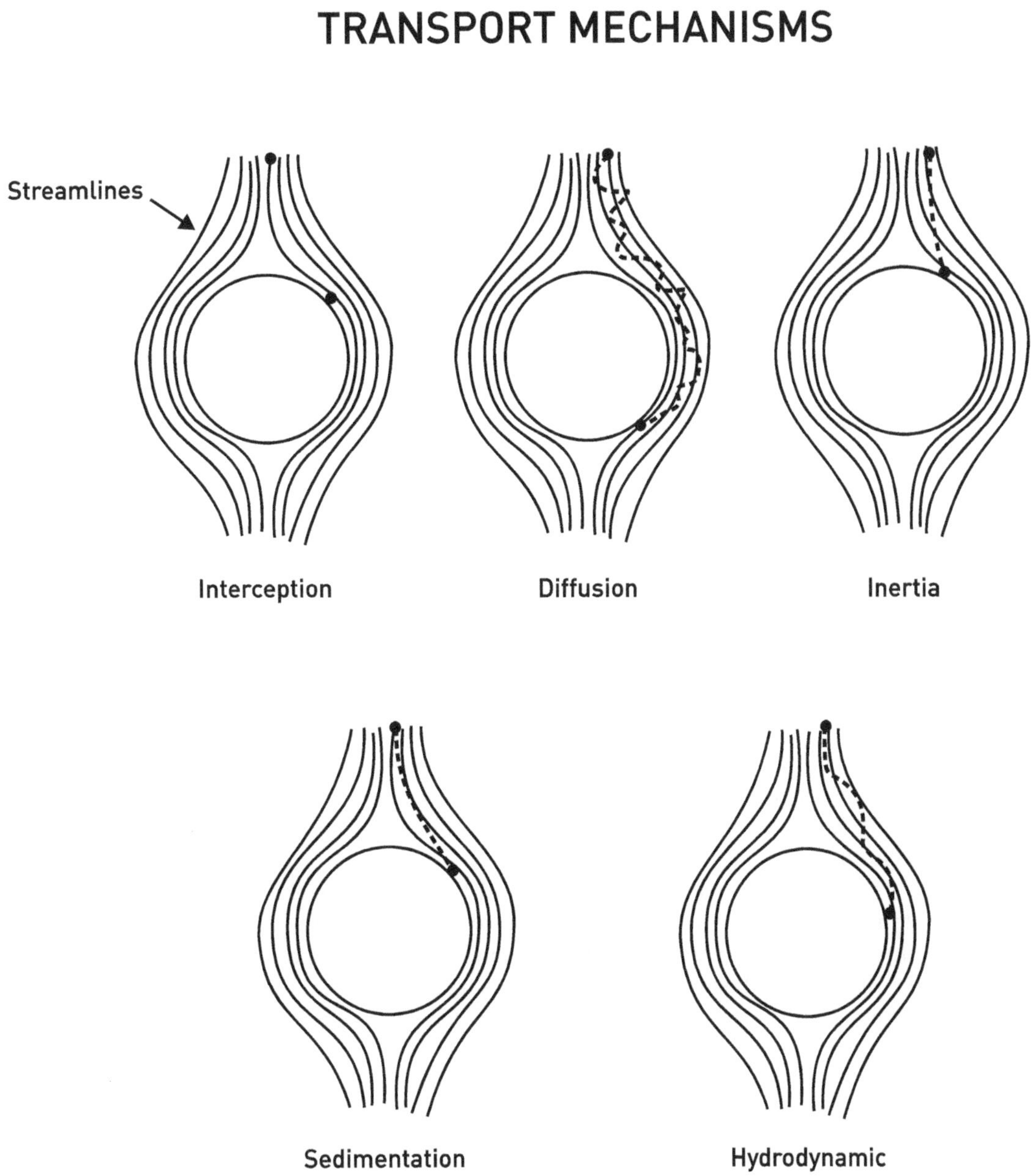

Source: Water Research Foundation

Figure 4-3 Particle transport and removal mechanisms in a granular filter

media. In this case, the particle inertia creates the media contact and subsequently removes the particle from the water.

Particles with density greater than that of water tend to be affected greatly by gravitational forces. These particles are removed through sedimentation. However, the importance of sedimentation in filtration diminishes with increased velocity through the filter, and inertia plays a greater role.

Hydrodynamic

Hydrodynamic transport and subsequent removal occur when asymmetrical particles tumble through the water, enhanced by the laminar flow. These rotations and translations occur because of a lack of balance in the hydrodynamic forces on the particle and will create media contact. When contacted with the filter media, particles can be removed from the water.

FILTER OPERATION AND MANAGEMENT

Coagulation and pretreatment processes prepare water for filtration. The final aspect of process management related to coagulation and filtration is management and control of the filter when it is placed into service after backwashing, during the filter run, and when the filter is backwashed at the end of the run. The benefits of performing coagulation and pretreatment carefully and correctly can be lost if a filter is not controlled and managed properly. This section presents information on managing filters for the purpose of attaining excellent filtered water quality.

Managing Filters During the Filter Run

Filters are typically operated in one of two modes. The more common mode of operation for modern granular media filters is constant rate, in which effluent flow control valves split flow between on-line filters or influent levels are used to manage effluent rates. The other mode of filtration management is declining rate, in which the level in the on-line filters is maintained constant but the rate in the filters that have the longest run time have lower filtration rates because of their increased dirty bed head loss.

During a filter run, filter management includes monitoring filtered water quality, monitoring head loss, and monitoring and controlling the rate of filtration. Instruments for monitoring the quality of filtered water are discussed in Chapter 9, including monitoring turbidity and particle counts. This chapter is focused on other operational details, including head loss and filtration rates.

Monitoring head loss is important because it enables operators to understand the status of each filter that is operating. If multiple filters reach terminal head loss in a short time period, removal of one or more filters for washing can result in the need to operate the remaining filters at higher rates, thus increasing head loss even more on the filters that have high head loss but remain in service. This puts strain on plant staff to perform backwashes in rapid succession.

Although many rapid-rate granular filters are operated as so-called constant-rate filters, in fact, filtration rates often are not constant throughout an entire filter run from start to finish. The need to increase or decrease water production and the need to remove filters from service for backwashing can result in rate changes to filters in service. Whereas decreasing the rate of filtration is unlikely to cause filtered water quality to deteriorate, increasing the filtration rate can do so. Cleasby et al. (1963) presented research results showing that filtered water quality could deteriorate when the filtration rate increased, with the extent of quality deterioration being greater when the rate increase occurred more rapidly; when the overall magnitude of the rate increase was greater; and when floc held in the filter bed was weak rather than strong. In addition, applying filtration rate increases can have a greater effect on filtered water quality when the filter is operating at high head loss, a condition indicative of filter pore spaces that are clogged with floc. This could lead to turbidity breakthrough, putting overall treatment performance at risk.

To counteract the potential for detrimental effects of filtration rate increases on filtered water quality, treatment plant staff should plan and manage filter operations to minimize the need for large or rapid increases in filtration rate. Monitoring filter effluent particle counts as well as filter effluent turbidity can provide an additional measurement to detect degradation of filter performance when rates are adjusted. Filters that are close to reaching terminal head loss may need to be taken out of service for backwashing before a general rate increase is applied to all operating filters. Because operating filters without rate changes is not a realistic expectation, filter aid polymers, which often are added just before filtration to strengthen floc, may be needed.

Backwashing Filters

Filter backwashing procedures influence filtered water quality. In general, filter backwashing is the reversal of water flow through the filter at a sufficient rate to expand the media bed such that captured solids can be dislodged and carried out of the filter cell. Effective removal of solids from the filter media can depend on the solids being captured and the filtration mechanisms capturing the solids. In some cases, the hydraulic shearing force of water alone is sufficient to remove captured solids from the filter bed. However, it is recommended that granular media filter facilities use surface wash or air scour to detach and remove the solids accumulated during filter runs. In Amirtharajah et al. (1991), a method is described for determining optimal backwash rates for filters using air and water backwash, including examples of how to apply the technique. The air that is introduced into the filter during the backwash must provide the porosity by expanding the bed so that the occluded matter is released from the filtering media, especially when it is deeply filtered. To determine whether the currently used backwash process is effective and efficient, several publications (e.g., *Filter Evaluation Procedures for Granular Media* [AWWA] and *Filter Maintenance and Operations Guidance Manual* [Water Research Foundation]) provide specific guidance for evaluating filter operation, testing methods, and interpretation of results. Seasonal water temperature variations normally require adjustment of backwash rate and possibly duration because of changing water viscosity. Warmer water expands the bed less than colder water does at the same backwash flow rate. In cold water conditions, a high backwash rate may lead to excess bed expansion and media loss. Backwash turbidity profiling helps one understand the optimal backwash duration. While a long backwash duration may yield a clean filter cell, it also may be excessive, resulting in lower net water production, increased cost for wash-water handling, and longer filter maturation times. Optimal filter backwashing retains some captured particles that assist in additional particle removal during the ripening phase of a filter run and can reduce the amount of time required for filter ripening. Monitoring the turbidity of backwash wastewater and correlating it with beginning of filter run performance can help identify the site-specific optimal solids retention. Removing all the solids during a backwash can lead to long ripening periods and increased vulnerability at the beginning of a filter run. A general turbidity target for backwash wastewater is around 10–15 ntu but should be optimized for each facility (Kawamura 2020). Floc retention analysis of the media determines the extent of solids captured within the filter media before backwash and remaining within the filter media after a backwash.

While a single improper or inadequate backwash may yield poor performance for the next filter run, long-term effects of inadequate backwashing can cause prolonged filtered water quality problems if mudballs form. Mudball formation within the filter media decreases the effective cross section of a filter bed, resulting in a higher filtration rate in the portions of the bed that are not clogged. Mudballs may also form vertical channels of high porosity within the filter bed, allowing water to pass without effectively filtering contaminants. One indication during backwash of mudball channeling is a high concentration of solids washing out immediately at the beginning of the cycle as backwash water passes through the higher-porosity part of the bed. The high solids washout is then followed by relatively clean backwash waste for the duration of the backwash as wash water continues passing through the vertical channel. Additional indicators include filter cracks, filter media pulling away from the walls, and boils or swirls during backwashing. Figures 4-4 through 4-6 present mudballs witnessed in granular media filters.

Source: Jim DeWolfe

Figure 4-4 Solids accumulation on the surface of filter media before backwashing

Another important aspect of filter backwashing that can negatively affect filtered water quality is the failure to backwash a filter that was in service for hours, removed from service, and then returned to service with floc accumulated in the filter bed. When floc is weak, a filter has high but not terminal head loss, or the rate of filtration is increased too quickly, restarting the dirty filter without backwashing puts great stress on the floc stored within the bed and can cause turbidity breakthrough. Restarting dirty filters may not be uncommon in small water systems where plant capacity is such that filters do not need to run continuously from return to service to attainment of terminal head loss or turbidity breakthrough, but doing so increases the likelihood of particle breakthrough and should be avoided if possible.

Source: Jim DeWolfe

Figure 4-5 Small mudball manually extracted from filter media bed

Source: Jim DeWolfe

Figure 4-6 Mudball forming channels in filter media bed

At filtration plants that recycle spent wash water within the plant, returning the spent wash water to the head of the plant in large flows while backwashing occurs can result in changes in flow through the plant, which would necessitate changes in coagulant feed rates to maintain a consistent dose. If recycling spent wash water changes the quality of source water, a different chemical regime might be needed as compared with that for treating source water with no wash water added. To avoid problems caused by large, intermittent flows of spent wash water, holding it in an equalization basin and slowing the return to the head of the plant to a uniform rate of flow can be helpful. Some equalization basins also allow settling and removal of captured solids to minimize reintroduction of pathogens. Some regulations limit recycle return rates to further mitigate process upsets.

Modified Backwash and Postbackwash Techniques for Improved Filter Performance

The tendency of filters to produce water with higher-than-normal turbidity after being returned to service has been extensively documented (Amirtharajah and Wetstein 1980). Deposited particles become collectors or part of the filter media after they have become attached to media grains or to previously attached particles. In addition, remnants of floc dislodged from filter media during filter washing but not washed out of the filter box can be discharged in filter effluent. Amburgey et al. (2004) have presented experimental data showing that floc remnants can become more electronegative as backwashing progresses and thus become less likely to reattach to media grains when filtration resumes after backwashing. Figure 4-7, adapted from Amirtharajah and Wetstein (1980), depicts a generalized particle breakthrough pattern over time after a backwashed filter is returned to service. The timescale is not uniform but is expanded to show details of filter behavior when the filter is returned to service and later is condensed to present hypothetical trends during the complete run and beyond.

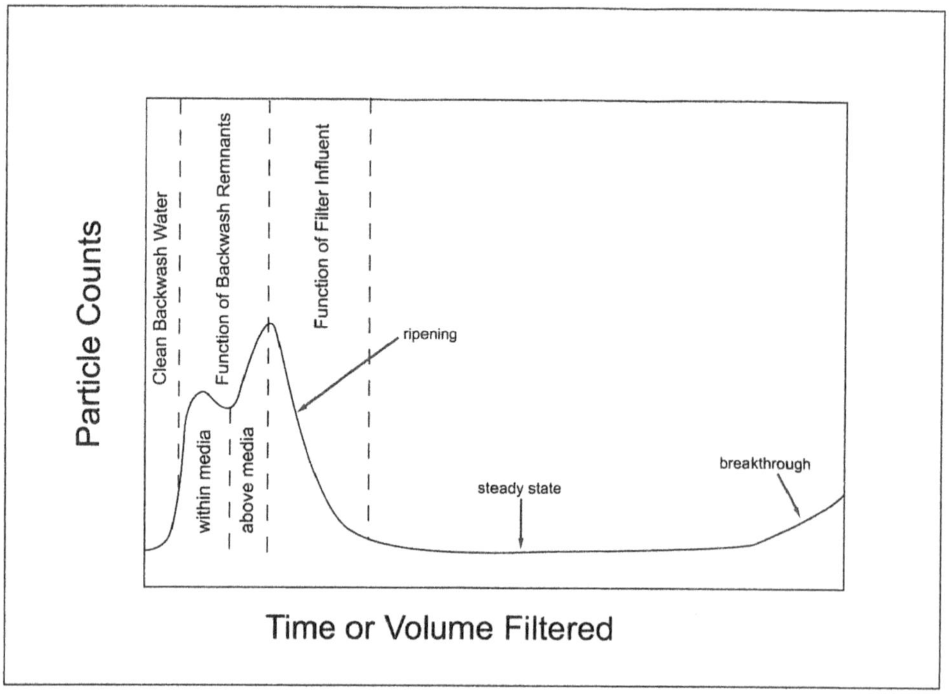

Source: Adapted from Amirtharajah and Wetstein 1980

Figure 4-7 Typical filter breakthrough curve

With standards on individual filter effluent turbidity in the operation of granular media filters, it is important to recognize the potential for particle breakthrough and to exercise appropriate operational procedures and costs to minimize it. There are several methods of minimizing particle breakthrough after a filter backwash. These may include those used during the active backwash process (reverse flow of water through the cell) such as adding chemical to backwash water and/or modifying backwash loading rates and backwash duration. Other methods are used at the conclusion of active backwash and may include addition of coagulant to the backwash water, extended terminal subfluidization wash (ETSW), addition of coagulant to filter influent, and filter to waste.

At some water utilities, a single approach of those listed here is used to help control the initial turbidity spike when a filter is returned to service after washing. Others use a combination of two or three techniques to enhance production of low-turbidity water when filters are returned to service. Currently, there is no way to estimate in advance the effect of the listed techniques for minimizing the initial turbidity spike, so a plant-specific trial-and-error approach is generally used to assess which technique might produce the desired filtered water quality. Some of the techniques require special equipment like chemical feed facilities for adding chemical to backwash water or filter-to-waste piping and valves for that method.

Adding coagulant to backwash water. Adding coagulant chemical or polymer to backwash water during the latter stage of filter washing is a technique that has been well established. The coagulant or polymer is used collect particles, some of which have sheared to a smaller size during backwash, for a more efficient capture at the start of the run. When this is done, shut off chemical addition before ending the filter wash so the wash water in the underdrain and piping does not contain added coagulant or polymer, as this water will be the first water produced when filtration resumes. Generally, the chemical used would be a positively charged coagulant. The appropriate chemical dosage may need to be determined in pilot plant or full-scale plant filtration trials.

ETSW. In ETSW, a subfluidizing rise rate is used to gently rinse backwash remnants (floc particles) out of the filter bed and the water above the filter bed before filter washing is concluded (Amburgey et al. 2003). The subfluidized rinse should be done for a time sufficient to displace water in the filter bed and the water between the top of the media and the water surface. The objective of the gentle rinse after a vigorous backwash is to wash particles out of the filter bed without dislodging additional particles from filter media. This has been proved to be effective in full-scale plants. Amburgey et al. (2003) reported that using a rise rate for the minimum fluidization velocity for the effective size (d_{10}) of the filter media produced substantially better results than using a minimum fluidization velocity for the d_{90} size of the media in the filter bed. They noted that when filter media could be seen at the end of the subfluidization wash (i.e., the turbidity of the water over the media was low), the subfluidization wash had met the objective of removing backwash remnants and providing for lower turbidity when the filter was returned to service. This may be a useful preliminary indicator that the procedure was effective for plant operators who use the ETSW. Results of the successful implementation of this backwash approach are presented in the case studies section.

Adding coagulant to filter influent. Adding coagulant or polymer to filter influent water as the filter box refills after backwashing has been demonstrated to work with metal coagulants and with cationic polymer. Dosage is determined on a trial-and-error basis, but anecdotal evidence in the industry suggests that the coagulant dosage (based on the volume of water in the filter box between the top of the media and the water surface) can be approximately one-third of the dosage used for coagulation in pretreatment. If this technique is used with cold water, consider the reaction time of the coagulant in cold water and avoid discharging unreacted coagulant (such as alum) to the clearwell. This technique has been applied successfully on Lake Michigan water using cationic polymer. Over-addition of coagulant can be detrimental to filter runs and effluent quality.

Filter to waste. Filter to waste is a widely practiced operational strategy to reduce the effects of particle breakthrough after backwash. Simply, it involves wasting poorer-quality water from the filters for a period of time after the filter is returned to service. This wasting is a means of operating the filter to allow for improving filtrate quality without sending filtered water to subsequent processes if it fails to meet the utility's quality goal. This method can involve a preset time based on experience that wasting should occur until particles reduce to acceptable levels, but the time to reach a desired water quality goal can vary with the filtration rate, backwash procedure, temperature, and coagulation conditions. Operating in filter-to-waste mode for an insufficient time can result in discharging water even though the filtered turbidity does not meet treatment plant goals, whereas operating filter to waste too long results in wasting water that should have been sent to subsequent processes. Therefore, the preferable approach is to use on-line turbidity analysis of the filter-to-waste water to enable operators to waste filtrate until the goal is attained. In some cases, up to several hours are necessary for wasting to achieve desired water quality and the opening and closing of valves can cause additional turbidity spikes because of rapid changes in flow. Controls should be provided to slowly and smoothly transition flow from filter to waste to filter effluent.

Water quality of and chemicals present in backwash water can have a significant effect on the filter ripening cycle. For example, some phosphates, fluoride, or disinfectant in the backwash water can have adverse effects. These effects are sometimes seen immediately or can build up over time.

Slow start. Starting a filter at a low filtration rate and gradually increasing the rate over time until the desired rate for operation during the run is attained is a procedure used at some plants. This approach requires filter rate control valves that can operate smoothly and with precise control, so they can be gradually opened a small amount, left in position for a few minutes, and then moved slightly again. In this manner, the filtration rate is gradually increased over time. This technique may be combined with filter to waste at plants with filter-to-waste piping that is inadequate for carrying the filter flow when the filter is operated in the upper range of its rate of flow.

Importance of Properly Managing Filter Operations

Careful management and operation of all aspects of water filtration and filter washing are part of a program to optimize filtered water quality. Quality optimization begins with optimizing coagulation and pretreatment processes, but it must continue through filtration for the maximum benefits of treatment to be attained. Additional information on topics presented in this portion of Chapter 4 can be found in *Water Filtration Practices* (Logsdon 2008).

BIOFILTRATION

Granular media filters can be operated biologically with the goal of biodegrading contaminants in addition to physically removing particles. Although biofiltration can be used to treat a range of water sources, including surface waters, groundwater, stormwater, and treated wastewater (i.e., reuse), the discussion in this section is limited to rapid-rate, aerobic gravity filters treating surface waters. As such, the operational and design principles presented in this section are only those specific to the addition of a biological process and should be considered secondary to the principles presented in the previous section for granular media filtration (i.e., biofilters should be designed and operated to also meet turbidity, hydraulic, and production goals). For more in-depth information on biofiltration, see AWWA Manual M80, *Biological Drinking Water Treatment*.

Introduction and Treatment Objectives

An increase in the application of biofiltration at surface water treatment facilities was observed after the implementation of the Stage 1 and Stage 2 Disinfectants and Disinfection Byproducts Rules. In response to these new regulations, many utilities implemented biofiltration to remove additional assimilable organic carbon (AOC) before distribution and thereby decrease chlorine demand, decrease formation of DBPs, increase control of AOC after ozone oxidation, and increase distribution system biological stability. Other common biodegradation goals for surface water treatment include removal of taste-and-odor compounds and manganese (Mn). Biofiltration is also effective for removal of a range of emerging and trace contaminants. More recently, utilities have reported a transition to more sustainable treatment as a goal for biofiltration implementation.

However, continued investigation will provide a better understanding of the role of microbial communities on water quality and minimize potential microbial risks. Recent studies have suggested that more research is necessary to understand variations in performance, byproduct formation, and microbial communities across seemingly similar treatment facilities. These examples highlight the importance of conducting site-specific testing and monitoring before and after implementation of biofiltration.

Fundamentals of Biofiltration

According to Brown et al. (2016), "Biofiltration is the operational practice of managing, maintaining, and promoting biological activity on granular media in a filter to enhance the removal of organic and inorganic constituents before treated water is introduced into the distribution system." Practically, biofiltration is implemented by suspending or significantly limiting chlorine (or other oxidant) feed to the influent of conventional gravity filters. Biofiltration may be implemented by relocating the point of primary disinfection downstream of filtration but is often implemented with other advanced treatment processes (e.g., upstream ozonation).

This operational change allows for natural attachment and growth of biofilms on media surfaces and in media pore spaces for the oxidation, reduction, and transformation of contaminants (Rittmann and Snoeyink 1984). In these aerobic systems, oxygen is used as an electron acceptor; thus, maintaining sufficient dissolved oxygen (DO) throughout the process is important. Contaminants are captured in biofilm and incorporated within microbial cells or are transformed via microbial processes. Key requirements for biofilters include a surface for attachment; DO; a continuous source of the nutrients nitrogen, phosphorus (P), and carbon; and sufficient contact time for biodegradation.

Process Design and Operation

Many utilities are operating biofiltration incidentally (i.e., they are operating conventional filters with minimal or no oxidant residual without a specific design or operational strategy aimed at optimizing biodegradation). Other utilities may purposefully choose to design their filters via a retrofit or during new construction to optimize biodegradation. Common modifications include the installation of air scour to improve backwashing and the replacement of underdrains to prevent clogging and/or allow for greater media depth and/or media change-out. Process design and optimization strategies to enhance contaminant removal performance while achieving hydraulic and turbidity removal goals are outlined in the following sections.

Pretreatment

Optimizing the influent water quality requires consideration of the source water quality and upstream treatment processes, which can be adjusted to optimize where and to what extent specific contaminants are removed. Optimization of coagulation, flocculation, and sedimentation processes may be achieved by adjusting coagulant doses and types and/or pH to change the concentration of particulate contaminants. For example, the run time of biofilters may be extended by lowering the influent turbidity and neutralizing the charge on particles to allow for minimizing or eliminating the use of a filter aid. Biofiltration may also allow for decreasing upstream chemical dosages. For example, in some situations, biofiltration may allow for reducing coagulation doses while achieving the same overall organic carbon removal (Upadhyaya et al. 2017), but this will need to be verified through pilot- or full-scale testing.

Upstream oxidation is a common strategy for increasing biodegradation of natural organic matter (NOM) into AOC and oxidizing dissolved metals. Under the right conditions, oxidation may convert dissolved metals into particulates that are easier to filter. In some cases, such as manganese oxidation, the resulting particles are highly negative in charge and may not filter easily. The use of a strong preoxidant ahead of biofiltration also breaks down complex organics into more readily biodegradable AOC. Higher concentrations of AOC compounds generally fuel bioactivity, resulting in greater biodegradation of contaminants. The most frequently used oxidant to promote biofiltration is ozone. Additionally, recent research (Lauderdale et al. 2018) has shown that low concentrations of preoxidants, such as chlorine, hydrogen peroxide, and potassium permanganate, can reduce the head loss accumulation caused by biomass growth in biofilters without affecting filtered water quality. However, high concentrations must be limited before biofiltration to prevent upsetting the balance of the microbial community.

Media Type

GAC, anthracite, and/or sand media are commonly used to promote biofiltration. Anthracite and GAC perform similarly in many situations (Upadhyaya et al. 2017). Anthracite is three to five times less expensive than GAC, does not abrade as quickly as GAC, and is not as susceptible to media loss during backwashing because of its greater density. However, several utilities have had GAC media in service for over 10 years without significant media degradation or loss.

Conversely, research has shown that GAC may support a more robust microbial community under stress conditions, such as low temperatures, chlorinated backwash water, or changing hydraulic conditions like adjustment of filter loading rates (Wang et al. 1995) because of its adsorptive properties and higher surface area for microorganisms to attach and grow. GAC also protects the microbial community by quenching disinfectants in the influent or backwash water. Although both bituminous and lignite GAC have been reported to perform similarly in biological applications (Najm et al. 2005), most utilities use bituminous GAC. Because of the limited resources available to generate GAC, carbon from other sources like coconut shell is becoming more popular.

Additionally, the adsorptive capacity of GAC provides a secondary buffer for removing contaminants. The ability to adsorb contaminants provides an initial removal mechanism after installation of new GAC while biological acclimation is in progress. Although adsorption of organic compounds will significantly decrease after the first three to six months of operation, the adsorption capacity for other microconstituents may extend much longer (Westerhoff et al. 2005). Furthermore, GAC adsorption sites can be naturally regenerated through desorption because of a reverse concentration gradient in the water or as a result of biological degradation (Dickenson et al. 2018).

Empty Bed Contact Time

The empty bed contact time (EBCT), or time the water is in contact with the microbial community on the filter media, is an indication of the potential for biodegradation (Zhang and Huck 1996). EBCT is calculated by dividing the media depth (e.g., feet) by the media hydraulic loading rate (e.g., gallons per minute [gpm] per square foot). (For example, (4-ft-deep media)/[(6 gpm/ft^2) × (1 ft^3/7.48 gal)] = 5.0-min EBCT.) Although EBCTs as low as 5 min may meet NOM removal goals, greater EBCTs may be required under some source water and treatment conditions or per local regulations. Additionally, greater EBCTs are often required for trace organic compounds (e.g., tastes and odors, endocrine disruptors, and pharmaceuticals), which are more resistant to biodegradation (Brown and Lauderdale 2006). If the filter box design provides flexibility for more media bed depth, additional filter media can be added to increase EBCT. Filter loading rates can also be decreased to increase EBCT, such as in winter when water demands tend to be lower and temperatures decrease, thus reducing the kinetics of biofiltration.

Filter Backwashing

Backwashing of biofilters follows the same principles and design criteria as conventional filters discussed earlier in this chapter, with a couple specific considerations with respect to biomass development on the filter media. Backwashing should balance removal of accumulated particles and biomass to maintain uniform hydraulic conditions and minimize head loss while maintaining a healthy biomass.

Although traditional backwashing strategies like surface wash and fluidization are often sufficient for biofilters, optimization of backwashing protocols is more critical in biofilters where biomass must be controlled in addition to accumulated particles. Most biofilters in the United States have a simultaneous air scour and fluidization backwash and use unchlorinated water (Evans et al. 2019a). The use of air scour in combination with backwash water (air–water wash) often improves backwashing by breaking up and removing excess biomass. Backwashing with an oxidant residual (e.g., 1 mg/L of chlorine or hydrogen peroxide) may also help remove excess biomass and minimize underdrain clogging (Lauderdale et al. 2018). However, backwashing with an oxidant residual across anthracite media may result in a significant reduction in the microbial community (e.g., Moll et al. 1999). Operational adjustments—such as to the fluidization rate, oxidant doses, or backwashing triggers—should be made on the basis of routine evaluations of the biomass and backwashing effectiveness. As with all filters, monitoring of differential pressure across underdrains and filter media is often helpful in optimizing backwash frequency for biofilters.

Shutdowns

As biofilters perform optimally under steady-state conditions and require a continuous source of DO, nutrients, and substrate, extended biofilter shutdowns like those caused by maintenance or low demand may result in a decline in performance after startup. If media are allowed to dry, the biomass may desiccate. Some utilities elect to leave biofilters submerged during a shutdown to prevent this. However, other utilities have observed excess algae growth in outdoor filters left submerged (Brown et al. 2016). DO concentrations may also significantly decrease, leading to anoxic or anaerobic conditions especially deeper within biofilters, which could significantly change the microbial community and/or lead to reduction of oxidized metals (Martin and Linder 2016). Some utilities have implemented a daily air scour to maintain aerobic conditions during shutdowns and/or to modify their backwash procedure during startup to allow for an extended backwash or filter-to-waste period. Recent research (Evans et al. 2020) has also shown that it may be preferable to avoid shutting down biofilters specifically in the winter when temperatures are low, as accumulated manganese may release after startup during lower temperature conditions. However, other utilities have taken biofilters offline for short durations (e.g., up to a month) with no observed effects during startup (Brown et al. 2016), suggesting site-specific monitoring of water quality and the biomass after shutdowns is required to tailor shutdown procedures.

pH Adjustment

The biofilter influent pH typically ranges from 6 to 9 and may affect contaminant removal, bioactivity (Upadhyaya et al. 2017), and the availability of phosphorus, a key nutrient, to microbial communities (Lauderdale et al. 2011). Bioactivity in the filter may affect filtered water pH. Coagulation, distribution system, and disinfection effects must be considered before making any full-scale changes to pH.

Nutrient Augmentation

Microorganisms use macronutrients like nitrogen and phosphorus for growth and microbial processes (LeChevallier et al. 1991). The addition of biologically available forms of these macronutrients (e.g., orthophosphate) to biofilters with measured nutrient limitations has provided observable hydraulic and potential water treatment benefits, but only at select locations (Stoddart and Gagnon 2017). If biofilter performance goals are not achieved and if monitoring demonstrates a nutrient limitation over multiple sampling events, testing of nutrient addition may be beneficial. Effects on the overall microbial community and distribution system (i.e., to ensure nutrient concentrations do not increase in the finished water) should also be considered.

Process Monitoring

Although there are many benefits of biofiltration, understanding whether biofilters are healthy and when operational changes are needed can be challenging. Biofilter process monitoring provides a continuous understanding of how biofiltration is benefitting overall treatment and plant operation. Before implementing biofiltration, biofilter performance goals documenting desired water quality improvements, production needs, and hydraulic limitations should be developed. These goals will provide a basis against which process monitoring data can be compared to determine whether process optimization is needed. Table 4-1 highlights common filter monitoring and control tools. Hooper et al. 2019 provides additional recommendations for biofiltration monitoring.

Table 4-1 Biofilter monitoring and control tools

Control	Operational	Water Quality	Biological
• Filter loading rate and/or EBCT • Backwash timing and/or rates • Filter run time • Oxidant dose, type, and feed location (ozone, chlorine, peroxide)	• Head loss and head loss profiling • Unit filter run volumes • Chlorine demand • Visual inspection	• General (e.g., pH, temperature, turbidity, nutrients) • Organic contaminant removal (e.g., dissolved organic carbon, ultraviolet absorbance at 254 nm, carboxylic acids) • Inorganic contaminant removal (e.g., iron [Fe], manganese) • DBP formation potential • Taste-and-odor compounds • Trace contaminants	• Adenosine triphosphate • Dissolved oxygen uptake • Biofilm formation rate • Extracellular polymeric substances • Microbial community analysis (e.g., polymerase chain reaction)

DBP—disinfection byproduct, EBCT—empty bed contact time

REFERENCES AND BIBLIOGRAPHY

Amburgey, J., A. Amirtharajah, B. Brouckaert, and N. Spivey. 2003. "An Enhanced Backwashing Technique for Improved Filter Ripening." *Journal AWWA*. 95(12):81-94.

Amburgey, J., A. Amirtharajah, M.T. York, B. Brouckaert, N. Spivey, and M.J. Arrowood. 2005. "Comparison of Conventional and Biological Filter Performance for *Cryptosporidium* and Microsphere Removal." *Journal AWWA*. 97(12):77-91.

Amirtharajah, A. 1991. "Optimum Backwash of Dual Media Filters and GAC Filter-Adsorbers with Air Scour." Denver: AWWA Research Foundation.

Amirtharajah, A., and D. Wetstein. 1980. "Initial Degradation of Effluent Quality During Filtration." *Journal AWWA*. 72(9):518-524.

Brown, J.C., and C.V. Lauderdale. 2006. "Efficient, Simultaneous Destruction of Multiple Drinking Water Contaminants Using Biological Filtration." *Florida Water Resources Journal*. 3:28-30.

Brown, J.C., G. Upadhyaya, J. Carter, T. Brown, and C. Lauderdale. 2016. "North American Biofiltration Knowledge Base." Project #4459. Denver: WRF (Water Research Foundation).

Cleasby, J.L., M.M. Williamson, and E.R. Baumann. 1963. "Effect of Filtration Rate Changes on Quality." *Journal AWWA*. 55(7):869-880.

Dickenson, E., E. Wert, E. Verdugo, K. Greenstein, J. Brown, G. Upadhyaya, P.G. Pope, O. Schneider, R. Marfil-Vega, and R.S. Summers. 2018. "Simultaneous Removal of Multiple Chemical Contaminants Using Biofiltration." Project #4559. Denver: WRF.

Evans, A., K. Onesios-Barry, K. Casteloes, J. Carter, C. Ha, L. Hakes, A. Stoddart, M. Earle, L. Anderson, N. Allward, G. Graham, and W. Knocke. 2020. "Optimizing Biofiltration for Improved Manganese Control under Cold-Water Conditions." Project #4749. Denver: WRF.

Evans, A.N., J.T. Carter, J. Brown, and C. Lauderdale. 2019a. "North American Biofiltration Practices." *AWWA Water Science*. 1(4):e1152. https://doi.org/10.1002/aws2.1152.

Evans, A., L.W. Stephens, and L. Terry. 2019b. "Harness Innovative Knowledge and Tools to Drive Biofilter Performance." *Journal AWWA*, 111:9: 98–101.

Evans, P.J., E.M. Opitz, P.A. Daniel, and C.R. Schulz. 2010. "Biological Drinking Water Treatment Perceptions and Actual Experiences in North America." Project #4129. Denver: WRF.

Hobeck, G.G., N.A. Clarke, and K.A. Dostal. 1962. "Effectiveness of Water Treatment Processes in Virus Removal." *Journal AWWA*. 54(10):1275-1292.

Hooper, J., K. Vickstrom, P. Evans, C. Alito, S. Black, and C. Lauderdale. 2019. "Guidance Manual for Monitoring Biological Filtration of Drinking Water." Project #4620. Denver: WRF.

Hozalski, R.M., S. Goel, and E.J. Bouwer. 1995. "TOC Removal in Biological Filters." *Journal AWWA*. 87(12):40-54.

Hozalski, R.M., and E.J. Bouwer. 2001. "Non-Steady State Simulation of BOM Removal in Drinking Water Biofilters: Model Development." *Water Research*. 35(1):198-210.

Juhna, T., and E. Melin. 2006. *Ozonation and Biofiltration in Water Treatment, Operational Status and Optimization Issues*. Utrecht, the Netherlands: Techneau.

Kawamura, S. 2000. *Integrated Design and Operation of Water Treatment Facilities*. John Wiley & Sons, Inc.

Kirisits, M.J., M.B. Emelko, and A.J. Pinto. 2019. "Applying Biotechnology for Drinking Water Biofiltration: Advancing Science and Practice." *Current Opinion in Biotechnology*. 57:197-204. https://doi.org/10.1016/j.copbio.2019.05.009.

Kohl, P.M., and D. Dixon. 2012. "Occurrence, Impacts, and Removal of Manganese in Biofiltration Processes." Project #4021. Denver: WRF.

Krasner, S., W. Mitch, P. Westerhoff, and A. Dotson. 2012. "Formation and Control of Emerging C-And N-DBPs in Drinking Water." *Journal AWWA*. 104(11):582-595.

Krasner, S.W., R. Shirkhani, P. Westerhoff, D. Hanigan, W.A. Mitch, D.L. McCurry, C. Chen, J. Skadsen, and U. von Gunten. 2015. "Controlling the Formation of Nitrosamines During Water Treatment." Project 4370. Denver: WRF.

Lauderdale, C., J.C. Brown, P.A. Chadik, and M.J. Kirisits. 2011. "Engineered Biofiltration for Enhanced Hydraulic and Water Treatment Performance." Project #4215. Denver: WRF.

Lauderdale, C., C. Alito, J. Hooper, K. Dowdell, and E. Wert. 2018. "Optimizing Biofiltration for Various Source Water Quality Conditions." Project #4555. Denver: WRF.

LeChevallier, M., W. Schulz, and R. Lee. 1991. "Bacterial Nutrients in Drinking Water." *Applied and Environmental Microbiology*. 57(3):857-862.

Martin, B., and K. Linder. 2016. "Filter Surveillance Improves Performance." *Opflow*. 42(11):8-9

Logsdon, G.S. 1979. "Water Filtration for Asbestos Removal." EPA-60/279-206. Washington: US Environmental Protection Agency.

Logsdon, G.S. 2008. *Water Filtration Practices: Including Slow Sand Filters and Precoat Filtration*. Denver: AWWA.

Logsdon, G., J. Symons, R. Hoye, and M. Arozarena. 1981. "Alternative Filtration Methods for Removal of *Giardia* Cysts and Cyst Models." *Journal AWWA*. 73(2):111-118.

Mallevialle, J., I.H. Suffet, and U.S. Chan, editors. 1992. *Influence and Removal of Organics in Drinking Water*. Boca Raton, Fla.: CRC Press.

Manem, J.A., and B.E. Rittmann. 1992. "Removing Trace-Level Organic Pollutants in a Biological Filter." *Journal AWWA*. 84(4):152-157.

Moll, D.M., R.S. Summers, A.C. Fonseca, and W. Matheis. 1999. "Impact of Temperature on Drinking Water Biofilter Performance and Microbial Community Structure." *Environmental Science & Technology*. 33(14):2377-2382.

Najm, I., M. Kennedy, and W. Naylor. 2005. "Lignite Versus Bituminous GAC for Biofiltration—A Case Study." *Journal AWWA*. 97(1):94-101. doi:10.1002/j.1551-8833.2005.tb10807.x

Patania, N., J. Jacangelo, L. Cummings, A. Wilczak, K. Riley, and J. Oppenheimer. 1995. "Optimization of Filtration for Cyst Removal." Project #703. Denver: AWWA and AWWA Research Foundation.

Rajagopalan, R., and C. Tien. 1976. "Trajectory Analysis of Deep-Bed Filtration with the Sphere-in-Cell Porous Media Model." *AIChE Journal*. 22(3):523-533.

Rittmann, B.E., and V.L. Snoeyink. 1984. "Achieving Biologically Stable Drinking Water." *Journal AWWA*. 76(10):106-114.

Stoddart, A., and G. Gagnon. 2017. "Water Quality and Filter Performance of Nutrient-, Oxidant- and Media-Enhanced Drinking Water Biofilters." *Environmental Science: Water Research & Technology*. 3(3):520-533.

Tobiason, J.E., J.L. Cleasby, G.S. Logsdon, and C.R. O'Melia. 2011 (6th ed.). "Granular Media Filtration." In *Water Quality & Treatment: A Handbook on Drinking Water* (J.K. Edzwald, editor). New York: McGraw-Hill and Denver: AWWA.

Tufenkji, N., and M. Elimelech. 2004. "Correlation Equation for Predicting Single-Collector Efficiency in Physicochemical Filtration in Saturated Porous Media." *Environmental Science & Technology*. 38(2):529-536.

Upadhyaya, G., J.C. Brown, A. Evans, J. Carter, C. Lauderdale, and O. Schneider. 2017. "Biofilter Conversion Guidance Manual." Project #4496. Denver: WRF.

Wang, J.Z., R.S. Summers, and R.J. Miltner. 1995. "Biofiltration Performance: Part 1, Relationship to Biomass." *Journal AWWA*. 87(12):55-63.

Wert, E., J.J. Neemann, D. Rexing, and R. Zegers. 2008. "Biofiltration for Removal of BOM and Residual Ammonia Following Control of Bromate Formation." *Water Research*. 42(1-2):372-378.

Westerhoff, P., R.S. Summers, Z. Chowdhury, and S. Kommineni. 2005. "Ozone-Enhanced Biofiltration for Geosmin and MIB Removal." Project #2775. Denver: WRF.

Young, G. 1961 (3rd ed.). *Whitten's Microbiology*. New York: McGraw-Hill.

Zearley, T.L., and R.S. Summers. 2012. "Removal of Yrace Organic Micropollutants by Drinking Water Biological Filters." *Environmental Science & Techology*. 46(17):9412-9419.

Zhang, S., and P.M. Huck. 1996. "Removal of AOC in Biological Water Treatment Processes: A Kinetic Modeling Approach." *Water Research*. 30(5):1195-1207.

Zhang, Z., L. Wang, and L. Shao. 2010. "Study on Relationship Between Characteristics of DOC and Removal Performance by BAC Filter." IEEE 2010 4th International Conf. Bioinformatics and Biomedical Engineering, Chengdu, China.

Chapter 5

Coagulation for Membrane Filtration

INTRODUCTION

Chapter Summary

- Membrane filtrate turbidity is not dependent on coagulation.
- Coagulation is used to remove natural organic matter for disinfection byproduct control, to reduce fouling, and to increase flux.
- Coagulation fundamentals (Chapter 1) still apply to membrane filtration applications.
- Coagulation with polyaluminum chlorides and aluminum chlorohydrate may result in less fouling than alum.
- Organic polymers can cause severe membrane fouling.
- Site-specific coagulation testing recommended.

One of the most important considerations for microfiltration (MF) and ultrafiltration (UF) membrane systems is pretreatment. Pretreatment of membrane systems may include adding coagulants, oxidants, polymers, and adsorbents such as powdered activated carbon (PAC) to the feedwater. Unlike granular media filtration, in which particles, including pathogens such as giardia and *Cryptosporidium*, are conditioned for removal through coagulation and adhere to media filters, membranes are an absolute barrier and will remove giardia and *Cryptosporidium* in the absence of a coagulant. However, the addition of coagulants may be needed if pretreatment of MF–UF is required for natural organic matter (NOM) removal or because of high turbidity of the source water. The benefits of pretreatment upstream of membrane filters include

- reduced membrane fouling,

- organic carbon removal,
- oxidation of iron (Fe) and manganese (Mn),
- removal of algae, and
- removal of taste-and-odor compounds.

Design and operational considerations of pretreatment and how pretreatment can affect performance of MF–UF membrane filters are discussed in this chapter. Low-pressure MF–UF membrane systems can be classified in many ways and are described in this chapter as hollow-fiber encased MF–UF and hollow-fiber submerged MF–UF systems. Different types of membrane materials are used in MF–UF membrane systems: polymeric and ceramic. This section will also include information about troubleshooting with recommended actions for variable feedwater quality, such as high turbidity and total organic carbon (TOC)–dissolved organic carbon (DOC), and process optimization to reduce energy consumption and improve permeability of MF–UF systems.

PRETREATMENT GOALS OF MF–UF SYSTEM

Particulate and Pathogen Reduction

For effective removal of turbidity, conventional filtration with granular media always requires chemical pretreatment (such as coagulant addition, flocculation, or clarification). Membrane system separation performance is independent of the variability of source water turbidity and therefore is more effective in reducing turbidity than granular media filters. The pores of MF–UF membranes are much smaller and generally uniform in size; therefore, they provide a predictable physical barrier to pathogens and particles in the feedwater. In MF–UF filtration, pathogen and particulate removal is independent of coagulant dose and effectiveness. However, membranes are prone to fouling, and the resulting membrane foulant layer can help improve finished water quality by further filtering out smaller particles. The foulant layer can also adsorb some larger organic materials that may normally pass through an MF–UF membrane. Generally, if the source water has an average turbidity of less than 10 ntu, the water may be applied directly to the membranes without clarification pretreatment. For source waters with high or variable turbidity, it may be more economical to provide pretreatment using coagulation, flocculation, and clarification upstream of MF–UF membranes to reduce the concentration of solids (AWWA 2016).

Total and Dissolved Organic Carbon Reduction

Optimizing the removal of TOC–DOC from source water by using coagulation pretreatment can help to meet disinfection byproduct (DBP) regulations. These compounds are generally in a dissolved state and will pass directly through MF–UF membranes without effective coagulation. The removal of NOM with coagulation–MF–UF is essentially the same as with coagulation and granular media filtration.

TOC can be characterized as either hydrophilic or hydrophobic in composition. The hydrophobic fraction of TOC is believed to contribute more to membrane fouling. This parameter can be quantified indirectly through monitoring specific ultraviolet absorbance (SUVA) as described in Chapter 1. For waters that contain high concentrations of TOC–DOC, pretreatment with coagulation, flocculation, and clarification can be effective in removing membrane fouling materials such as the hydrophobic fraction of TOC (AWWA 2016).

Iron and Manganese Removal

Many groundwaters have elevated concentrations of dissolved iron and manganese, which must be removed before distribution. Removal of dissolved iron and manganese is dependent on oxidation of these species so that they are converted to particulate form and can be removed in subsequent clarification and/or filtration processes. Oxidation happens with physical aeration or with the addition of chemical oxidants, such as permanganate, chlorine, or ozone. Aeration and chlorine are usually more effective for iron oxidation, while permanganate is more effective for manganese oxidation. For manganese removal in most conventional water treatment plants, chlorine is added upstream of the filters to create a continuously regenerated manganese dioxide (MnO_2) coating on anthracite, sand, and manganese dioxide filters. With MF–UF membranes, removal is achieved only through straining, and manganese removal may be less effective than with oxidation plus adsorption filtration. When considering MF–UF membrane filtration for a surface water treatment plant with manganese as a constituent of concern, extended pilot testing with all the potential source water conditions is highly recommended to determine chemical oxidation requirements for precipitation of the manganese. In some cases, oxidation plus adsorption filtration may be required upstream of MF–UF membranes.

Algae Removal

In addition to causing taste-and-odor issues in source water like reservoirs and lakes, the occurrence of algal blooms can cause serious problems in water treatment processes. Although MF–UF membranes can completely remove algae cells, membrane fouling can occur because of the presence of algae cells themselves and their extracellular organic matter (EOM). Presence of EOM in source water is an important operational issue that needs to be considered when evaluating challenges with MF–UF membranes. Pretreatment of MF–UF membranes using coagulation, flocculation, and sedimentation is an effective option for the removal of algae and EOM, ultimately improving membrane performance.

Taste-and-Odor Removal

The presence of algal byproducts such as geosmin and 2-methylisoborneol (MIB) can cause taste-and-odor issues. The addition of PAC upstream of MF–UF membranes can be effective in the removal of geosmin and MIB. Pilot- or full-scale testing is recommended to ensure that the additional suspended solids load from the PAC does not adversely affect membrane flux or fouling rates. If sufficiently high PAC doses cannot be used upstream of the membranes, upstream clarification for PAC removal, downstream granular activated carbon use, and oxidation processes may be options for treating taste-and-odor compounds.

TYPES OF MEMBRANE SYSTEMS

In this section, various types of MF–UF membrane products (polymeric and ceramic membranes) will be discussed in addition to MF–UF membrane process configurations (encased and submerged). Regardless of the type of membrane product or process configuration, pretreatment is one of the most important considerations for membrane systems. Coagulation is used not only to achieve set water quality goals but also to create a feedwater to the membranes that helps minimize fouling.

Over the years, the water industry has researched the effect of different pretreatment technologies on water quality and MF–UF performance. This research will continue, as membranes are a viable filtration alternative in drinking water plants. There are distinct differences in the design parameters of coagulation and the selection of a coagulant when used as pretreatment to membranes. For example, coagulation is necessary to minimize turbidity, sometimes in combination with a clarification process, upstream of media filters to avoid reduction in filter run time and turbidity breakthrough. For MF–UF membrane plants, the conditions of coagulation can be vastly different because the membranes are an absolute barrier, and turbidity can be captured on the membrane surface with no concern about turbidity breakthrough. With membrane systems, however, coagulation with or without a clarification process can affect the membrane system performance in terms of fouling and loss of membrane-specific flux. Another complicating factor is that membrane system configurations and membrane fiber packing densities differ between suppliers, and they are constantly being upgraded and changed. This means that the pretreatment regimen optimized for one system might not be as effective for another system. As a result, pilot testing with the selected coagulant is always recommended.

MF–UF membrane systems are offered with warranties and sometimes performance guarantees. As a result, there are stated requirements in the contractual agreements for pretreatment, operation, and data storage for maintaining the warranty and/or performance guarantee on the system. Before making changes to pretreatment in an MF–UF membrane treatment plant, it is recommended to check with the membrane system supplier and confirm the effect of such a change on the system's performance, as well as any contractual obligations.

Commercially available polymeric MF–UF membrane products include polypropylene (PP), polyvinylidene fluoride (PVDF), polyether sulfone (PES), and polyvinylpyrrolidone (PVP) blends. Each material varies in fouling resistance and chemical resistance. Of these, PES–PVP blends and PVDF offer the highest oxidant tolerance. In addition to polymeric-based membranes, several ceramic MF–UF membrane products are commercially available. When compared with polymeric-based membranes, ceramic membranes have higher chemical stability, higher permeability and flux, longer service life, and microbial fouling resistance. However, their production cost is significantly higher, which is reflected in the purchase price. There are several commercially available membrane geometries, including hollow fiber, tubular, and flat sheet. Hollow-fiber membrane refers to the configuration of pumping through the membrane fiber, allowing the membranes to be backwashed, which controls particulate and organic fouling. Among the polymeric MF–UF membrane products, hollow-fiber PVDF membranes will be the subject of focus in this section, as they are the most common in drinking water treatment.

There are two general process configurations for MF–UF systems: encased (or pressurized) and submerged (or immersed). Figure 5-1 shows process flow diagrams for direct filtration to encased and submerged membranes compared with conventional treatment to granular media filtration. Pretreatment will affect each of these MF–UF membrane systems differently, and each membrane system will have a different chemical compatibility and removal efficiency for contaminants. Some of the most important considerations for pretreatment for MF–UF membrane systems include (see the "Membrane Pretreatment Chemicals" section for a detailed discussion)

- coagulants type and pH,
- coagulant aid polymers,
- oxidant and adsorbent types,
- flocculation time and mixing energy, and
- a preclarification system.

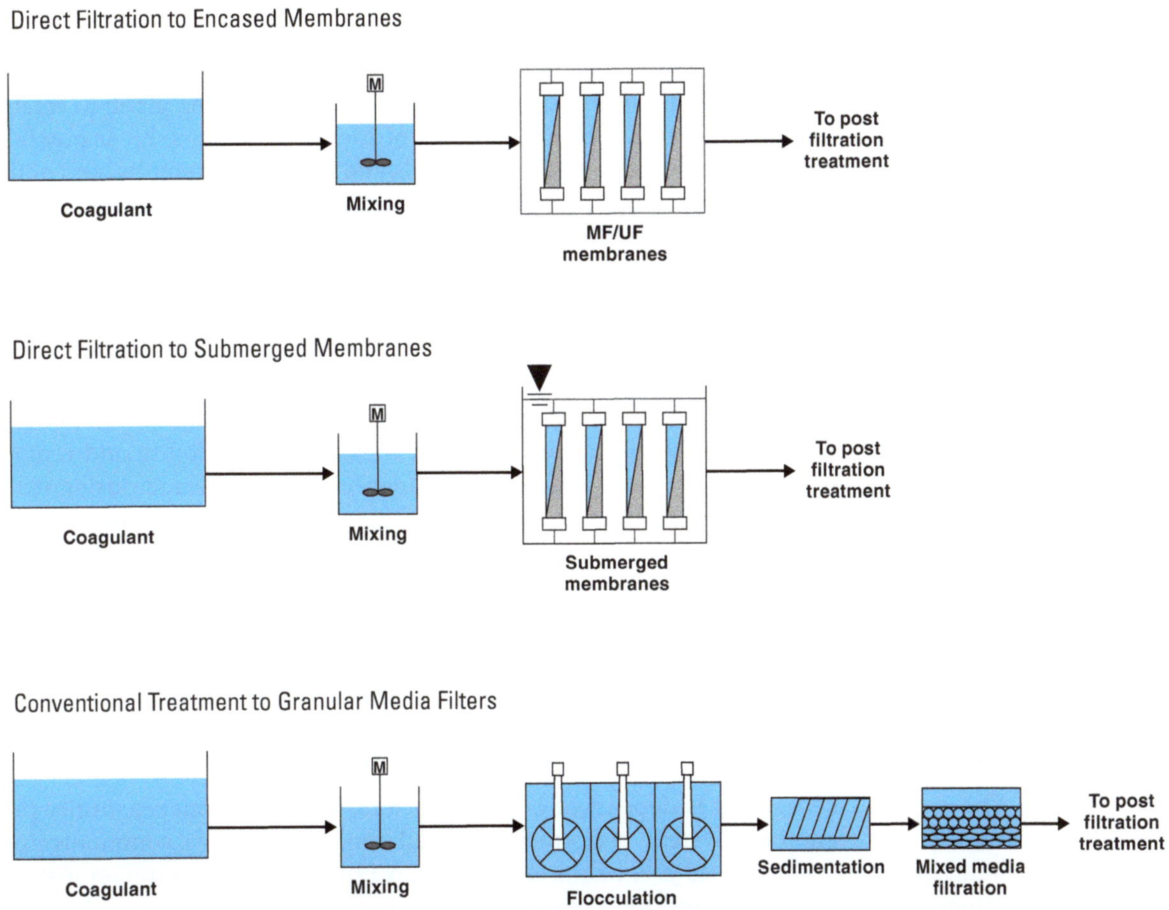

Figure 5-1 Process flow diagrams for direct filtration options and conventional pretreatment

Hollow-Fiber Encased and Submerged PVDF Membrane Systems

The installation of hollow-fiber membrane systems as a filtration alternative to conventional granular media filters has some special considerations for coagulation as pretreatment. The question of whether clarification is needed upstream of hollow-fiber membrane systems is common, but in practice, there are systems with and without clarification. There have been many pilot-scale systems and full-scale installations in which the pretreatment has not been optimized for the membrane system, and this has resulted in issues with operation. Each type of hollow-fiber membrane system is unique in terms of the membrane fiber materials and properties, the fiber packing density, and the hydraulic flow patterns for filtration and backwashing. These characteristics in combination with source water quality, the type and concentration of organics present, and removal goals might dictate whether clarification is needed for sustained operation and thorough cleaning of the membrane by backwashing. Submerged hollow-fiber membrane systems can typically tolerate a higher solids loading rate in the tank and thus not require upstream clarification. Submerged membrane fibers still require adequate backwashing and cleaning, so it is important to monitor the performance of the system over time.

Ceramic Membranes

Ceramic membranes are newer to the drinking water industry than polymeric membranes; therefore, there are fewer full-scale installations. More attention is being given to ceramic membranes because of their sustainability (upward of 20-year membrane life is possible), ability to be cleaned with a wider range of chemicals and chemical strengths, and ability to accommodate higher-pressure (5 bar or 72.5 psi) backwashes. The conditions for coagulation upstream of ceramic membranes are site specific and may be for DBP formation potential reduction and to reduce fouling.

MEMBRANE PRETREATMENT CHEMICALS

Coagulants and pretreatment chemicals will affect each type of membrane system differently. Each membrane system will have a different chemical compatibility and removal efficiency for the contaminants. It is critical to coordinate chemical type and dosing with a membrane system supplier before using such pretreatment mechanisms because an unapproved chemical may void a membrane warranty.

The following subsections describe two types of coagulants in general use for water treatment: metal-salt coagulants and organic polymers. They also describe the chemical compatibility of commonly used pretreatment chemicals, including oxidants and PAC, and the required mixing time and energy that is suitable for inline coagulation.

Type of Coagulants

The choice of a coagulant is dependent on many factors. Cost, performance, utility preference, and membrane system performance are all considered. Most commonly used alum-based and ferric-based coagulants can be chosen, but a membrane supplier may recommend a particular type for optimized membrane performance. Unlike granular media filtration, the performance of the membrane process for particulate removal is not dependent on the surface charge of the resulting floc particles, and therefore, coagulant chemistry is less important. Coagulant chemistry does affect the removal of NOM and contribute to membrane fouling. Proper coagulant selection, dose, and pH must be considered before a full-scale implementation. Bench- or pilot-scale testing is often required to determine appropriate coagulation conditions.

The choice of the coagulant is linked to water quality goals. For example, if a certain level of DOC removal is required, one coagulant may perform better than another for the water type. The pH of coagulation is an important parameter not only to control the floc development and organics removal but also to minimize the possibility of residual metals from the coagulants from passing through the membrane in a dissolved form. Dissolved aluminum (Al) can also pass through an MF–UF membrane if the pH and mixing of the coagulation step are not adequate. This can also happen for iron, but aluminum has a secondary regulatory limit in drinking water (0.2 mg/L) that must be monitored closely.

Aluminum-based coagulants. Aluminum-based coagulants include aluminum sulfate (alum), polyaluminum chloride (PACl), and aluminum chlorohydrate (ACH), which is a very high-basicity PACl. As described in Chapter 1, alum is more effective at pH lower than 7. PACl and ACH are preferred over alum because they can be effective even at pH values greater than 7.5. If alum is fed upstream of MF–UF membrane systems, the pH should be 6.5–6.7 to minimize the solubility of aluminum in water (AWWA 2018). Alum can cause more fouling than PACl and ACH. This should be taken into account when alum is selected.

Ferric-based coagulants. Ferric-based coagulants include ferric chloride ($FeCl_3$) and ferric sulfate ($Fe_2(SO_4)_3$). Ferric-based coagulants are less soluble than aluminum-based coagulants and can be used over a much wider pH range (5.0–8.5) in the feedwater. The dosage for $FeCl_3$ depends on the quality of the feedwater—specifically, alkalinity and the TOC concentration.

Oxidants

Depending on the constituents found in the source water, oxidation may be required to precipitate dissolved ions before membrane filtration. Oxidants such as chlorine dioxide, potassium permanganate, and ozone can also be used for the removal of metals. Applying oxidant before membrane filtration is a complex subject not covered in this manual. If a chemical oxidant is used, then the compatibility of the MF–UF membrane with the proposed chemical oxidant must be checked with the supplier. Oxidants can permanently alter the chemical composition and physical integrity of some membranes. If the membrane type has a low resistance to chemical oxidants, the oxidant residual must be removed from the feedwater before the membranes. PVDF membranes have higher tolerance to oxidants, while PP membranes generally have lower tolerance to oxidants.

Coagulant Aid Polymers

Coagulant aid polymers are another category of pretreatment chemical that can be used for removal of turbidity. It is important to verify type and dose with membrane supplier to see if the polymer can be used in pretreatment. For conventional pretreatment plants, polymers are often used for enhanced clarification, particularly with sand ballasted sedimentation. Polymers are also used in backwash treatment, in which treated backwash water is returned to the head of the water plant. Both processes can cause polymer to be introduced to the membrane stream. Applying polymers directly to polymeric MF–UF membranes can pose significant risk, especially if cationic and nonionic polymers are used. Improper use of polymers can cause significant fouling, resulting in rapid loss of plant capacity. The fouling can be very difficult to remove, requiring extended cleaning of the membrane system. Therefore, it is important to coordinate with the membrane system supplier the polymer type and dosing before using them ahead of MF–UF membranes. Membrane system suppliers generally prefer to evaluate polymers individually and review project-specific details such as feedwater quality and clarifier type. This allows greater input into how these polymers might best be incorporated into MF–UF membrane system and overall plant optimization.

Powdered Activated Carbon

Adsorbents such as PAC can be effective in DOC and DBP removal, as well as taste-and-odor control. Most frequently, PAC is used in conjunction with a clarification process but can also be added directly to some membrane systems. PAC dosing must be coordinated with the membrane supplier because PAC can be abrasive and damage membranes. It may be necessary to limit PAC dosing and frequency of application to avoid damaging membranes prematurely. Ceramic membranes may be advantageous for use with PAC because of their much higher abrasion resistance.

If PAC is used ahead of the clarification process, select a carbon product that settles well. In this case, jar testing can be used to confirm adequate dosing and performance of PAC. There is always a risk of PAC carryover during clarifier upsets, which can increase the amount of PAC reaching the membranes. Membrane performance may need to be modified by having more frequent backwashes.

There are various risks to be aware of if PAC is planned to be dosed directly upstream of membranes. PAC retained in membrane modules will rapidly consume chlorine during the cleaning process. The chlorine residual concentration needs to be monitored carefully to ensure proper concentration throughout the cleaning process. Another concern is membrane abrasion, especially if PAC builds up within the modules. Different types of carbons behave differently from an abrasion perspective, which is why discussing options with a membrane supplier is important.

Inline Coagulation Mixing Time and Energy

For inline coagulation applications, it is important to consider the mixing energy and flocculation time that is required for adequate floc development for membrane filtration. Floc development that is optimized for MF–UF membrane filtration often requires less time and mixing energy than for conventional media filters. Mixing is required to create a pin floc that is suitable for inline coagulation with MF–UF membranes.

Clarification Before MF–UF Systems

Clarification reduces the solids loading to MF–UF membrane systems and reduces the opportunity for fouling. By providing an effluent that is low in turbidity, MF–UF membranes can operate at an increased flux with reduced cleaning frequency, ultimately reducing capital and operating costs. Variations on conventional pretreatment include high-rate clarification using plate or tube settlers, sludge blanket clarifiers, and dissolved air flotation (DAF). MF–UF membrane compatibility with any upstream treatment clarification processes that use polymers should be confirmed before design by evaluating similar full-scale plants and pilot testing if necessary. DAF is commonly used when algae counts are elevated in a source water. Alum and ferric chloride can also be used, but they require optimization of the clarification process to reduce the risk of carryover and membrane fouling.

OPERATIONAL OPTIMIZATION

This section describes techniques that can be implemented to optimize coagulant feed systems upstream of MF–UF membrane systems. This section also includes a troubleshooting guide with recommended actions for variable feedwater quality and process optimization to reduce energy consumption and improve permeability of MF–UF systems.

Optimization Goals

To optimize the coagulation process upstream of MF–UF membranes, there should be selected goals, and these goals should be part of the standard operating procedures and the operations and maintenance manual. Objectives of MF–UF membrane optimization can include

- TOC removal requirements,
- iron and manganese removal,
- point-of-entry DBP sampling,
- increased permeability,
- pressure decay, and
- decreased cleaning intervals.

Control of coagulant dose, mixing energy, and time are important operational parameters for good performance of an MF–UF membrane system. Mixing is a critical step to create a pin floc that is easily removed by the membranes if direct filtration is used. On the basis of real-time feedback for water quality parameters, such as SUVA and TOC, coagulant dose can be optimized to reduce membrane fouling and improve hydraulic throughput.

Methods to Improve Coagulant Control

Control of coagulant dosing can be challenging in source waters with seasonal or other variations in water quality. In source water with relatively stable SUVA, dosing can be controlled by on-line monitoring of ultraviolet (UV) absorbance in the incoming source water. UV has been shown to correlate well with the presence of humic substances (Köhler et al. 2016), which make up a large portion of material removed during coagulation. In this instance, the use of optical sensors for coagulant dose control could be a viable method of process control. In recent UF membrane pilot studies on various sources of surface water, automatic coagulant dosing based on on-line measurement of UV absorbance was successfully applied to optimize aluminum coagulation dose (Keucken et al. 2017). Careful cross calibration of raw and filtered samples is recommended when using on-line sensor data for process optimization, even in low-turbidity water.

Methods to Improve Hydraulic Throughput

Although particle removal by MF–UF membranes is independent of source water quality and pretreatment conditions, membrane productivity can be reduced by water quality variations. For example, if source water turbidity increases, the operator may need to reduce the membrane flux or decrease the time intervals between backwashing and chemical cleaning operations. The operator should be aware of the effect of source water or pretreatment changes at the facility, as well as the options available for addressing these changes. In some cases, a significant increase in turbidity in feedwater for a 24-h period will have little or no effect on membrane operations. However, if there is a significant increase in turbidity for more than 48 h, the flux should be decreased or the backwash conditions revised to maintain the desired time interval between chemical cleaning operations. The operator should always evaluate the effect on plant capacity before revising any operating set points.

Methods to Remove Fouling Related to Coagulation

When using coagulation ahead of low-pressure membranes, the particle charge of the resulting floc particles plays an important role in fouling (McLeod 2014). Precipitated floc particles that possess a cationic charge will be attracted to the membrane surface. This attraction will cover the surface area of the membrane element quickly, resulting in reduced ability of water to pass through the element. Figure 5-2 depicts how coagulation at higher pH ranges results in anionic floc charge, which has been shown to reduce fouling on membranes.

Figure 5-3 shows accumulation of floc material over the first two-week period of operation after a routine chemical cleaning. This accumulation is directly related to the pressure increase of transmembrane pressure. No pH or particle charge correction was used, and the source water pH for this source was low (approximately 7.0). Measuring the precipitated charge with a zeta potential analyzer during this period of operation showed a cationic floc charge (McLeod 2014). Figure 5-3 also shows clean membrane elements after the second two weeks of operation under the same conditions with particle charge correction using a controlled caustic addition to increase pH. Zeta potential measurements during this operating period depict a slightly anionic charge measurement. The lack of accumulation of particulates is a result of "neutralizing" or altering the precipitated particle charge that emulates the charge of the membrane element.

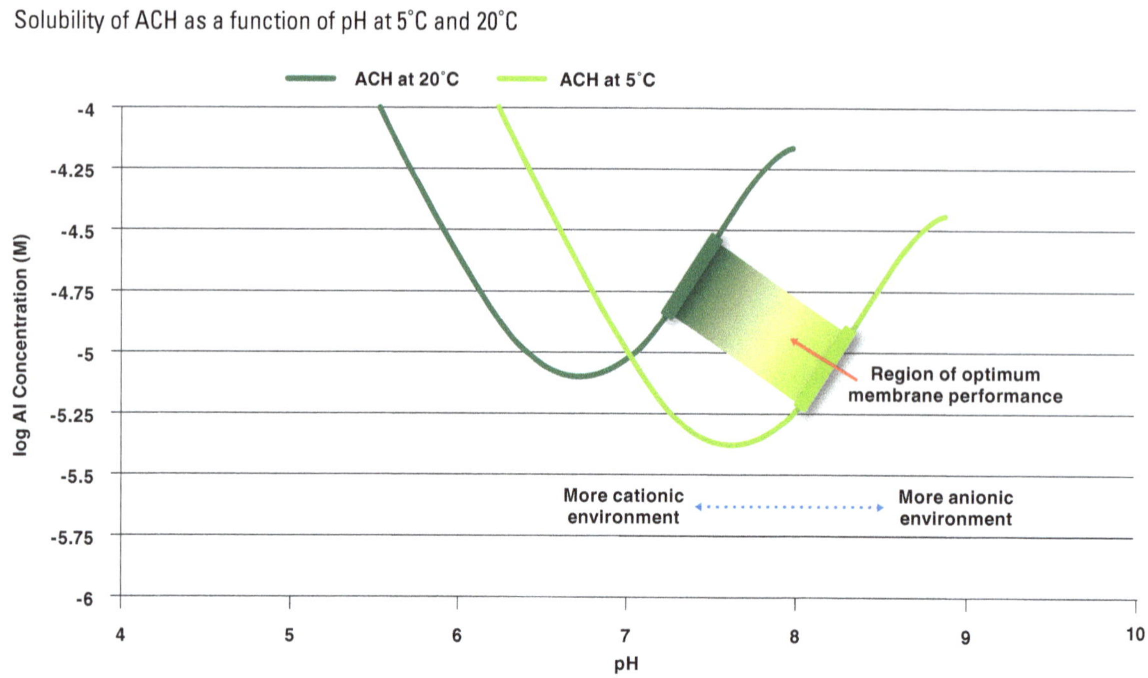

Source: ACH solubility data adapted from Pernitsky and Edzwald 2003

ACH—aluminum chlorohydrate

Figure 5-2 Region of pH range for optimal membrane performance

(a) Accumulation of floc material on membranes

(b) Clean membrane elements with particle charge correction using caustic addition

Source: City of Berthoud, Colo.

Figure 5-3 Comparison from Carter Lake Filter Plant, Berthoud, Colo.

REFERENCES AND BIBLIOGRAPHY

AWWA. 2018 (1st ed.). M62, *Membrane Applications for Water Reuse*. Denver: AWWA.

AWWA. 2016 (2nd ed.). M53, *Microfiltration and Ultrafiltration Membrane for Drinking Water*. Denver: AWWA.

Bolto, B.A. 1995. "Soluble Polymers in Water Purification." *Progress in Polymer Science*. 20(6):987-1041.

Choi, K.Y., and B.A. Dempsey. 2004. "In-line Coagulation with Low-Pressure Membrane Filtration." *Water Research*. 38(19):4271-4281.

Frenkel, V., and G. Best. 1998. "Effective Colour and TOC Removal from Drinking Water by Microfiltration." *Environmental Science & Engineering Magazine*. May:50.

Judd, S.J., and P. Hillis. 2000. "Optimization of Combined Coagulation and Microfiltration for Water Treatment." *Water Research*. 35(12):2895-2904.

Kim, H.C., J.-H. Hong, and S. Lee. 2006. "Fouling of Microfiltration Membranes by Natural Organic Matter After Coagulation Treatment: A Comparison of Different Initial Mixing Conditions." *Journal of Membrane Science*. 283(1-2):266-272.

Köhler, S.J., E.E. Lavonen, A. Keucken, P. Schmitt-Kopplin, T. Spanjer, and K.M. Persson. 2016. "Upgrading Coagulation with Hollow-Fibre Nanofiltration for Improved Organic Matter Removal During Surface Water Treatment." *Water Research*. 89:232-240.

Keucken, A., G. Heinicke, K.M. Persson, and S.J. Köhler. 2017. "Combined Coagulation and Ultrafiltration Process to Counteract Increasing NOM in Brown Surface Water." *Water*. 9(9):697.

Loi-Brügger, A., S. Panglisch, P. Buchta, K. Hattori, H. Yonekawa, Y. Tomita, and R. Gimbel. 2006. "Ceramic Membranes for Direct River Water Treatment Applying Coagulation and Microfiltration." *Water Supply*. 6(4):89-98.

McLeod, G.A. 2014. "Mitigating Low-Pressure Membrane Fouling by Controlling the Charge of Precipitated Floc Particles." *Journal of Membrane and Separation Technology*. 3(4):213-218.

Pernitsky, D.J., and J.K. Edzwald. 2003. "Solubility of Polyaluminum Coagulants." *Journal of Water Supply: Research and Technology–AQUA*. 52(6):395-406.

This page intentionally blank.

Chapter 6

Coagulation and Filtration in Wastewater, Water Reuse, Seawater, and Stormwater Treatment

Chapter Summary

- Key coagulation principals apply to wastewater, seawater, and stormwater the same way they do to potable water production with some modifications.
- Coagulation and filtration can enhance the removal of specific nutrients or deleterious compounds such as phosphorous or suspended particles.
- Coagulation is a vital pretreatment process for seawater reverse osmosis (RO) systems.

Coagulation plays an important role in the removal of particulate matter, including microorganisms, from water. Removal of other contaminants such as arsenic, iron (Fe), manganese (Mn), and dissolved organic matter is achieved via coprecipitation, sorption, or other physiochemical mechanisms. Chemical coagulation has also proved useful in pretreatment of wastewater, recycled water, seawater, and stormwater.

The same fundamental principles outlined in previous chapters apply to these applications. The vast majority of suspended particles will carry a negative charge and will require neutralization by a coagulant. Dissolved organic material, although potentially different from surface water natural organic matter (NOM), will exert a coagulant demand that will affect the required coagulant dose. Ultimately, the type of coagulant and the coagulation pH will need to be selected so that near-neutral floc particles are produced, allowing clarification and granular media filtration processes to work properly.

Similar to drinking water treatment, identifying matrix conditions (e.g., pH) and determining the proper chemicals and dosages for coagulation are necessary steps for the treatment of wastewater, recycled water, seawater, and stormwater. Coagulation and filtration treatment of these matrices is similar to drinking water treatment through the creation and growth of floc with subsequent removal of solids through clarification and/or filtration processes. To maintain the careful process control over chemical coagulation and subsequent filtration, various on-line monitoring devices and monitoring procedures are used for the treatment of wastewater, recycled water, seawater, and stormwater.

COAGULATION AND FILTRATION IN WASTEWATER TREATMENT

Phosphorus Removal

Coagulation is commonly practiced in municipal wastewater treatment plants (WWTPs) for the chemical precipitation of phosphorus. Phosphorus is essential to the growth of biological organisms, including algae. To control eutrophication in the receiving water, maximum effluent concentration (or mass loading) of total phosphorus (TP) is mandated in discharge permits of many wastewater treatment facilities. Untreated municipal wastewater may contain 4–15 mg/L of phosphorus as P (Tchobanoglous and Burton 1991). Bacteria used in conventional biological treatment and enhanced biological phosphorus removal processes consume this vital nutrient, and when excess biomass is wasted, a net reduction of phosphorus concentration occurs. To achieve additional removal, chemical coagulants such as metal salts and lime are added before primary and secondary sedimentation to precipitate phosphate. At these additional locations, competing reactions necessitate dosing higher than stoichiometrically calculated values. Biological treatment, coagulation, and sedimentation processes can reduce secondary effluent phosphorus concentrations to 0.3–1.0 mg/L as P, which may be insufficient to meet the facility discharge permit—further treatment is required.

Postsecondary coagulation (i.e., postprecipitation) involves the addition of metal salts to secondary effluent to remove chemical precipitates in either tertiary clarifiers or, more commonly, in effluent filters. Polymers are also used to bind chemical precipitates, organic matter, bacteria, and other insoluble solids into larger, more rapidly settleable or filterable flocs. The use of coagulants and polymer is dependent on their need to result in removing contaminants to meet the facility's discharge permit limits (e.g., for concentration of total suspended solids [TSS] or phosphorus)—filtration without chemical addition may be sufficient.

Metal salts used to coagulate secondary effluent to produce phosphate include aluminum sulfate (alum), sodium aluminate, polyaluminum chloride (PACl), ferric chloride ($FeCl_3$), and ferric sulfate ($Fe_2(SO_4)_3$). Several factors affect chemical choice, including initial phosphorus concentration, required effluent phosphorus concentration, pH, alkalinity, chemical cost, injection location, sludge storage and handling capacity, sludge disposal cost, and treatment process compatibility. As an example, Table 6-1 presents results from testing performed at a municipal water resource recovery facility (WRRF) (Narasiah et al. 1991) that identified the facility-specific requirements for alum addition in secondary wastewater at various TP removal efficiencies and pH. These ratios are presented as Al.

Table 6-1 Facility-specific molar ratio of Al:P for TP removal

Efficiency (% Removal)	pH 7.5	pH 6.5	pH 5.5
90	2.47	2.07	1.94
95	3.21	2.64	2.42
99	8.72	4.02	3.62

Table 6-2 Common modern-day wastewater filters

Common Name	Operation	Filtering Media, Typical	Fluid Direction Through Media	Media Depth, Typical	Figure
Conventional deep bed*	Semicontinuous	Sand	Downward	72 in.	–
Conventional	Semicontinuous	Anthracite on sand	Downward	24 in. anthracite + 12 in. sand, 36 in. total	N/A
Traveling bridge	Continuous	Sand	Downward	11 in.	6-1
Moving bed*	Continuous	Sand	Upward	72 in.	6-2
Cloth media	Continuous	Pile cloth	Outside-in, horizontal	±5⁄32 in.	6-3
Compressed media	Semicontinuous	Polyethylene woven balls	Upward or downward	Variable	6-4
Disc	Continuous	Woven polyester or stainless steel	Inside-out, horizontal	N/A	6-5
Membrane	Semicontinuous	Hollow microfibers	N/A	N/A	–

N/A—not applicable

* May be configured as a denitrification filter to also biologically remove nitrate

Secondary biological treatment processes typically consume alkalinity and may depress pH because of nitrification. Without pH control, a postprecipitation coagulation process may be inefficient and further reduce pH to the point of effluent permit violation. Jar testing, sampling, and laboratory analysis of both secondary and tertiary effluent are essential to properly select chemicals and respective doses. Relative to coagulation in primary or secondary wastewater processes, chemical dosage rates are closer to stoichiometric-calculated values. Coagulant dose rises nonlinearly relative to increased percentages of phosphorus removed and varies because of the facility-specific chemical application process.

Tertiary Treatment

It is common to add a filtration step at the end of conventional WWTPs when conventional secondary wastewater treatment cannot reliably meet facility effluent quality requirements. Usually, it is efficient and economical to treat and remove most contaminants before filtration, allowing the secondary clarified wastewater to then be "polished" by the filter. As an example, a typical average influent concentration of TSS to a municipal WRRF is approximately 200 mg/L (Tchobanoglous and Burton 1991), whereas the secondary clarification treatment objective is less than 30 mg/L (USEPA 2010), necessitating greater than 85-percent removal. Should the discharge concentration regulatory requirement for TSS be 5 mg/L (as is common for fragile receiving water bodies in the United States), the filtration processes must reliably ensure the remaining 12.5 percent of original TSS concentration be removed.

Many different types of filters are used at WRRFs. Granular (e.g., sand, anthracite), cloth, compressed, and microscreen media or membranes are used in different configurations and flow directions. Filters may be semicontinuous (i.e., requiring offline backwashing) or continuous and include those identified in Table 6-2. A coagulation step may not be needed for disc, compressed media, cloth media, and membrane filter applications, but for granular media filter applications, coagulation is generally required to ensure effluent suspended solids limits are met.

Each of the following must be carefully considered to properly select a wastewater filter technology:

- treatment capability,
- required pretreatment,
- hydraulic capacity,
- solids loading capacity,
- backwash rate,
- facility backwash treatment capacity,
- footprint,
- maintenance requirements (e.g., preventive, corrective, predictive),
- materials of construction, and
- installed cost.

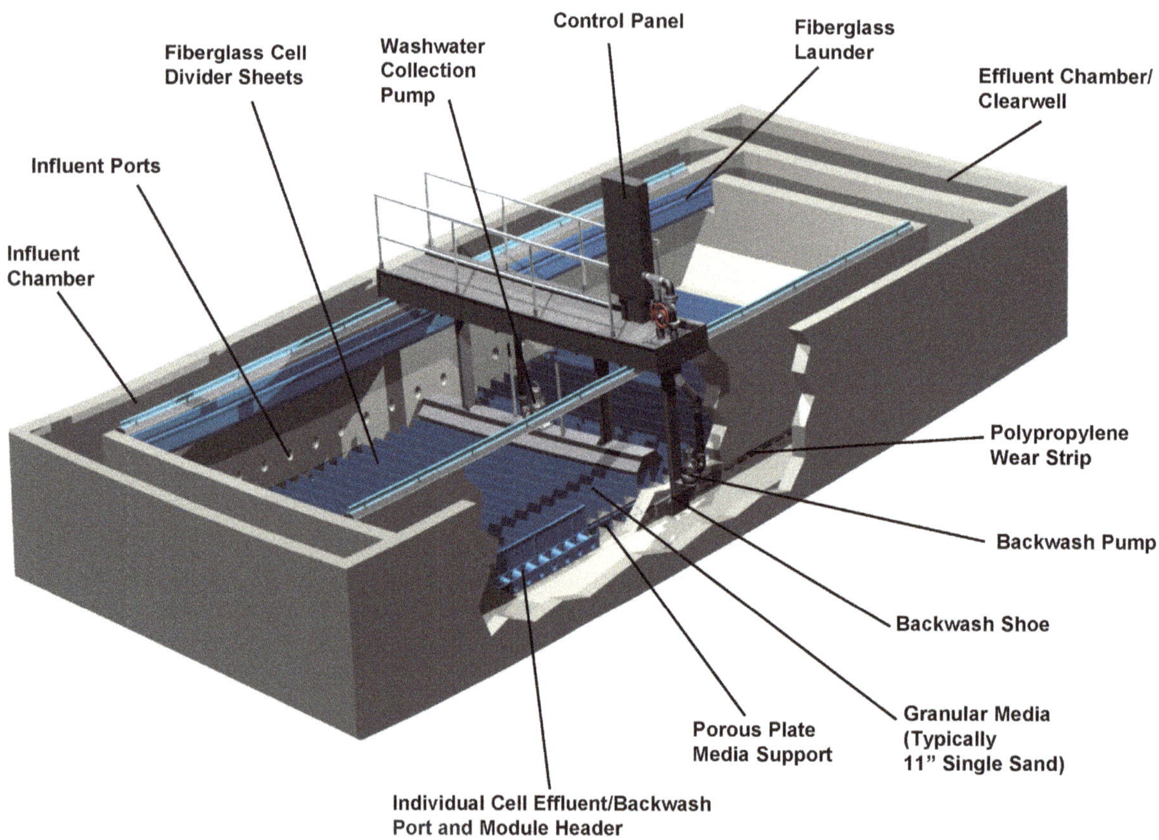

Source: Aqua Aerobic Systems

Figure 6-1 Traveling bridge filter

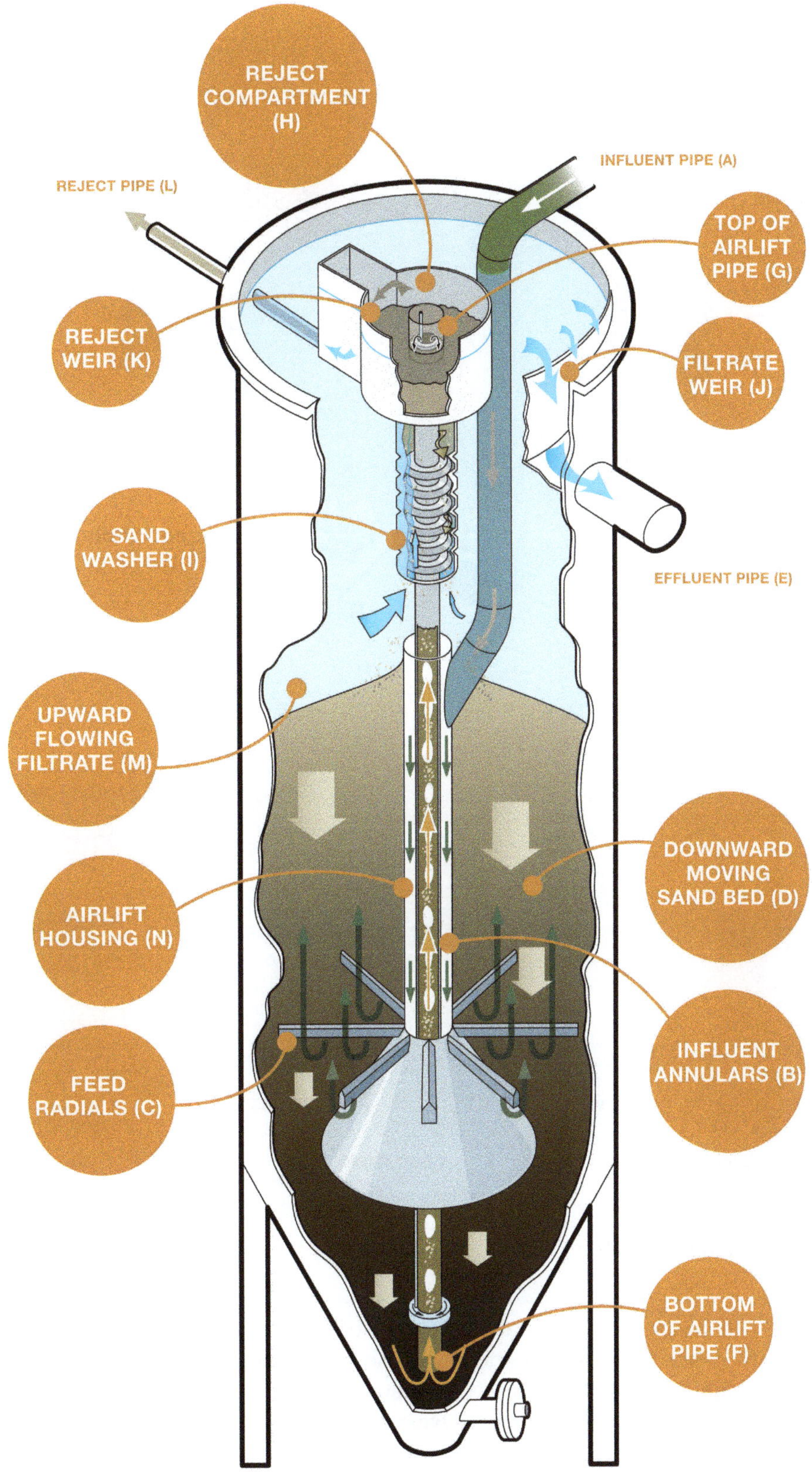

Source: Parkson Corporation

Figure 6-2 Moving bed filter

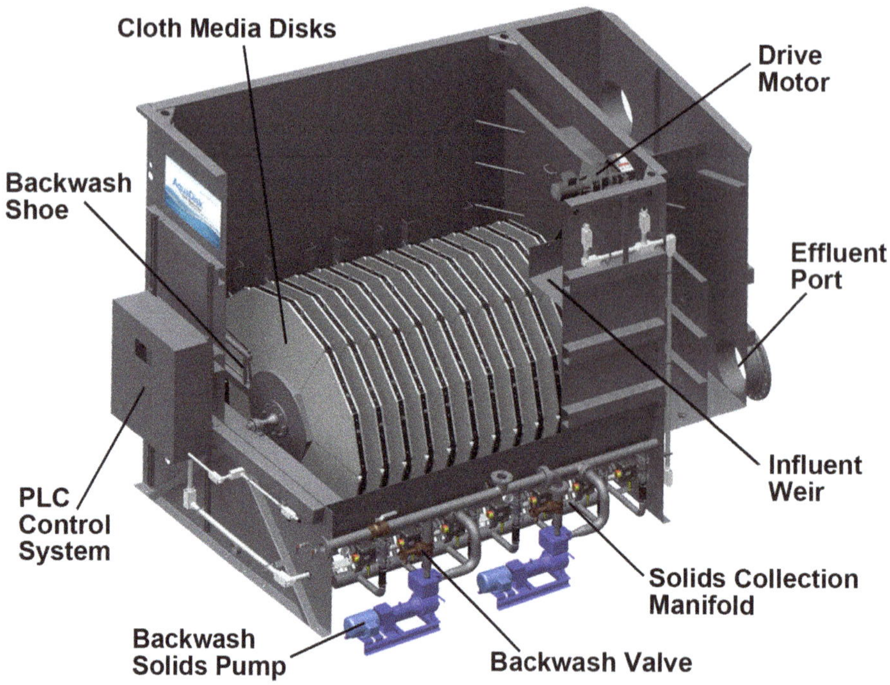

Source: Aqua Aerobic Systems

PLC—programmable logic controller

Figure 6-3 Cloth media filter

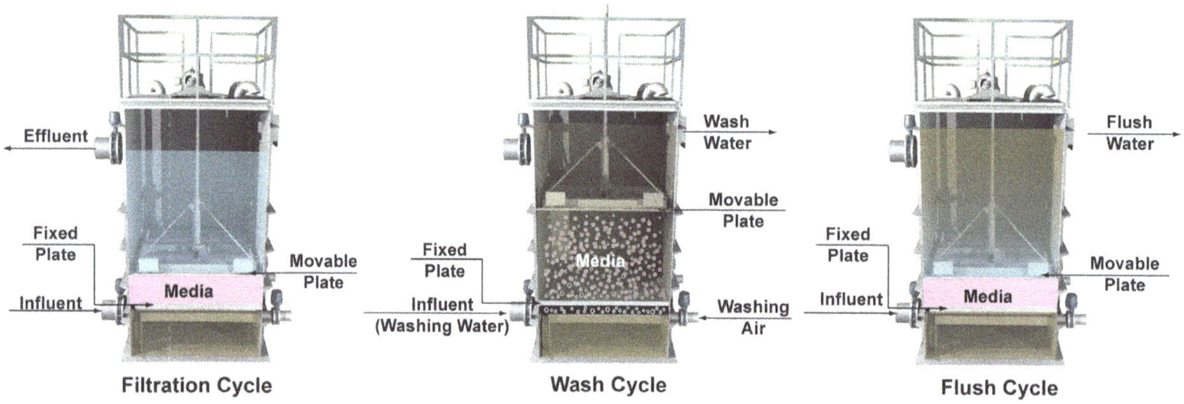

Source: Schreiber

Figure 6-4 Compressed media filter

In general, for flow rates in excess of 20 million gal/day (mgd), the lower relative maintenance requirements and backwash rates of conventional downflow granular filters outweigh less advantageous considerations (e.g., footprint). However, non-granular media filters are often chosen for wastewater applications over a wide range of flow rates because of the typical advantages related to continuous operation, solids loading capacity, footprint, and installed cost. Postsecondary membrane filtration is chosen when treatment performance is paramount, as the installed, operational, and maintenance costs are often much more expensive than other filtration technologies.

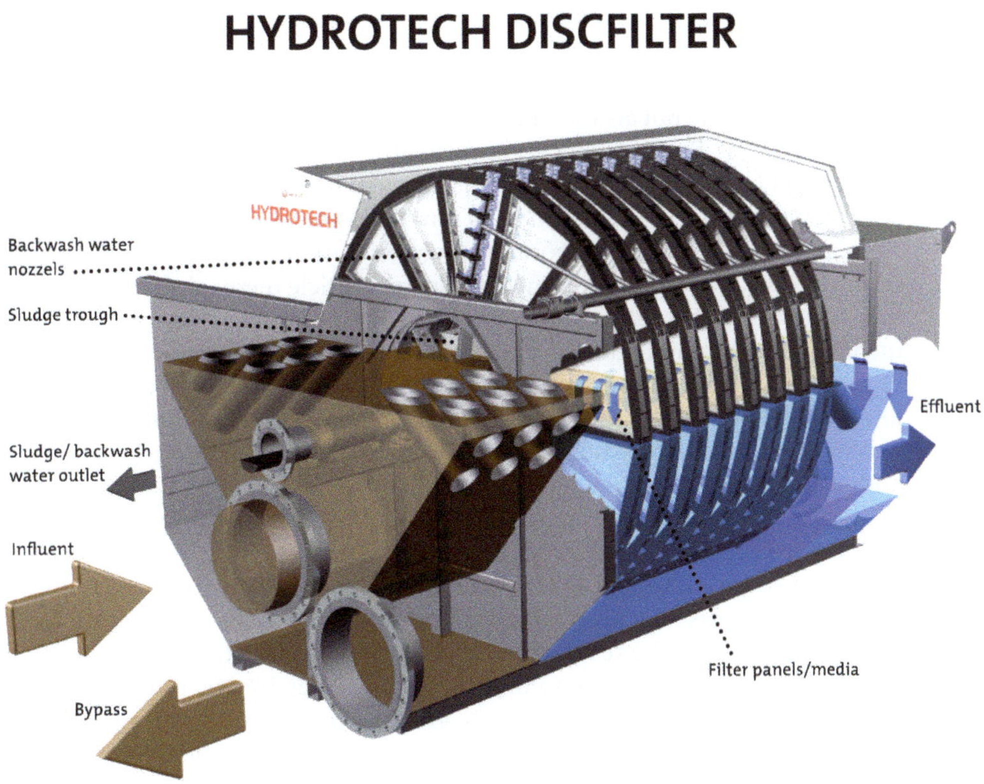

Source: Veolia Water Technologies

Figure 6-5 Disc filter

Comparison with Drinking Water Treatment

Similar coagulant chemicals are used for drinking water as well as postsecondary wastewater treatment; however, treatment objectives are different. Unlike for drinking water treatment, chemicals not certified by NSF International may be used for wastewater treatment. The primary correlation between drinking water and wastewater is tertiary treatment in which TSS removal requires different chemical application methods and dosage than turbidity removal for water treatment.

In secondary treated wastewater, a strong correlation exists between TSS concentration and turbidity (Tchobanoglous and Burton 1991):

$$\text{TSS, mg/L} = (2.3 \text{ to } 2.4) \times (\text{Turbidity, ntu}) \quad \text{(Eq 6-1)}$$

WWTPs that discharge into receiving waters with high assimilative capacity may have a TSS effluent requirement of 30 mg/L, which would correlate to 13 ntu. This result is two orders of magnitude higher than the treatment goal of many drinking water treatment facilities.

Another correlation may be drawn in secondary treated wastewater: Suspended solids are typically 3–5 percent phosphorus. As an example, an effluent TP concentration requirement may be 0.3 mg/L for a sensitive receiving water body. As such, to meet this requirement, effluent must contain less than 6–10 mg/L TSS. As shown above, this concentration range correlates to a turbidity as low as 2.5 ntu, which is well above regulatory requirements for drinking water treatment.

As discussed in Chapter 4, granular media filters use numerous mechanisms, including straining but also sedimentation, interception, and diffusion, to remove particles less than 10 μm. Specific filter conditions may also allow contaminant removal through biological metabolism by biofilm that is attached on the media surface. However, removal of relatively large (i.e., greater than 10 μm) wastewater secondary solids is typically sufficient to meet TSS and phosphorus concentration requirements in tertiary effluent. Thus, reliance on straining as the sole mechanism for particle removal allows for filter technology to be used for wastewater that cannot be used in drinking water treatment. In Table 6-2, media depth is minimal for several technologies. Because minimal-depth filters are predominantly used on flows less than 20 mgd, reject volumes from routine backwashing are relatively unimportant. However, using these types of filters for higher-capacity wastewater treatment facilities creates significant retreatment or reject volumes as well as added and more complex maintenance.

As described in previous chapters, should solids be captured only at the top of a granular media bed, rapid head loss would necessitate overly frequent backwashing. Pretreatment processes must be optimized to remove relatively large particles and allow filtration mechanisms to occur vertically through the media bed depth. Case study 4 in Chapter 10 presents the results of bench- and pilot-scale coagulation and filter testing to improve filter performance at Central San Wastewater Treatment Plant in Martinez, Calif.

Deeper designs of relatively large granular media (e.g., sand with 1.5–2.0-mm effective size) compensate for straining potential and produce acceptable wastewater filter effluent. Alternately, use of two or three stratified layers of media with increasing density and decreasing size (from top to bottom) removes larger solids near the bed top and smaller solids further down.

For capital cost reasons, many WRRFs do not have dedicated postsecondary unit process structures and equipment for coagulation and flocculation. Instead, chemicals are often dripped into secondary clarifier effluent launders or filter influent headboxes, injected into the filter influent piping, or added in other convenient locations. Thus, mixing and flocculation occur in piping and filter structures—not necessarily efficient, but often economic, given the variable necessity and requirements of chemical addition.

Monitoring and Process Evaluation

For non-reuse wastewater treatment facilities, monitoring of the coagulation and filtration processes is often limited. At some WRRFs, on-line TSS or turbidity sensors are used before filtration to alert operators of secondary clarification process problems, which may result in solids overload of the filters. Coagulant dosage may be controlled by facility flow rate, optimally after secondary clarification. Monitoring of orthophosphate is critical for controlling coagulant dosing in phosphorus removal applications. With no automated dosage control, costly overapplication and through-filter metal leakage may occur.

Postfiltration process evaluation is typically limited to filter effluent turbidity (often recorded as suspended solids). Analysis of composite sampling of facility effluent is performed a few times per week, according to the facility's permit requirements. Samples may be tested for TSS and phosphorus within a few hours to provide process evaluation.

Another very common wastewater effluent permit parameter is five-day biological oxygen demand (BOD_5). This parameter is defined as the milligrams of oxygen needed to break down the organic matter contained in a liter of water over five days. Most of the effluent BOD from a well-operated treatment facility is in particulate form. Therefore, to provide quicker process evaluation, facility personnel may draw a correlation over time between BOD_5 and TSS from respective composite samples. It is possible that permitted

BOD_5 effluent concentration may require TSS concentrations to be lower than its specific permitted effluent concentration. Attention must be paid to the parameters that control overall process operation.

COAGULATION AND FILTRATION IN WATER REUSE

Nonpotable Reuse Applications

Nonpotable reuse is a general term for water that is used more than one time before it passes back into the water cycle—for example, wastewater that has been treated to a level that allows for its reuse for a beneficial purpose such as irrigation; industrial processes; heating, ventilation, and air-conditioning; or toilet flushing.

Coagulation and filtration treatment for water reuse depends on the application and water quality requirement of the water to be treated. The removal of residual particulate matter remaining after conventional wastewater treatment (i.e., secondary effluent) is required because it does not remove all suspended solids (Khan 2013). Particulate matter and suspended solids contribute to turbidity, which could be associated with undesirable chemical contaminants or pathogens (bacteria, viruses, and protozoan cysts and oocysts). At a minimum, tertiary treatment is required in most applications. Tertiary treatment for recycled water production may include filtration only or a combination of coagulation, flocculation, and filtration.

To protect the public against the adverse health effects associated with pathogens, the quality of recycled water should meet federal, state, provincial, or local regulatory requirements. For drinking water treatment using conventional filtration (or direct filtration), the Surface Water Treatment Rule requires that the turbidity limit of the combined filter effluent does not exceed 0.5 ntu in 95 percent of measurements each month and 5 ntu at any time. In comparison, the turbidity of recycled water produced by tertiary treatment using granular media filter should not exceed 5 ntu more than 5 percent of the time within a 24-h period and 10 ntu at any time.

Potable Reuse Applications

Potable reuse is a general term for advanced treated water that is used to augment drinking water supplies (WateReuse 2015). Potable reuse covers both indirect and direct potable reuse. Coagulation and filtration treatment for water reuse depends on the application and water quality requirement of the water to be treated. Similar to nonpotable reuse, potable reuse applications require the removal of residual particulate matter. Additionally, the presence of particulates can negatively affect the performance of downstream treatment processes such as reverse osmosis (RO) and disinfection.

For potable reuse applications, various forms of advanced water treatment (AWT) technologies are used in combination with either environmental buffers or engineered buffers to ensure that all necessary public health requirements are met to allow the water to be used as source water for drinking water treatment plants. AWT technologies related to filtration include microfiltration (MF)–ultrafiltration (UF) (with or without coagulation) and RO. An alternative AWT technology that provides a potential substitute for RO, especially when brine discharge is not practical, is an ozonation process followed by biologically active filter (BAF) (USEPA 2017). More discussion on BAF can be found in Chapter 4.

Comparison with Drinking Water Treatment

Similar to drinking water treatment, coagulation and filtration are used for particle removal in water reuse applications. Particle removal serves dual purposes: removal of microbial agents including pathogens and those associated with particulates, colloids, or organics and improvement of the effluent quality for disinfection and other subsequent treatment processes. For tertiary treatment and AWT plants using MF or UF, they typically use filtration without coagulant, allowing dissolved organic constituents to pass through unchanged. Membrane filters can remove dissolved organic material when used in conjunction with coagulation.

Like drinking water, removal effectiveness is highly dependent on operating conditions and the feedwater quality. Rapid changes in flow or feedwater quality can result in high levels of particulate breakthrough in a media filter because of temporary nonoptimal coagulant doses. Lower concentrations of organics and other particulates after filtrations can reduce the demand for chemical oxidants in disinfection; additionally, lower turbidities can improve ultraviolet (UV) disinfection by increasing the UV transmittance and allowing more efficient pathogen inactivation.

Monitoring and Process Evaluation

Rigorous process control and monitoring protocols are required for potable reuse to ensure continuous public health protection. The purpose of monitoring is to assess process performance and trigger alarms if there is a change in normal operating conditions from a pathogen or chemical constituent outlook. On-line real-time and offline monitoring of reused water is essential to protecting public health.

AWT processes are fully capable of producing a comparable, if not superior, water quality when compared with existing drinking water supplies. However, the response time for failure mitigation potable reuse applications is inevitably shorter in potable reuse schemes. Therefore, accurate and robust monitoring and process control technologies are fundamental for the operations and performance of AWT processes.

Monitoring must focus on mitigating the risks to public health through application of critical control points (CCPs) for both pathogen and chemical control (USEPA 2017). CCPs are locations where essential unit processes at AWT facility occur, and powerful continuous monitoring techniques can evaluate process performance and provide fail-safe control. Fail-safe is not intended to mean failure proof, but it means that the system will revert to a safe condition if a critical component fails.

For direct potable reuse applications, there are more stringent monitoring practices that are set by ANSI/AWWA Standard G485, Direct Potable Reuse Program Operation and Management. For example, process control monitoring and CCP set points and alarms must be established to ensure the treatment system is reliably meeting performance goals and must allow adequate response time. Also, on-line analyzers and instruments used to monitor CCPs, treatment processes, and water quality parameters must be calibrated and checked for proper functioning at prescribed intervals on the basis of manufacturer recommendations.

COAGULATION AND FILTRATION IN SEAWATER TREATMENT

Desalination with RO

Seawater desalination with RO is one of the most common treatment processes that is used for the production of drinking water. Pretreatment of seawater, including coagulation, clarification, and filtration (either granular media or membrane), is required to protect the thin composite membranes from damage as a result of fouling. RO membrane fouling can cause significant decrease in the flow of permeated water. This reduced flow rate must be overcome by increasing the working pressure, resulting in increased energy and cost of operation. Seawater contains, in addition to dissolved inorganic salts, many types of organic compounds. These organic compounds include polysaccharide, amino acid, and lipids, which are byproducts from the metabolisms of microorganisms, such as plankton and marine bacteria. Seawater also contains NOM consisting of compounds that can form complexes with metals.

Coagulation for seawater treatment is usually accomplished by the addition of iron salts such as ferric chloride ($FeCl_3$) or ferric sulfate ($Fe_2(SO_4)_3$). For effective coagulation and subsequent flocculation of seawater, pH should be between 6.8 and 7.2. However, coagulation using iron salts at the natural pH of seawater (typically around pH 8.2) may nonetheless reduce colloids and NOM and produce good flocs with subsequent removal in filtration process. The high salinity of seawater has some important implications for coagulation chemistry.

For seawater treatment, aluminum salts such as aluminum chloride (alum) and PACl are not recommended to be used as coagulants, as the high salinity results in high aluminum solubility. See Chapter 1 for additional details.

Polymers can be used as an aid in seawater treatment for the formation of flocs and their subsequent removal in filtration process but are avoided in many instances because of concerns with RO membrane fouling.

It is important to evaluate key operational parameters in coagulation and filtration for seawater treatment. Jar testing is a convenient method to evaluate parameters such as pH, coagulant dose, mixing value (i.e., *G* value), and contact time because of changes in these parameters from the higher density of seawater.

Filtration for pretreatment of seawater RO is very similar to drinking water and wastewater. The quality control of filtered water is carried out by monitoring the silt density index (SDI_{15}), modified fouling index ($MFI_{0.45}$), turbidity, and residual Fe^{3+} concentration. Filters used in pretreatment of seawater may be gravity or pressure filters with anthracite and sand granular media or MF or UF.

In seawater subject to algae blooms, treatment is typically preceded by pretreatment with dissolved air flotation (DAF), which is well suited to float low-density particles.

Comparison with Drinking Water Treatment

The fundamental differences between seawater and other source water (e.g., reservoirs, rivers, groundwater) reside in the high salinity of seawater and differences in its organic compounds and algae. As a result, the type of coagulant and its dose can be different for seawater pretreatment compared with treatment of other source water for the production of drinking water. In the coagulation and flocculation of seawater, the coagulant used is an iron salt, with $FeCl_3$ or $Fe_2(SO_4)_3$.

The required mixing intensity in the flocculation process is different in seawater treatment. The required intensity varies depending on the quality of seawater and the coagulant used. Hydrolysis of the coagulant used (usually ferric chloride) is slower when the salinity is higher, and the flocculation takes longer.

In coagulation–filtration treatment applications in which a clarification process is used before a filtration process, flocs formed in seawater are usually removed using a DAF process.

Monitoring and Process Evaluation

Monitoring of key water quality parameters is necessary for process control in seawater RO pretreatment.

Some of these parameters may be measured using on-line instrumentation (e.g., total organic carbon [TOC] analyzer); however, others must be analyzed in a laboratory using samples collected from various sampling points. Table 6-3 provides a summary of water quality parameters that are typically monitored in seawater RO pretreatment.

COAGULATION AND FILTRATION IN STORMWATER TREATMENT

Objectives

While separate stormwater systems provide much needed flood control and reduction, the collection and discharge of stormwater runoff into natural waterways results in a decrease in water quality. To reduce the negative effects of this practice, regulators have developed receiving water quality standards, similar to those for wastewater discharges. To maintain these standards, governmental agencies mandate treatment requirements for stormwater that is discharged from these systems.

The primary objective in the treatment of stormwater, similar to that of wastewater treatment, is to reduce negative effects on the receiving water body to protect public health, the environment, and quality of life. Stormwater runoff pollutants include suspended solids, nutrients, metals, bacteria, and hydrocarbons. They are introduced to the

Table 6-3 Water quality parameters monitored in seawater RO pretreatment

		Sample Analysis*,†	
Parameter	Sampling Location	On-line	Laboratory
Turbidity	Feed seawater, after sedimentation or DAF, before and after filtration	✓	✓
Suspended solids	Feed seawater, after coagulation, after sedimentation or DAF, before and after filtration	N/A	✓
TOC	Feed seawater, after sedimentation or DAF, before and after filtration, and after RO	✓	✓
pH	Feed seawater, after coagulation, after sedimentation or DAF, before and after filtration, and after RO	✓	✓
ORP	Feed seawater, after coagulation, after sedimentation or DAF, before and after filtration, and after RO	✓	✓
Conductivity	Feed seawater, after coagulation, after sedimentation or DAF, before and after filtration, and after RO	✓	✓
Particle count	Feed seawater, after sedimentation or DAF, before and after filtration	✓	✓
Residual coagulant (Fe^{3+})	After sedimentation or DAF, after filtration, RO feedwater	N/A	✓
Boron	Feed seawater (rarely measured), after RO	✓	✓
Redox/ORP	Feed seawater and feed RO	✓	✓
Free chlorine	Before and after feed seawater chlorination, before RO feed	✓	✓
SDI_{15}‡	Feed seawater, after filtration, and RO feedwater	✓	✓
$MFI_{0.45}$§	Feed seawater, after filtration, and RO feedwater	✓	✓
Oil and grease	Feed seawater, after coagulation, after sedimentation or DAF, after filtration	N/A	✓

DAF—dissolved air flotation, MFI—modified fouling index, N/A—not applicable, ORP—oxidation–reduction potential, redox—oxidation–reduction, RO—reverse osmosis, SDI—silt density index, TOC—total organic carbon

* Other analytical methods may be found in the *Standard Methods for the Examination of Water and Wastewater*, 24th Edition (2022).

† Standard analytical methods for some parameters are not suitable for seawater samples.

‡ The determination of SDI must be carried out according to ASTM Standard D4189-07, Standard Test Method for Silt Density Index (SDI) of Water.

§ The determination of $MFI_{0.45}$ must be done according to ASTM Standard D8002-15, Standard Test Method for Modified Fouling Index ($MFI_{0.45}$) of Water.

watershed from a variety of sources, such as soil erosion, fertilizers, animal waste, and automobiles.

The approach traditionally implemented to treat stormwater runoff is through the installation of best management practices (BMPs), which are measures, both structural and nonstructural, to reverse the effects of development on hydrology and water quality. Examples of commonly used structural BMPs include bioretention facilities, engineered wetlands, sand filters, and retention ponds or lakes. These systems help treat the stormwater before it enters the storm sewer network. The removal mechanism of these BMPs includes sedimentation, filtration, and infiltration.

These treatment processes perform best when implemented in a series, or a "treatment train." A treatment train approach incorporates two or more BMPs at a time. BMPs based on physical removal processes (to remove larger solids and floatables) implemented together with BMPs that promote biological and chemical processes (to remove suspended and dissolved pollutants) produce robust treatment.

Coupling coagulation and filtration processes has been found to be successful in the treatment of suspended and other dissolved pollutants such as nutrients, bacteria, metals, and hydrocarbons. The most commonly used coagulant for stormwater treatment is aluminum sulfate (alum). Other coagulants include ferric chloride, chitosan, polyacrylamide, and a mixture of lanthanum and bentonite clay. When added to water, alum produces the precipitates aluminum hydroxide ($Al(OH)_3$) and aluminum phosphate ($Al(PO)_4$). Dissolved phosphorus is removed directly through the chemical reaction that produces $Al(PO)_4$. All other contaminants are removed through the sorption to the $Al(OH)_3$ precipitate. Typical alum dose ranges from 5 to 10 mg/L. Table 6-4 outlines the pollutant removal ability of alum to treat stormwater.

Jar and pilot testing is important to determine coagulation parameters, such as the coagulant dose, settling time, and the type of demand buffer application to control the pH. The volume of alum floc that will be produced can also be calculated during this step. The volume anticipated will dictate required storage and maintenance frequency. Because the coagulation process must occur in an environment where the stormwater has enough time to interact with the coagulant and settle the floc, the application of coagulation to treat stormwater runoff is best suited for areas with existing lakes and ponds.

Table 6-4 Percentage removal efficiencies of alum-treated stormwater runoff

Parameter	Without Alum (24 H of Settling)	Alum Dose *mg/L as Al*		
		5	7.5	10
Dissolved organic nitrogen	20	51	62	65
Particulate nitrogen	57	88	94	96
Total nitrogen	20*	65*	71*	73*
Dissolved orthophosphate	17	96	98	98
Particulate phosphorus	61	82	94	95
Total phosphorus	45	86	94	96
Total suspended solids	70	95	97	98
Biological oxygen demand	20	61	63	64
Total coliform	37	80	94	99^+
Fecal coliform	61	96	99	99^+

* Dependent on the type of nitrogen species present

Options for coagulant addition include mechanical injection into the stormwater system before entering the pond or manual application by boat throughout the pond. If using a mechanical system, the dosing system and pond outlet should be a minimum of 30.5 m (100 ft) apart to ensure proper mixing and settling time. Chemical injection pumps for coagulant dosing are coupled with flowmeters to apply a dose based on flow rate. During the chemical dosing process, it is important to ensure that the precipitation process is supported through mixing. Maintaining a pH of 5.5–7.5 is also critical to ensure aquatic organisms are protected during the treatment process. To maintain a stable pH, a buffer (such as sodium hydroxide) can be applied. Similar to the coagulant dosing system, the buffer may be applied using mechanical methods that monitor the flow rate and apply the chemical at the dosage determined in the laboratory testing process.

Removal of floc that accumulates on the bottom of a pond can be performed using mechanical dredging methods. To make removal easier, a sump area may be constructed within the pond where most of the material will be deposited. A common disposal method of pond solids is into the sanitary sewer system.

Immediately following coagulation, stormwater may be filtered to remove most of the remaining suspended particles. There are a variety of filtration BMPs that may be implemented, such as bioretention and sand filters. While these BMPs provide a filtration mechanism, they do not rely on the coagulation process to be effective. Tables 6-5 and 6-6 outline the inflow and outflow characteristics of stormwater runoff for bioretention and media filter BMPs, respectively.

Bioretention BMPs are depressed vegetated areas that treat the stormwater runoff through infiltration. They are known for their ability to treat suspended solids. During the design process, the volume, geometry, drain time, vegetation, and media selection will drive decisions regarding the inlet, outlet, underdrain, and irrigation infrastructure. An impermeable liner should be added so that the surrounding infrastructure is not affected. If the system is to be designed and implemented without a liner, a geotechnical engineer should perform an investigation to negate adverse effects. Drought tolerant vegetation should be planted to reduce long-term irrigation demands. Vegetation root depths should not exceed liner depth.

Sand or media filter BMPs are a type of infiltration BMP in which the stormwater runoff is routed through a sand bed and out of an underdrain. The design considerations are similar to bioretention BMPs, as they both rely on infiltration to treat the stormwater. Similar to bioretention facilities, the volume, geometry, drain time, and media selection will drive decisions regarding the inlet, outlet, underdrain, and irrigation infrastructure. The primary difference is that sand filters are not meant to support vegetation. Special consideration should be made to provide energy dissipation at the inlet of the sand filter to ensure that erosion does not occur.

The application of chemical coagulation and filtration of stormwater runoff is beneficial for the secondary treatment of dissolved and fine suspended particles. While the cost, engineering, and effort of coagulation exceed many other treatment options, its ability to

Table 6-5 Median average removal efficiency of bioretention BMPs

Parameter	Inflow *mg/L*	Outflow *mg/L*
Total nitrogen	1.46	1.15
Total phosphorus	0.13	0.13
Total suspended solids	44.6	12.9

Table 6-6 Median average removal efficiency of sand filter BMPs

Parameter	Inflow *mg/L*	Outflow *mg/L*
Total nitrogen	1.51	0.63
Total phosphorus	0.20	0.11
Total suspended solids	44.0	8.0

remove target contaminants in a controlled way may be desirable for municipalities interested in meeting regulated effluent requirements.

Comparison with Drinking Water Treatment

Primary differences in stormwater treatment compared with drinking water are the treated water quality goals. The objective of stormwater effluent from the coagulation and filtration process is to have a quality that supports the environment, public health, and recreation uses. As a result, special attention must be placed on understanding the byproducts produced during the coagulation process. While alum treatment has been found to produce nontoxic byproducts, laboratory tests should be performed before wide-scale implementation to understand effluent water quality.

However, similarities do exist. Despite the variance of source water between drinking water and stormwater, similar contaminants are present. Thus, the same chemical processes may be used. Additionally, like in drinking water treatment, water reuse, and wastewater treatment, sand and media filtration is used to filter coagulated and flocculated particles.

Monitoring and Process Evaluation

Following construction, a monitoring and process evaluation plan may evaluate performance, prioritize maintenance, and document the effectiveness of stormwater treatment measures. Specific plans must define the treatment objectives and performance criteria and outline the monitoring procedures. Because factors could influence the treatment process, including the design elements, watershed conditions, and maintenance activities, qualitative and quantitative metrics must both be included in the monitoring and process evaluation plan.

The National Stormwater BMP database has provided guidance for monitoring and reporting BMP functions. In addition to details of the general watershed, BMP location, and monitoring equipment, a monitoring plan must require evaluation and reporting of site-specific precipitation, flow, and water quality. Precipitation information is important to evaluate the BMP's treatment ability during varying storm intensity. Flow data and water quality data, specific to the BMP of interest, may be used to calculate BMP removal efficiencies.

REFERENCES

ANSI (American National Standards Institute)/AWWA. 2018. G485. Direct Potable Reuse Program Operation and Management. Denver: AWWA.

APHA, AWWA, and WEF (American Public Health Association, American Water Works Association, Water Environment Federation). 2023 (24th ed.). *Standard Methods for the Examination of Water and Wastewater.* W.C. Lipps, E.B. Braun-Howland, and T.E. Baxter, editors. Washington, D.C.: APHA, AWWA, and WEF.

ASTM International. 2007. D4189. Standard Test Method for Silt Density Index (SDI) of Water. West Conshohocken, Pa.: ASTM International.

ASTM International. 2015. D8002. Standard Test Method for Modified Fouling Index (MFI-0.45) of Water. West Conshohocken, Pa.: ASTM International.

AWWA. 1999 (5th ed.) *Water Quality & Treatment: A Handbook of Community Water Supplies.* New York: McGraw-Hill.

Droste, R.L., and R.L. Gehr. 1997. *Theory and Practice of Water and Wastewater Treatment.* Hoboken, NJ Wiley.

Geosyntec Consultants and Wright Water Engineers. 2011. *International Stormwater Best Management Practices (BMP) Database Pollutant Category Summary: Solids (TSS, TDS and Turbidity).* Greenwood Village, CO: Geosyntec Consultants.

Khan, S. 2013. Drinking Water Through Recycling: The Benefits and Costs of Supplying Direct to the Distribution System—A Report of a Study by the Australian Academy of Technological Sciences and Engineering (ATSE)." Melbourne, Australia: ATSE.

Lefebvre, E., and B. Legube. 1993. "Coagulation–Flocculation by Ferric Chloride of Some Organic Compounds in Aqueous Solution." *Water Research.* 27(3):433-447.

Narasiah, K.S., C. Morasse, and J. Lemay. 1991. "Nutrient Removal from Aerated Lagoons Using Alum and Ferric Chloride – A Case Study." *Water Science & Technology.* 23(7-9):1563-1572.

Tchobanoglous, G., and F.L. Burton. 1991. *Wastewater Engineering: Treatment, Disposal, and Reuse.* New York: McGraw-Hill.

USEPA. 2017. *2017 Potable Reuse Compendium.* EPA-CDM CRADA 844-15. Washington: USEPA.

USEPA. 2010. "Chapter 5. Technology-Based Effluent Limitations." In *NPDES Permit Writers' Manual.* EPA-833-K-10-001. Washington: Water Permits Division, Office of Wastewater Management, Office of Water, USEPA. www.epa.gov/sites/production/files/2015-09/documents/pwm_chapt_05.pdf (accessed May 13, 2019).

USEPA. 1999. *Preliminary Data Summary of Urban Storm Water Best Management Practices.* EPA-821-R-99-012. Washington: Office of Water, USEPA.

WateReuse. 2015. *Framework for Direct Potable Reuse.* WateReuse Project 14 -20. Alexandria, VA.

Chapter 7

Jar Testing and Bench Testing

INTRODUCTION

When evaluating process changes to modify or optimize water treatment, it is neither practical nor desirable to test the changes at full scale. Jar testing provides a valuable tool for realistically simulating coagulation, flocculation, and sedimentation at a full-scale treatment plant without requiring the quantity of source water, chemicals, and water disposal capacity that would be required for full-scale testing. Jar testing approximates, at bench scale, the conventional treatment steps that occur in the treatment train of the full-scale plant. In general, the jar test method is capable of sufficiently duplicating chemical treatment processes; however, there are occasionally limitations to duplicating more advanced physical treatment processes at bench scale.

Jar testing may be conducted to evaluate different facets of water

Chapter Summary

- Jar testing is only useful to the extent that it can simulate full-scale plant conditions. Even when theoretical conditions are matched closely in the jar test procedure, the clarified or filtered water quality rarely matches full-scale implementation, and there is often a need to adjust the test parameters to match jar test results to full-scale observations.
- A jar test protocol should be developed before any testing is conducted or any experimental data are collected.
- Successful interpretation of jar test results is possible only when adequate data have been collected and recorded.
- Although jar testing often focuses on coagulation and filtration, other treatment processes can be evaluated at bench scale, including oxidation, softening, and adsorption.

treatment. Tests can be performed to evaluate the effects of changes in chemical dosages and points of application; choose alternative coagulants; add polymeric coagulant aids; implement alternative preoxidation strategies; vary mixing intensities and times; and change overflow rates on the removal of particles, natural organic matter, or other water quality parameters of concern. Although there are many potential facets of mixing and flocculation that can be evaluated through jar testing, chemical doses are the most directly transferrable data from bench to full scale.

This chapter covers jar and bench testing for many common treatment processes. Jar testing can be conducted for several other processes not included in this chapter, but these processes have specific testing requirements that are better described in manufacturer operations and maintenance (O&M) manuals for those specific technologies.

Types of Jar Testing

Bench-scale evaluation of sedimentation. Bench-scale evaluation of sedimentation is, perhaps, the most common form of jar testing practiced at water treatment . The objective of this testing is typically to optimize coagulation and flocculation to minimize settled water turbidity. This is commonly performed in response to changes in source water quality (e.g., temperature, turbidity, total organic carbon [TOC]). Depending on the variability of the source, jar testing to optimize coagulation and flocculation for sedimentation may be required anywhere from seasonally to several times per day.

The most common independent variable to evaluate during jar testing of sedimentation is chemical dose. In this scenario, multiple test jars would be run side by side, each with a different dose of the plant's primary coagulant and/or coagulant aid polymer. Alternate coagulants or polymers may also be evaluated. In this scenario, the controlled variable would be the rapid mix and flocculator speeds, which would be set to simulate the full-scale treatment plant. After mixing and flocculation, turbidity samples would be drawn from the jars at set intervals to allow for the construction of a turbidity versus settling time plot.

If the full-scale treatment plant is designed to allow for the easy adjustment of rapid mix or flocculator speed, jar testing may also investigate altering mixer speeds to optimize coagulation. This is typically not performed as frequently as adjustment to coagulant dosage, but some facilities may adjust mixing parameters on a seasonal basis to account for changes in water temperature and/or viscosity. In this case, multiple rounds of jar testing are recommended. The first would focus on optimizing chemical dosages at a set mixing condition; once the chemical dosages are set, they will be held as the control variable while rapid mix and/or flocculator speeds are varied between jars. Even though changes in rapid mix or flocculation conditions may lead to changes in settled turbidity, those changes may not always result in a discernable effect on the overall filter performance. Therefore, it is recommended that jar test results be verified at full scale, using full-scale unit filter run volumes (UFRVs) as the key performance indicator to determine whether the changes were beneficial.

Bench-scale evaluation of filtration. Most source waters require chemical pretreatment to produce a filterable floc, so filtration testing is usually conducted along with coagulation and flocculation jar testing, either with an intermediate sedimentation step or without (also known as "direct filtration").

An exception is bench-scale testing of biofiltration. Bench-scale testing is often used to explore certain aspects of biofiltration, such as the removal of specific compounds like 2-methylisoborneol and geosmin (Elhadi et al. 2006) or compounds shed by biofilters (Anderson et al. 2008). In these instances, the feed to the biofilters is often tap water that has been dechlorinated and supplemented with nutrients and/or compounds being assessed for removal.

One area in which bench-scale filtration testing is used in an academic setting is membrane filtration. Such tests are used to assess specific operational parameters of membrane treatment, such as flux decline. Allgeier and Summers (1995) provide a protocol for a rapid bench-scale membrane test, whereas DiGiano et al. (2000) present protocols for two alternative bench-scale filtration tests: a batch recycle membrane test and a batch internal recycle membrane test. Although bench-scale membrane testing methods are well established, they are not often used in a utility setting.

Bench-scale evaluation of filtration following sedimentation. Together, sedimentation and filtration are the primary mechanisms for solids removal in water treatment. Although both processes can be tested at the bench-scale separately, if the goal is to evaluate overall removal of solids and/or specific contaminants in water treatment, it can be beneficial to combine these processes during bench testing. The specific test arrangement should be defined by the test goals. If the goal is to predict filtered water quality and continuous-flow testing is not required, batch testing of filtration following sedimentation can be accomplished simply by filtering jar test supernatant through paper, cellulose nitrate, or glass-fiber filters to simulate full-scale filter performance (Weiss et al. 2003, Hering et al. 1997, Hudson and Wagner 1981). For testing in which the evaluation of a continuous flow system is desirable, some researchers such as Bundy et al. (2007) constructed flow-through treatment systems including coagulation, flocculation, sedimentation, filtration, and disinfection. Such systems essentially function as very small pilot-test apparatuses and should be operated as such.

Bench-scale evaluation of direct filtration. Direct filtration can be a difficult process to optimize because, unlike sedimentation, jar tests cannot rely on settled water turbidity for a signal to indicate optimal coagulant dosing. Instead, it is necessary to filter the water following pretreatment (coagulation and flocculation) to assess performance efficacy. Although some researchers have approximated direct filtration performance using standard filter paper (Wagner and Hudson 1982), it is more common for researchers to use bench-scale granular media filters to assess filter effluent quality from direct filtration processes (Graham et al. 1992, Tanaka and Pirbazari 1986, Scheuch and Edzwald 1981). A step-by-step procedure for conducting a filter index test using filter paper is provided in Appendix A. Alternatively, there are commercially available bench-scale systems that combine jar testers with small-diameter (~2-in. diameter) filter columns that can be loaded with granular media to evaluate jar testing, as shown in Figure 7-1. When using such systems, testers should take care to maintain similitude to the full-scale system to the extent possible, minimize wall effects, and recognize limitations between the volume of water that can be flocculated in the jar compared with the volume of downstream filter column.

Many full-scale direct filtration facilities favor zeta potential testing for evaluating coagulation effectiveness. Zeta potential testing analyzes the particle charge in colloidal suspensions. As zeta potentials approach zero, the interparticle repulsive forces that maintain a stable colloidal suspension decrease, which allows particles to approach each other and agglomerate into stable flocs. Bench-scale zeta potential testing can be used to optimize pretreatment in direct filtration facilities, which in turn will improve filter productivity and filter effluent quality (Pernitsky et al. 2011).

Bench-scale evaluation of ballasted flocculation processes. Jar test procedures used for the evaluation of sedimentation can be modified to evaluate ballasted flocculation through the incorporation of microsand or glass beads during the rapid-mix/ flocculation step (Desjardins et al. 2002, Smith and Edwards 2002). Ballasted flocculation jar testing can be used to optimize the dosages of coagulant, polymer, and/or ballast material used by the process.

In addition to assessing settled water turbidity, it may be important to evaluate filterability of the settled water. Desjardins et al. (2002) found that although coagulant dosage and pH are the most important parameters controlling settled water quality, filterability of the settled water was sensitive to the ratio of ballast to coagulant and polymer.

Bench-scale evaluation of dissolved air flotation plants. Bench-scale testing of dissolved air flotation (DAF) is similar to that of ballasted flocculation in that it largely mirrors jar testing of sedimentation, with a modified process. For DAF testing, a source of

Source: Phipps & Bird

Figure 7-1 Example of combined jar and filtration testing apparatus

compressed air, a saturator, and appurtenances to distribute air-saturated water to the jars is required. See Figure 7-2 for an example of a DAF jar test apparatus.

Testing can be performed using standard jar test equipment that is modified to be dosed with the air-saturated water (Jarvis et al. 2009, Bunker et al. 1995), or a taller graduated cylinder can be modified to accommodate the coagulation–flocculation and flotation steps (Gonzalez Galvis 2019, Teixeira et al. 2010). In addition to the standard parameters evaluated during jar testing for sedimentation (e.g., coagulant and/or polymer dose, retention time, mixing *G*-values), parameters associated with the flotation process should be considered, including flotation time, saturator pressure, and recycle ratio.

PREPARING FOR A JAR TEST

A jar test protocol should be developed before conducting any testing or collecting any experimental data. While establishing a realistic and useful jar test protocol requires time and effort, a well-planned and organized approach for jar testing will save time in the long run by reducing guesswork and identifying the critical resources and information needed to ensure that a jar test yields useful data. For more information on preparing for a jar test,

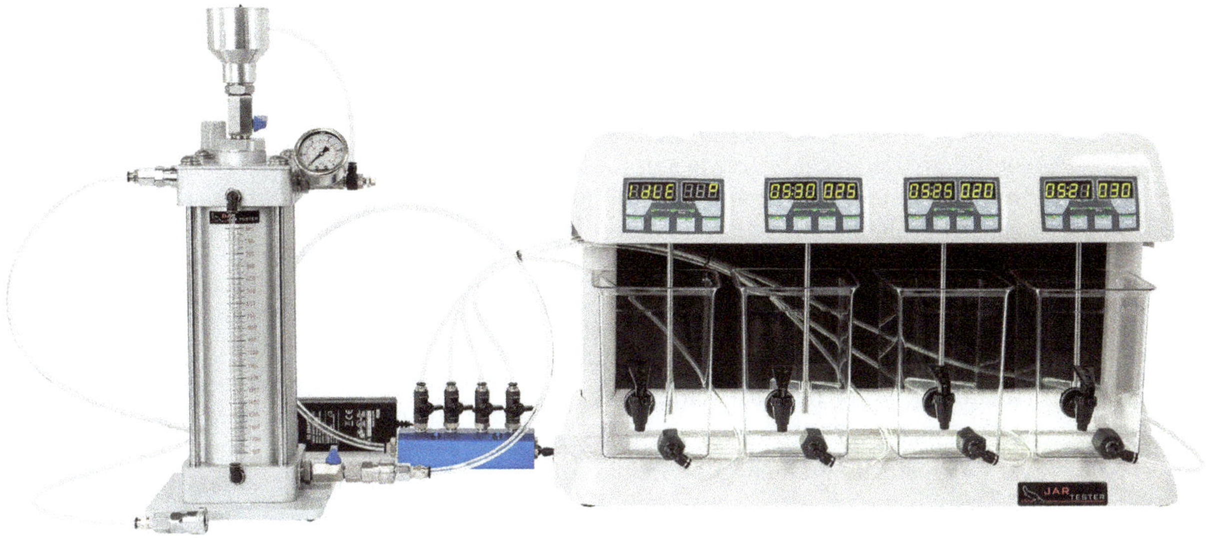

Source: Platypus

Figure 7-2 Example of bench-scale DAF testing apparatus

see M12, *Simplified Procedures for Water Examination,* or one of the readily available generic jar testing standard operating procedures (SOPs).

Defining Study Goals

Jar tests can be conducted for several reasons, so it is critical to define the objective of the test before beginning. The most common use of the jar test is for day-to-day process control, but it may also be used for other reasons. For example, goals might include

- evaluating alternative chemicals such as coagulants, pH controls, preoxidants, coagulant aids, or adsorbents;
- optimizing the coagulant conditions (type, dose and pH) for charge neutralization and turbidity removal;
- optimizing the coagulant conditions for TOC removal; and
- evaluating physical modifications, such as varying mixing intensities and points of chemical application.

Water quality parameters that are commonly used to assess treatment performance include

- turbidity and color, before and after jar testing;
- dissolved organic carbon (DOC) concentration or ultraviolet (UV) absorbance at 254 nm (UV254);
- disinfection byproducts (DBPs) and/or DBP formation potential;
- iron (Fe) and manganese (Mn) concentrations, before and after jar testing;
- taste and odor of the treated water;
- sludge characteristics;
- algae concentrations;
- residual aluminum (Al) levels;
- coagulation pH; and
- floc charge measurements (zeta potential or streaming current).

Correlating to Full-Scale WTP

Jar testing is only useful to the extent that it can accurately simulate full-scale plant conditions. Doing this requires knowledge of the hydraulic characteristics of the full-scale treatment steps: rapid mixing (also called flash mixing), flocculation, and clarification, which will be defined by the basin geometry and flow characteristics of the full-scale plant. To achieve similitude between the jar testing and full-scale treatment, it is necessary to know the following key parameters:

- velocity gradient-mixing intensity in the rapid mix and flocculation basins,
- retention times in the rapid mix and flocculation basins, and
- surface loading rate (SLR) of the sedimentation basin.

> NOTE: Even when theoretical conditions are matched closely in the jar test procedure, there is often a need to adjust the test parameters to match jar test results to full-scale observations. Customizing a jar test procedure so that it can yield results indicative of plant performance is an iterative process. Theoretical parameters can be used as a starting point and then adjusted until the full-scale plant results are adequately simulated by the jar test. An example of this is presented as a case study, "Jar Test Calibration," in Chapter 10.

Other full-scale treatment parameters may be relevant depending on the goals for the jar testing. For example, if jar testing is being used to evaluate time-dependent reactions, it would be necessary to know the actual retention time in the full-scale treatment basins to accurately model the influence of time on the results.

As described in Chapter 2, the intensity of mixing in rapid mixing and flocculation processes is generally quantified by the velocity gradient, *G*, which has units of s^{-1}. For jar test results to be relevant to the full-scale treatment plant, the *G*-values during jar testing should correspond to those in the treatment plant. The velocity gradient varies significantly with water temperature (viscosity) independent of the mixing-device speed. Accordingly, jar tests should be conducted with water temperatures the same as or as near as possible to the plant water temperature conditions to simulate plant conditions. In practice, this can be accomplished by using water baths or storing water samples in temperature-controlled refrigerators set to match water temperatures in the full-scale plant. In all cases, the temperature of the sample water during jar testing must be recorded.

Initial mixing (rapid mix). When a jar test is performed, the rapid mixing time should be equivalent to the retention time of the mixing chamber in the treatment plant, and *G*-values should match the full-scale conditions, if possible. Typical mixing intensities for a well-designed flash-mix basin range from 300 to 1,000 s^{-1}. In a typical jar test, initial mix durations of 30–60 s are used, with paddle speeds of 100–300 revolutions per minute (rpm). If rapid mixing in the plant is done by static mixers, inline mixers, or chemical induction mixers, simulation of the full-scale mixing conditions may not be possible using typical jar test equipment.

Flocculation. In jar testing, the retention time and *G*-values of the flocculation basin should correspond to those in the WTP. If the plant performs tapered flocculation in different compartments, it would be necessary to vary the mixing speeds over the course of the jar test, with the more intense mixing steps occurring first followed by the less intense mixing.

When using jar tests to evaluate flocculation, in jar tests, flocculation is done on a batch basis for a specific retention time. At full scale, on the other hand, retention times for floc particles in continuous-flow flocculation basins will vary over a range because of the short-circuiting that inevitably occurs in those basins. If experience shows that the results of jar testing diverge from those seen at full scale when evaluating flocculation at the same retention time, it may be necessary to develop an empirical correction to the jar test flocculation time to account for the short-circuiting that will occur at full scale.

Velocity gradient. The design documents or O&M manual for mixing equipment should provide data for actual in-plant velocity gradients for the rapid-mix and the flocculation basins. If the information is not readily available, use the procedure for calculating velocity gradient described in Chapter 2.

Detention times. The theoretical detention or retention time is defined by the following equation:

$$\text{Detention Time} = \frac{\text{Tank Volume}}{\text{Flow rate}} \tag{Eq 7-1}$$

For example, the detention time of a tank with the dimensions 7 ft (2.14 m) × 6 ft (1.83 m) × 10 ft (3.06 m) (depth × width × length) operating at a flow rate of 105 gpm (23.9 m^3/h) would be as follows:

$$\frac{7\text{ft} \times 6\text{ft} \times 10\text{ft}}{105\,\frac{\text{gal}}{\text{min}}} \times 7.48\frac{\text{gal}}{\text{ft}^3} = 30\text{ min} \tag{Eq 7-2}$$

$$\frac{2.14\text{ m} \times 1.83\text{ m} \times 3.06\text{ m}}{2.39\,\frac{\text{m}^3}{\text{h}}} \times 60\frac{\text{min}}{\text{h}} = 30\text{ min} \tag{Eq 7-3}$$

As shown, the fixed basin dimensions and the flow rates through each basin can be used to readily calculate the theoretical detention time through a treatment process. However, short-circuiting in the process units can cause the effective detention time through a plant's basin to be significantly lower than the calculated theoretical detention times. Special studies may need to be performed to characterize the effective detention time. Finally, make sure to account for not only the size of each basin but also the number of basins in service when dividing the volume by the flow rate.

Sedimentation. When jar testing is used to evaluate sedimentation, a key test objective is to optimize floc development and settleability. This assumes type of sedimentation in the sedimentation process is flocculant settling (type II), as described in Chapter 3. Also, because the depth of the jar apparatus is limited, it assumes the settling velocity of the flocs remains constant, which may not be the case for deep sedimentation basins or unusually configured sedimentation basins.

The sedimentation basin overflow rate, or surface loading, determines the length of time that the water is allowed to settle in a jar before a sample is taken. The time of sedimentation in the jar test for turbidity removal may be as low as 5 min.

The contact time in the full-scale basin may range from 2 to 4 h. Thus, the settling time needed in the jar test can be a fraction of that in the full-scale treatment process. Short holding times for turbidity removal are not adequate for assessing chemical reactions that may be occurring in the basin, such as formation of DBPs or for evaluating the efficacy of powdered activated carbon (PAC). If the purpose of the bench-testing procedure is to determine a reaction time-dependent parameter during the treatment process, it will be necessary to hold the jars for the actual retention time in full-scale process basins rather than to take samples at shorter times related to the basin overflow rate.

Sedimentation is evaluated through jar testing by measuring the turbidity of the treated water, after coagulation and/or flocculation mixing has stopped, from samples drawn from a port in the jar that is submerged below the water surface. Standard jars used in jar tests have sample ports that are 10 cm below the water surface when using the full capacity of the jar (i.e., 2 L in a 2-L jar). For these jars, dividing the settling distance (i.e., 10 cm) by the sample time (i.e., duration between cessation of mixing and drawing of sample) gives the settling velocity in terms of cm/min. This settling velocity can then be converted to the equivalent SLR, commonly expressed in either gallons per day per square foot (gpd/ft^2) or gallons per minute per square foot (gpm/ft^2), through basic unit conversion:

$$\frac{\text{Settling Distance}}{\text{Sample Time}} = \frac{\text{Volume} \div \text{Time}}{\text{Area}} = \frac{\text{Flow Rate}}{\text{Surface Area}} = \text{SLR} \qquad \text{(Eq 7-4)}$$

or

$$\frac{\text{cm}}{\text{min}} = \frac{1{,}440\ \text{min}}{\text{day}} \times \frac{1\,\text{gal}}{3{,}785\ \text{cm}^3} \times \frac{929\ \text{cm}^2}{1\,\text{ft}^2} = \frac{\text{gpd}}{\text{ft}^2} \qquad \text{(Eq 7-5)}$$

When evaluating sedimentation during jar testing, the standard practice is to collect several samples from each jar at set time intervals that correspond to the SLRs of interest (i.e., the SLRs in the full-scale treatment process or alternate rates that are being evaluated). The conversion factors in Table 7-1 can be used to convert the settling velocities observed in jar testing to their equivalent SLRs.

Jar testing is particularly useful when evaluating how modifying coagulation and/or flocculation chemistry will affect sedimentation. An example is illustrated in Figure 7-3, which shows settled water turbidities recorded from two jars—one that represents standard full-scale coagulation and one that includes the addition of a coagulant aid polymer. The plant's maximum SLR was 1,800 gpd/ft^2, which corresponds to a settling velocity of 5 cm/min or a settling time of 2 min in a standard 2-L jar with a sampling port 10 cm below the water surface. Additional samples were collected at 5 and 10 min, which correspond to SLRs of 706 and 353 gpd/ft^2, respectively. This jar test clearly indicates that the coagulant aid polymer has minimal effect when plant flows are low (settling time of 10 min, corresponding to a 353-gpd/ft^2 loading rate), but as SLRs increase (either because plant flows increase or because sedimentation basins are taken out of service), the polymer is more effective in reducing settled water turbidity.

As described in Chapter 3, some clarification processes use inclined tubes or plates to reduce the surface area required for clarification. The tubes or plates effectively reduce the distance required for particles to settle before they contact a surface and are removed from suspension in the treated water. Typically, tube or plate settlers used for clarification of drinking water are arranged for countercurrent flow, as indicated in Figure 7-4, with water flowing up through the tubes or plates and the settled solids sloughing downward because of gravity.

Table 7-1 Settling velocity conversion factors for clarification basins

Units for Settling Velocity	Multiply by	Equivalent Surface Loading Rate
cm/min	353.4	mgd/ft^2
cm/min	$3.53\ 10^{-4}$	gpd/ft^2
cm/min	0.245	gpm/ft^2
m/h	0.623	gpm/ft^2
in./min	0.409	gpm/ft^2

gpd—gallons per day, gpm—gallons per minute, mgd—million gallons per day

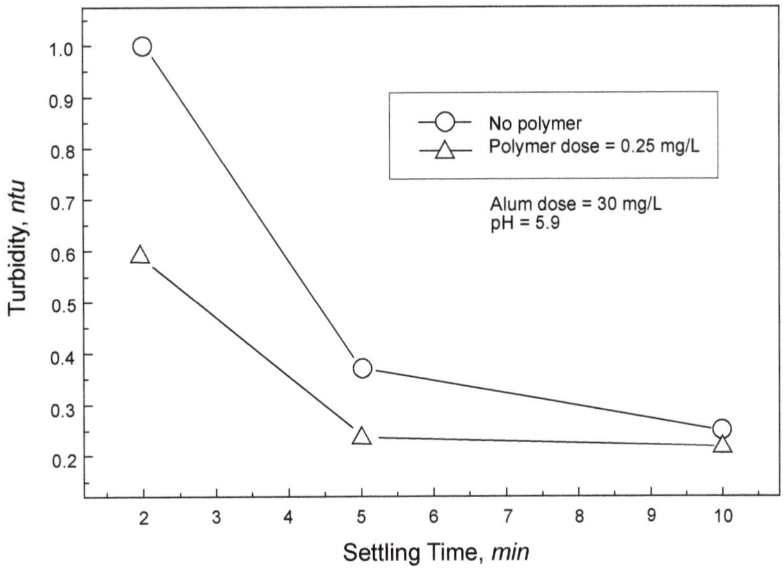

Source: AH Environmental Consultants 1997

Figure 7-3 Example of the use of jar tests: settled turbidity versus settling time

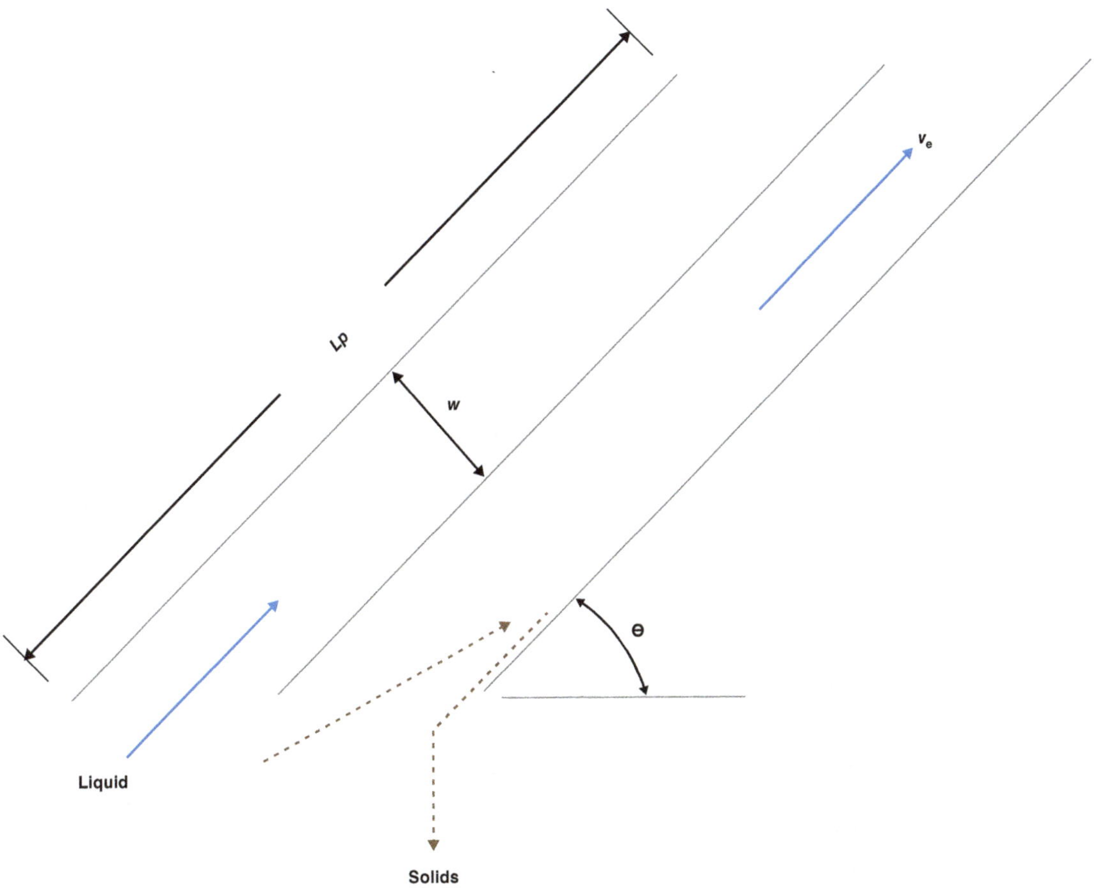

Source: Damon Roth

Figure 7-4 Flow geometry for countercurrent plate or tube settler

When using jar testing to evaluate performance of plate or tube settlers, the relevant parameter is settling velocity. The minimum velocity required for particle removal is given by Eq 7-6:

$$\nu \geq \frac{\nu_\theta w}{L_p \cos\theta + w \sin\theta} \qquad \text{(Eq 7-6)}$$

Where: ν = the minimum settling velocity
ν_θ = the water velocity between surfaces
w = the perpendicular distance between surfaces
L_p = the length of the settling surface
θ = the angle of the inclined plates or tubes

If many plates or tubes are used (as is typical in a plate or tube settler), the water velocity between surfaces, ν_θ, can be estimated using Eq 7-7:

$$\nu_\theta \geq \frac{Q}{Nwb} \qquad \text{(Eq 7-7)}$$

Where: Q = the flow through the unit
N = the number of channels made by plates or tubes
b = the tube or plate width

If the observed settling velocity of the flocs in the jar tests exceeds this minimum settling velocity, they are expected to be removed from suspension through the settler; if the observed settling velocity is below the calculated minimum settling velocity, some flocs may remain in the settled water.

Filtration. Although jar testing is widely applied for evaluating coagulation and flocculation chemistry for turbidity removal via sedimentation, extending jar testing to bench-scale filtration testing is much less common because granular media filtration is difficult to simulate accurately at bench scale. Various techniques have been used in practice, including filtering the water through Whatman No. 40 filter paper or small-diameter media filters. Although the use of filter paper does not represent the filtration mechanisms in a granular media filter, it will remove particles larger than the 8-μm opening size of the filter paper.

As discussed in Chapter 1, measurements of floc charge (streaming current or zeta potential) from jar tests can be used to give a reasonably good estimate of whether a certain set of coagulation conditions will result in good full-scale granular media performance. The correlation between jar test performance and full-scale plant performance can be improved if operators have been able to determine floc charge ranges that correspond to good filter performance at their plant. If such a relationship has not been established, a zeta potential measurement of –5 mV or greater (i.e., closer to zero) has been shown to be a good target value, as shown in Table 1-2.

Bench-scale testing of membrane filtration, although not often applied in a utility setting, is also an option (DiGiano et al. 2000). Membranes can be tested either as hollow-fiber filters (Biscardi and Duranceau 2017, Jang et al. 2005, Park et al. 2002) or as flat-disc filters (Choi and Dempsey 2005, DiGiano et al. 2000). Because membrane filtration tests are often focused on fouling at the membrane surface and the associated rise in transmembrane pressure, wall effects in filtration cells used for membrane filtration testing are often less impactful to the parameters of interest.

Chemicals and points of application. Another planning step involves defining which chemicals to use in the jar test and the point of application. If the objective of the test is to assist in day-to-day plant optimization, then the chemicals that are currently used in the treatment process should be used in the jars and in the same order that they are used in the plant. If alternative chemicals are to be evaluated, these must be used in the jars. The operator should have data sheets for each chemical with information about the chemical formula, specific gravity, percent weight, viscosity for liquids, solubility for solids, and safety information. If the chemical being evaluated in a test has never been used in the process, the previous information should be supplemented with data for unit cost, shipping, and storage requirements, as well as a current safety data sheet.

Current treatment performance data. To determine how well the jar testing conditions or procedures simulate conditions in the full-scale facility, the operator should obtain treatment plant performance data. This information will be compared with that from the jar tests conducted under the same conditions. Water quality data from the plant's settled water are generally used, such as the settled water turbidity, pH, and TOC or UV absorbance values.

Equipment

Jar testers. Jar test equipment consists of jars to hold the water, the impeller, a mechanism to turn the impeller, and laboratory equipment to analyze the results. The following sections describe each of these items and the role of each in the jar test procedure. There is a large amount commercially available jar test equipment. An example of a common jar test apparatus is shown in Figure 7-5. Typically, these devices include

Source: Phipps & Bird

Figure 7-5 Example of a jar test unit

four or six jars with sample ports and paddle mixers and can be preprogrammed to stir at particular speeds for particular amounts of time, simulating the coagulation, flocculation, and sedimentation processes.

Types of jars. Various types of jars have been used, including cylindrical 1- and 2-L glass beakers. The most commonly used jars today are 2-L square beakers. The square plastic beakers provide better mixing conditions than round beakers; therefore, the remainder of this section will focus on these containers. Desirable attributes for the jars include transparency, volume markings, appropriate sampling point, ease of cleaning, and durability.

Stirrer. A stirring mechanism turns the impellers. Two basic types of stirring mechanisms are used: gear-driven and magnetically driven units. A gear-driven unit has a variable-speed motor that turns four to six gears and stirrer shafts. The motor(s) and gears are located above the jar test containers so that the shafts for the impellers extend down into each jar. Units are available that can be controlled from about 10 to 300 rpm. This is the type used in the standard jar test apparatus shown in Figure 7-5.

The magnetically driven unit works on the same principle as a magnetic stirrer plate. The paddle contains a magnet that turns as the metal under it rotates. Either type of unit may be used for jar testing. The magnetically driven unit has the advantage of providing open space above the jars to add chemicals. It is also possible to construct a jar stirrer using one or more variable-speed mixers. Units can be purchased that operate over a broad rpm range and allow for an intensive rapid mix. They offer the advantage of testing different mix intensities simultaneously because the stirrers are individually controlled.

Impellers. Several types of impellers are available: paddle type, turbine, marine, and axial flow. Ideally, the impeller that best simulates the full-scale process should be used. Interchanging impellers require a stirrer mechanism equipped to allow the replacement of the shaft or impeller.

Magnetic stirrers are available only with flat paddle impellers and would be difficult to retrofit with a different impeller type. They can, therefore, effectively simulate the action of a paddle, walking-beam, or flat-blade turbine types of mixing devices.

Standard top-mounting stirrers are generally provided with flat paddles. However, the shaft and impeller can be changed to allow testing of different impeller types. Figure 7-6 shows values for a flat paddle impeller. Values for testing with magnetic impellers or marine paddle impellers can be found in Appendix B.

Water bath. A water bath is a tank in which the jars sit during the jar test as water circulates around them to maintain the correct water temperature. If the testing is conducted in an external laboratory, the water bath temperature can be maintained using a circulating unit with an integrated digital temperature controller. However, if testing is conducted at the full-scale facility's laboratory, the flow from the sample tap can be directed to a water bath to maintain the jar temperature at the water temperature of the full-scale process. This can be particularly useful when optimizing performance under current conditions, particularly when evaluating chemical reactions that require matching chemical reaction times. However, the use of a water bath is not usually required unless the treated water is very cold, and the jar test is to be done in the warm laboratory.

A simple test determines whether a water bath is necessary: Take a sample of the source water and immediately run a jar test; take another sample of the same water, let it warm up a few degrees, and then conduct the same test and note whether the results differ. Cold waters are frequently difficult to treat because lower temperatures result in water that is more viscous, is higher in density, and has slower chemical reaction times for flocs to form. Therefore, even a slight temperature increase could make treatment easier than in the full plant, which causes misleading results.

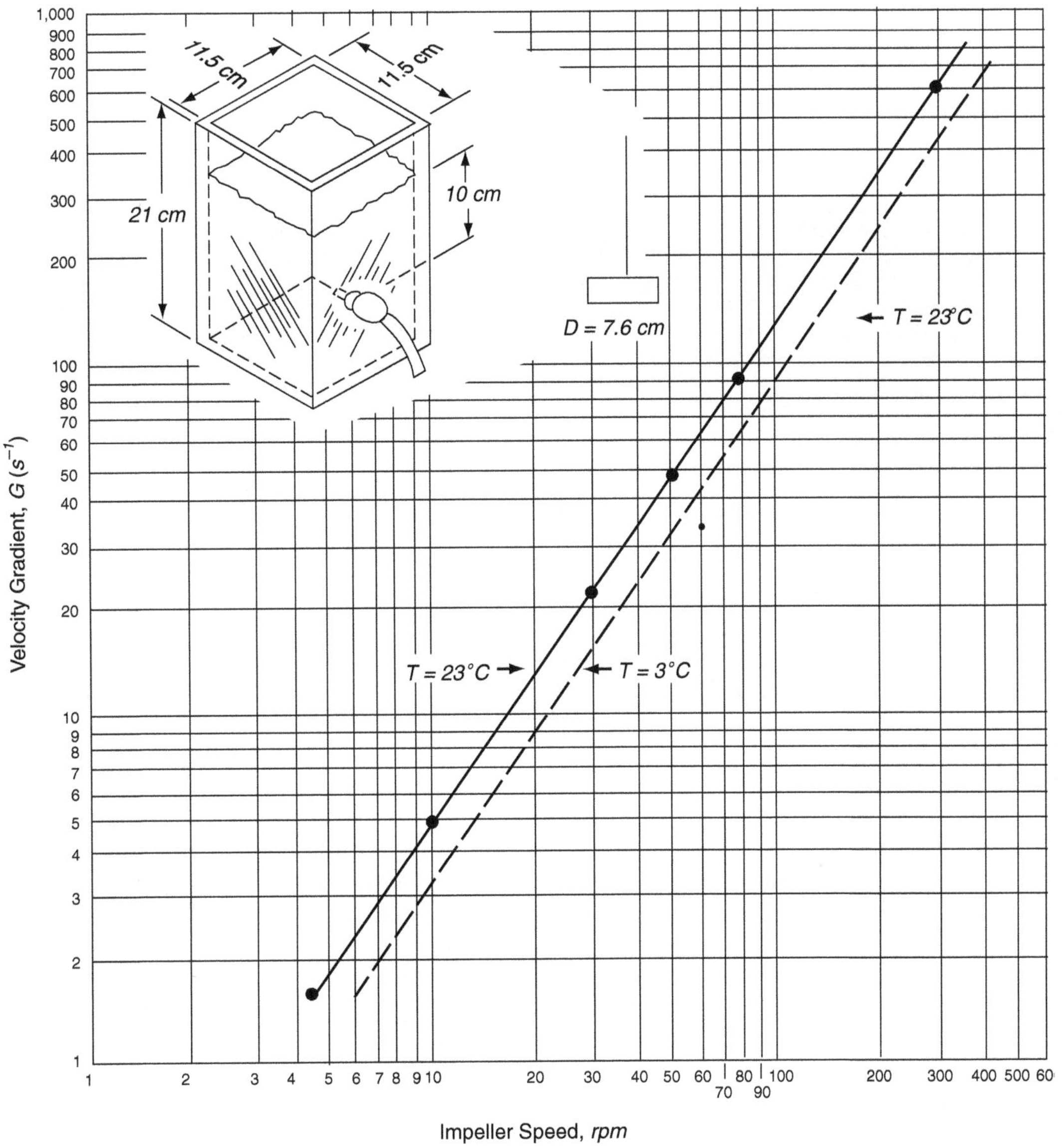

Source: Cornwell and Bishop 1983

Figure 7-6 Laboratory *G* curve for flat paddle

Analytical equipment and laboratory ware. A variety of laboratory equipment is required to prepare standardized solutions and analyze water samples. The most essential analytical equipment includes a turbidimeter, pH meter, and thermometer. Regulatory requirements may require operators to perform more advanced analytical analyses, which may call for the use of a TOC analyzer or UV spectrophotometer. If the analytical equipment is not available, the operator must make arrangements with a commercial laboratory for needed tests and obtain suitable sample containers. Other equipment frequently used to assess the results of jar tests includes zeta potential meters, which are used to evaluate the coagulant dose required for charge neutralization, and particle counters, which are used to assess particle removal through treatment.

An assortment of pipettes, burettes, syringes, graduated cylinders, volumetric flasks, and a laboratory weight scale may also be required to make the necessary solutions. If

Source: Zone 7 Water Agency

Figure 7-7 Wooden holder for six dosing cups: small plastic cups fit into each hole drilled in the wood

chemicals are to be added without dilution, a micropipette is required. Many operators use polytetrafluoroethylene septa (small, flat disks originally intended as liners of glass sample bottles) for holding minute volumes of neat chemical until ready for use. When needed, they are simply tossed into the jar along with the neat chemical, where they quickly sink to the bottom as the chemical is being mixed into the water.

Tools for simultaneous chemical addition. When comparing various chemical dosages in a jar test, it may be necessary to add the chemicals to all jars simultaneously. Under most conditions, manually adding the chemical quickly to each jar is acceptable. However, when small time changes are important, it can be advantageous to be able to dose all jars at exactly the same time. For example, if a test involves a short mix time of just a few minutes, and it takes 1 min to dose all six jars, the first jar dosed will have a significantly longer reaction time than the last jar. In such cases, operators have built devices such as the cup holder (Figure 7-7) and the septa bar (Figure 7-8) for holding small cups filled with the premeasured chemicals. Such devices allow dosing of all jars simultaneously. Alternatively, sample collection times can be staggered with chemical feed to ensure that each jar has the same reaction time between chemical feed and sample collection.

Chemicals

The chemicals used for pretreatment processes generally fall into one of five categories:

- Coagulants
- Coagulant aids
- pH controllers
- Oxidants–disinfectants
- Adsorbents (e.g., activated carbon)

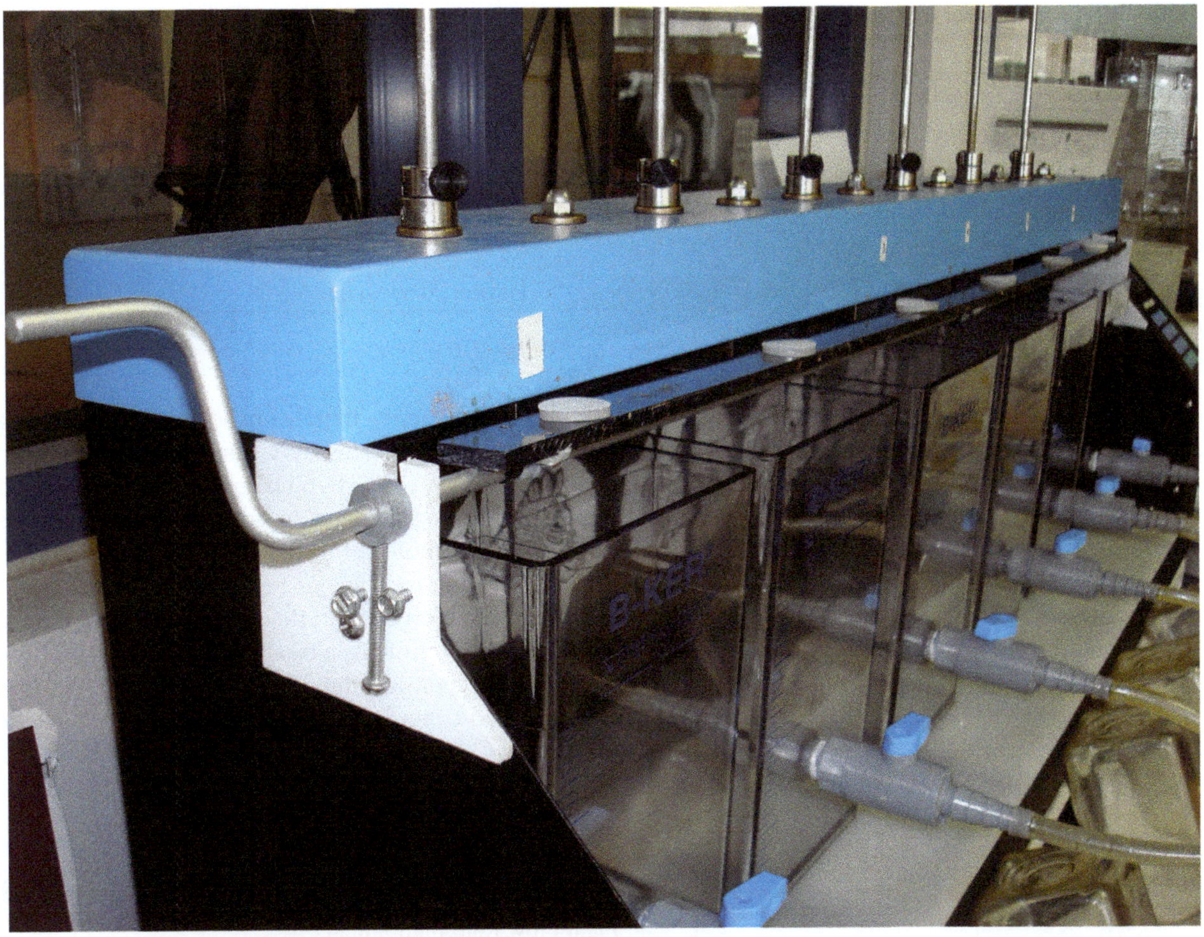

Source: Alameda County Water District

Figure 7-8 Septa bar: metal bar to hold multiple septa over jars until ready to dump septa into all jars simultaneously

In addition to those listed, other chemicals may be needed during and after the jar test. Some of the chemicals listed are available as pure substances of uniform quality. Depending on their origins, others may vary in chemical composition because of their complexity, causing differences in effectiveness. To achieve useful results, jar testing must use treatment chemicals that are being used or will be used at the full-scale plant and are certified for use in drinking water treatment (NSF/ANSI 60 2005).

Coagulants. Coagulants and how they work are discussed in detail in Chapters 1 and 2. The most commonly used coagulants include

- aluminum salts (alum),
- iron salts (ferric chloride [$FeCl_3$] and ferric sulfate [$Fe_2(SO_4)$]), and
- polymeric inorganic coagulants (partially neutralized metal salts such as polyaluminum chloride [PACl]).

In addition to those listed, organic polymers can also be used as coagulants. However, use of organic polymers as the primary coagulant is less common; as described in Chapter 2, they are typically used as coagulant aids. Therefore, organic polymers are discussed later in this section under the heading "Coagulant Aids."

Expressing coagulant concentrations. When a jar test is performed, the correct dosages of chemicals must be applied. The operator must understand whether the applied dosage is stated "as product," "as ingredient," or some other terminology. Additionally,

Table 7-2 Factors for determining molar metal dosage

Coagulant	Factor
Dry alum ($Al_2(SO_4)_3 \bullet 14H_2O$)	3.365
Liquid aluminum product (alum, PACl, etc.)	Percent Al_2O_3 by weight ÷ 5.1
Dry ferric chloride (anhydrous, $FeCl_3$)	6.165
Dry ferric sulfate (anhydrous, $Fe_2(SO_4)_3$)	5.004
Dry ferrous sulfate (anhydrous, $FeSO_4$)	6.579
Liquid iron product (ferric sulfate, polyferric chloride, etc.)	Percent Fe by weight ÷ 5.58

$Al_2(SO_4)_3$—aluminum sulfate, H_2O—water, PACl—polyaluminum chloride

some coagulants (particularly alum) have different active metal concentration, depending on the waters of hydration attached to the compound. This may result in differences in percentage concentration for different products, so each chemical should be considered separately. One way to avoid confusion is to express dosages or concentrations in terms of the molar amount of active ingredient, i.e., the moles of iron or aluminum per liter of solution. This convention is of even greater importance for jar tests comparing different coagulants. Equimolar dosages of aluminum and iron contain the same number of atoms. To determine the molar metal dosage applied at the full-scale treatment plant, data for the plant flow rate, bulk chemical feed rate, and the chemical data sheet are needed. The data sheets for most liquid aluminum- or iron-based coagulants report values for the specific gravity and percentage aluminum oxide (Al_2O_3) or percentage Fe. To convert to molar metal dosage, apply the formula:

$$\text{Dosage}, \frac{\mu\text{mol}}{\text{L}} = \frac{\text{Chemical Feed Rate} \frac{\text{lb}}{\text{day}}}{\text{Plant Flow Rate, mgd} \times 8.34} \times \text{Factor} \qquad \text{(Eq 7-8)}$$

The "molar metal factor" depends on the coagulant used, as shown in Table 7-2. The molar metal dose at the full-scale plant obtained in this way can then be used as a baseline value for the jar test.

Example: A treatment plant feeds 5,000 lb (2,270 kg) of liquid alum per day at a flow rate of 5 mgd (19 mL/d). According to the data sheet, the product contains 8.3 percent Al_2O_3 by weight. Therefore, the molar-equivalent dosage equals the following:

$$\frac{5{,}000}{(5 \times 8.34)} \times \frac{8.3}{5.1} = 195\ \mu\text{M}\left(\frac{\mu\text{mol}}{\text{L}}\right) \text{ as Al} \qquad \text{(Eq 7-9)}$$

If the operator were to test ferric chloride as an alternative coagulant, a comparable dosage to the currently applied alum would be 195 μM as Fe. The 195 μM as Al corresponds to 58 mg/L as alum; ferric chloride at a dosage of 58 mg/L as Fe would be five times the molar Al concentration. Figure 7-9 illustrates the relationship of molar metal concentration to alum or iron concentration in mg/L.

Making up stock solutions from the dry chemical. During jar testing, it is often inconvenient to feed a dry chemical to the jars. To test dry metal salts, stock solutions should be made up in accordance with Table 7-3 to achieve the stock solution concentrations specified.

After preparing the stock solutions specified in Table 7-3, the appropriate volume of stock solution can be added to the test jars to achieve the target dose. For example, adding 1 mL of alum solution prepared as specified in Table 7-3 to a 2-L jar results in a dose of 5 mg/L as alum or 17 μM as Al.

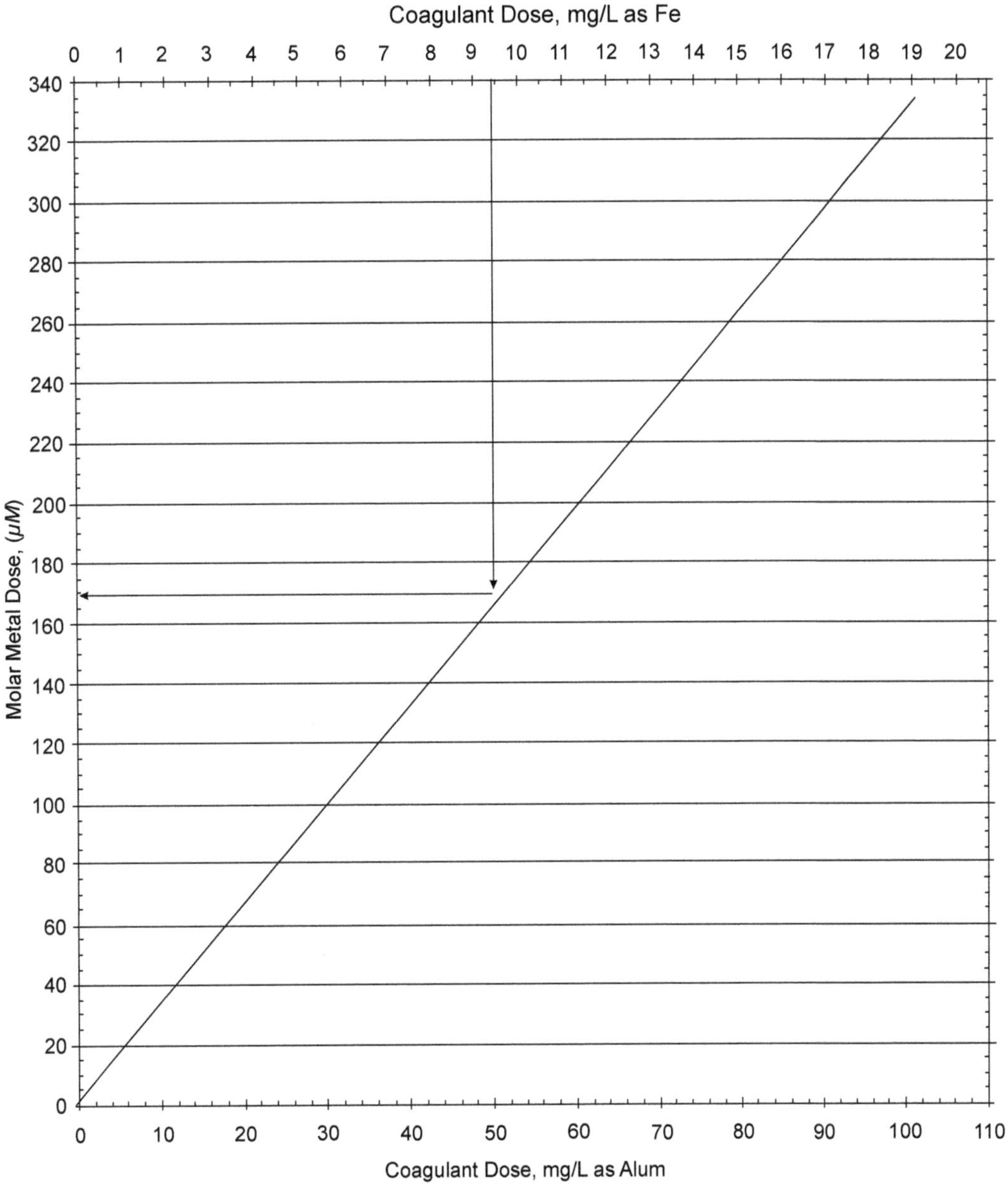

Figure 7-9 Expressing coagulant doses in molar metal concentrations

Table 7-3 Stock solution concentrations achieved by dissolving specified mass of dry chemical into deionized water and diluting to 1 L

Chemical	Mass of Dry Chemical *g*	Stock Solution Concentration *mg/L as product*	Stock Solution Concentration *Percentage active product*
Alum ($Al_2(SO_4)_3 \bullet 14H_2O$)	10	10,000 mg/L as alum	0.17% as Al_2O_3
Ferric chloride ($FeCl_3$, anhydrous)	2.93	1,000 mg/L as ferric	0.10% as Fe^{3+}
Ferric sulfate ($Fe_2(SO_4)_3$, anhydrous)	3.57	1,000 mg/L as ferric	0.10% as Fe^{3+}

Working with liquid coagulants. There are two approaches to dosing coagulants when they are in liquid form. In many cases, liquid coagulants can be added directly to test jars with a micropipette to achieve the target coagulant dose. However, if a micropipette is not available, a stock solution can be created that results in even numbers to get a set dose in the jar. For example, to prepare a solution that results in a 5-mg/L dosage for each 1 mL added to the 2-L jar, the following calculations apply:

$$\begin{array}{c}\text{Stock Solution}\\ \text{Concentation in Percent}(\%)\end{array} = \frac{\text{Desired Dosage, }\dfrac{\text{mg}}{\text{L}} \times \text{Volume of Jar, mL}}{\text{mL of Stock Solution Added to Each Jar} \times 10{,}000\left(\dfrac{\text{mg}}{\text{L}}\text{ to Percent Conversion}\right)} \quad \text{(Eq 7-10)}$$

For this example,

$$\frac{5\dfrac{\text{mg}}{\text{L}} \times 2{,}000\,\text{mL}}{1\text{ mL} \times 10{,}000} = 1.0\text{ Percent Stock Solution} \quad \text{(Eq 7-11)}$$

Therefore, if a 1.0 percent stock solution is made, every 1.0 mg/L of this solution added to the 2-L jar results in a dosage of 5.0 mg/L of the product being tested.

The product strength, which is the mass of the active ingredient per gallon, also must be known. This value, in lb/gal, is often provided by the chemical supplier for each load of material delivered. If it was not provided directly, it can be calculated from the specific gravity of the material and the percent active ingredient, both of which are generally provided by the manufacturer. Specific gravity can be verified by using a hygrometer, or it can be measured directly using a graduated cylinder and a scale. Some manufacturers will provide tables relating the specific gravity to the percentage of active ingredient for a particular coagulant. The following equation is for the product strength, in lb/gal:

$$\text{Strength, }\frac{\text{lb}}{\text{gal}} = \text{Percent Active Ingredient} \times \text{Specific Gravity} \times 8.34\frac{\text{lb}}{\text{gal}} \quad \text{(Eq 7-12)}$$

When the product strength and the desired percent stock solution to be prepared are determined, the following equation can be used to determine amount of neat product (mL) to add to the deionized water to make the working solution:

$$\text{Volume of Coagulant, mL} = \frac{\%\text{ Stock Solution} \times \text{Flask Volume, mL} \times 8.34\dfrac{\text{lb}}{\text{gal}}}{100 \times \text{Product Strength, }\dfrac{\text{lb}}{\text{gal}}} \quad \text{(Eq 7-13)}$$

Example: A water utility is preparing to evaluate dosages of alum in increments of 5 mg/L (i.e., dosages of 5, 10, 15 mg/L, etc.) in 2-L jars. In this case, using alum with a specific gravity of 1.22 and 48 percent active ingredient, the calculations to make 500 mL of solution for the jar tests are as follows:

$$\text{Strength, }\frac{\text{lb}}{\text{gal}} = 0.48 \times 1.22 \times 8.34\frac{\text{lb}}{\text{gal}} = 4.88\frac{\text{lb}}{\text{gal}} \quad \text{(Eq 7-14)}$$

$$\text{Volume of Alum, mL} = \frac{1.0 \times 500 \text{ mL} \times 8.34 \dfrac{\text{lb}}{\text{gal}}}{100 \times 4.88 \dfrac{\text{lb}}{\text{gal}}} = 8.5 \text{ mL} \qquad \text{(Eq 7-15)}$$

Therefore, add 8.5 mL of neat liquid alum to the 500-mL volumetric flask and fill to the mark with deionized or distilled water. Adding 1 mL of the resulting stock solution to a 2-L volumetric flask would equate to an alum dose of 5 mg/L, making it easy to dose the alum using a 1-mL pipette to achieve the target doses.

Stock solutions are typically only prepared using simple hydrolyzing metal salts such as alum, ferric chloride, or ferric sulfate. Prehydrolyzed metal salts such as PACls should be dosed neat because they lose their effectiveness when diluted; this should be verified with the coagulant supplier if in doubt. Dilutions of ferric and alum are typically shelf stable. However, dilutions of PACl or other coagulants should be prepared daily.

In lieu of diluting liquid coagulants into a stock solution, it is often possible to add the necessary volume of neat coagulant to test jars with a micropipette, an example of which is shown in Figure 7-10. However, care must be exercised when pipetting viscous liquids.

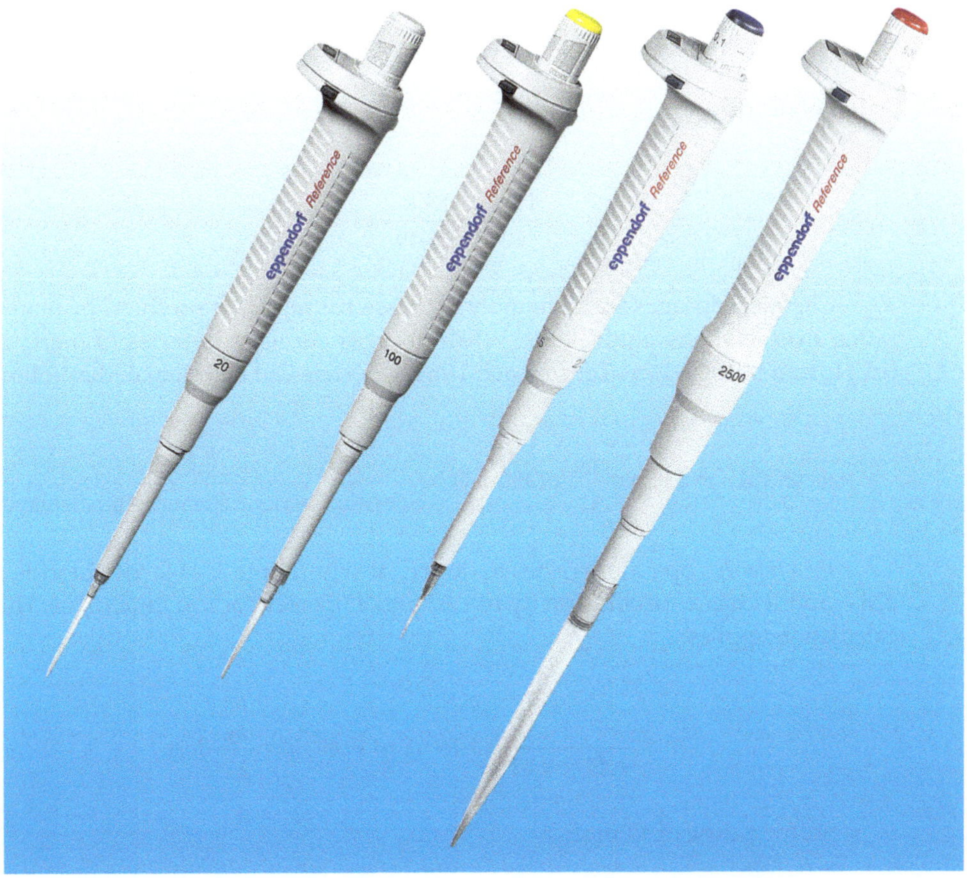

Source: Eppendorf North America

Figure 7-10 Examples of micropipettes capable of dispensing 0.1–2,500 μL

The amount in microliters (μL) to add to a 2-L jar can be calculated from the percentage Fe or percentage Al_2O_3 content and the specific gravity of the coagulant as follows:

For aluminum products:

$$\text{Volume, μL} = \frac{\text{Dose, μM} \times 5.10 \times \text{Volume of Jar, L}}{\text{Percent } Al_2O_3 \times \text{Specific Gravity}} \times \text{Dilution Factor} \qquad \text{(Eq 7-16)}$$

For iron products:

$$\text{Volume, μL} = \frac{\text{Dose, μM} \times 5.59 \times \text{Volume of jar, L}}{\text{Percent Fe} \times \text{Specific Gravity}} \times \text{Dilution Factor} \qquad \text{(Eq 7-17)}$$

Example: A water utility tests two alternative coagulants, a polymeric ferric coagulant (specific gravity of 1.49, 12.27 percent Fe) and aluminum chlorohydrate (ACH) (specific gravity of 1.34, 23.5 percent Al_2O_3). The plant currently uses ferric chloride at a dosage of 9.4 mg/L as Fe, which corresponds to a molar metal concentration of 168 μM. To evaluate the effectiveness of these chemicals at equimolar concentrations, the following amounts must be added to a 2-L jar:

ACH:

$$\frac{168 \times 5.1 \times 2}{23.5 \times 1.34} \times 1 = 54 \text{ μL} \qquad \text{(Eq 7-18)}$$

Polyferric:

$$\frac{168 \times 5.59 \times 2}{12.27 \times 1.49} \times 1 = 103 \text{ μL} \qquad \text{(Eq 7-19)}$$

Chemical dosages are commonly expressed as mg/L "as product" by many operators, rather than as molar equivalents. When dosages are used in terms of mg/L, the amount (μL) to add to the 2-L beaker for each 1-mg/L dosage can be calculated as follows:

$$\text{Amount(μL) to add per } \frac{\text{mg}}{\text{L}} \text{ of Product} = \frac{2}{\text{Specific Gravity} \times \text{Percent Active Ingredient}}$$

Example: An operator wishes to dose a jar with 17 mg/L of ferric chloride as product. This product has a specific gravity of 1.47 and 43 percent active ingredient. The following calculation applies:

$$\frac{2}{1.47 \times 0.43} = 3.16 \text{ μL of } FeCl_3 \text{ per } \frac{\text{mg}}{\text{L}} \text{ dose} \qquad \text{(Eq 7-20)}$$

So, for a dose of 17 mg/L,

$$3.16 \text{ μL} \times 17 = 53.7 \text{ μL} \qquad \text{(Eq 7-21)}$$

This amount can be dispensed directly onto the septa using a micropipette.

Polymeric inorganic coagulants. The effectiveness of dry and liquid coagulants varies with dilution, and highly diluted stock solutions may deteriorate over time. Thus, new stock solutions must be prepared every day. Polymeric inorganic coagulants such as PACl can degrade very quickly and should not be diluted but used neat.

Coagulant aids. In a most general sense, a coagulant aid is any substance used in conjunction with a primary coagulant, such as alum, to assist coagulation. The most common coagulant aids are synthetic organic polymers. Polymers are chains of small subunits or monomers, and they may contain ionizable groups as discussed in Chapter 1.

Like metal salts, polymers may be supplied either as a dry chemical or as an emulsion (i.e., liquid chemical). Polymer dosages are typically based on the mass of neat product that is applied, although the amount of active polymer may vary between products. Dry polymers typically contain up to 90 percent active polymer, whereas emulsions typically range from 20 to 55 percent active polymer. Unlike metal salts, polymers require an activation process in addition to dilution. Dry polymers must be rehydrated, whereas emulsion polymers require mixing with high shear to invert the beads of hydrogel that maintain the stability of the polymer in the liquid emulsion. For dry polymers, the rehydration process may take multiple hours, whereas the time frame for inverting emulsion polymers is in the order of minutes. For both types of polymers, providing additional time to further dissolve the polymer and allow the polymer chains to unfurl will increase the efficiency of the product.

Preparation and use of coagulant aids. Polymers are available as liquid, emulsion, and dry powder products. The following paragraphs include instructions for preparation and use, which vary for these types, in cases in which specific dilution instructions are not provided by polymer suppliers. When possible, follow the specific dilution instructions provided by the polymer manufacturer or vendor.

Liquid and emulsion polymers:

1. Add 200 mL of distilled water to a clean 1-L volumetric flask.
2. After shaking the product container vigorously, weigh out 2.0 g of the polymer product onto an aluminum or plastic weighing dish.
3. Using a squeeze bottle of distilled water, rinse all the polymer into the volumetric flask.
4. Cap and shake vigorously for at least 1 min.
5. The strength of the inverted polymer emulsion will be 1.0 percent, or 10,000 mg/L.
6. Fill the volumetric flask to the 1-L mark with distilled water to dilute the inverted emulsion.
7. Cap and shake for at least 1 min.
8. The strength of the diluted stock solution will be 0.2 percent, which is the equivalent of 2,000 mg/L or 2.0 mg/mL.
9. Therefore, 1.0 mL of the stock solution added to a 2-L jar will be equivalent to a dose of 1.0 mg/L.
10. Most polymer stock solutions of this strength will degrade within 24 h.

Dry polymers. The preparation of dry polymers is like that described for liquid and emulsion polymers. However, after adding the polymer to the 1-L volumetric flask, drop in a magnetic stir bar, place the flask on a magnetic stirrer, and mix at medium to medium-high speed for at least 2 h. After the polymer has completely dissolved in solution, remove the stir bar and proceed with dilution.

Polymers are expensive, but they usually work well at low concentrations. The maximum dosage should be less than 2–3 mg/L, and dosages as small as 0.05 mg/L may prove effective as aids for coagulation or flocculation and even as low as 0.005 for filter aids. Note the maximum permissible dosage listed on the NSF International certification and ensure that this is not exceeded.

pH Control. Control of pH during jar testing is important for several reasons:

- Simulating existing conditions
- Evaluating alternatives for optimal DOC reduction
- Evaluating alternatives for optimal turbidity removal
- Minimizing coagulant residual in the distribution system

If the coagulation reaction using aluminum or iron salts as the primary coagulants occurs at nonoptimized pH conditions, the quality of the treated and filtered water can be degraded by elevated concentrations of dissolved aluminum or iron. Particularly when alum is used, the metallic salt coagulants are generally more susceptible than the polymers to loss of effectiveness under nonoptimal pH conditions. Both alum and ferric coagulants will lower the pH of the water, especially when the buffering capacity of the water is low.

Often, pH is controlled only by the dosage of the coagulant applied. If the WTP operates this way, jar tests should also be conducted the same way to find the optimal chemical dosage for routine treatment of water.

Sometimes, however, the properties of source water cause optimal treatment to occur at a pH significantly different from that obtainable from the coagulant alone. A jar test series will quickly demonstrate this condition. In some cases, cost-effective treatment would adjust the source water pH with another chemical (acid or base) to achieve the best pH for the coagulant of choice. Again, the jar test is perhaps the most valuable tool for rapidly determining the best combination of coagulants and other chemicals to achieve the most cost-effective and process-efficient reaction pH.

Adjusting pH during a jar test can be a hectic task involving measurements of and adjustments to pH in six jars during a 1-min rapid mix. The easiest and generally most accurate method is to predetermine the required acid or base dose by conducting a titration. This is done by taking a small sample of source water, say 100–200 mL, and adding the coagulant dosage equivalent to that which will be added to the 2-L jar. Place the sample on a standard stirrer and titrate with acid or base, recording the dosage to reach the desired pH level. A single titration can determine the acid or base dosages needed to reach different pH levels. With this information, the proper volume can be premeasured and fed to the jar during, before, or after coagulant addition. The pH at the end of the rapid mix phase still should be measured and recorded. Remember, however, that the pH should be measured both before and after the jar test. In some waters, the phenomenon of pH increase throughout the jar test is caused by carbon dioxide (CO_2) off-gassing.

Common pH-adjusting chemicals. Although WTPs have a variety of chemicals that are used to adjust pH in a full-scale facility, jar test pH is typically adjusted with sodium hydroxide and sulfuric or hydrochloric acid for convenience. For lime-softening application, lime cannot be replaced with sodium hydroxide.

If an alternate base is used in place of lime, once the dosage of the base has been established, the following conversion factors can be used to adjust values if the WTP and the jar test series use different products by converting to mg/L as calcium carbonate ($CaCO_3$) and then back to any other base:

- Unslaked lime (quicklime): mg/L CaO × 1.78 = mg/L $CaCO_3$
- Hydrated lime: mg/L $Ca(OH)_2$ × 1.35 = mg/L $CaCO_3$
- Soda ash: mg/L Na_2CO_3 × 0.94 = mg/L $CaCO_3$
- Sodium hydroxide: mg/L NaOH × 1.25 = mg/L $CaCO_3$

Other chemicals. Successful simulation of full-scale plant conditions requires the addition of any other chemicals normally added during rapid mix, flocculation, or sedimentation at the same dosages in the jar tests. Chlorine, chlorine dioxide, ozone, and potassium permanganate or sodium permanganate are the chemicals most used as primary disinfectants and/or oxidants. These or other additional chemicals may affect treatment performance. For example, fluoridation chemicals can complex with alum if fluoride is added in pretreatment, which results in the need to add more alum to attain proper coagulation. Some phosphate chemicals that are used for corrosion control can cause increased filtered water particle counts and turbidity when filters are returned to service if filters are backwashed with water containing these chemicals (Amburgey et al. 2004).

Adsorbents. Activated carbon is commonly used to adsorb color-causing organic compounds, taste-and-odor compounds, and synthetic organic chemicals in drinking water treatment. PAC is typically used in conjunction with the coagulation process. PAC dosages can range between 1 and 100 mg/L depending on the type and concentrations of organic compounds present. Because of their adsorbing capabilities, PAC can be potentially used for removing organics associated with TOC.

As described previously, a successful jar test should implement the addition of chemicals used in the full-scale treatment process. For this reason, PAC can be potentially implemented on routine jar tests. The operators can also test the effect of PAC dose on the removal of taste-and-odor compounds with threshold odor number testing or sending samples to a laboratory for analysis. Note the maximum permissible dosage listed on the NSF International certification and ensure that this is not exceeded during the application of adsorbents.

Preparation and use of adsorbents. PAC is available as dry powder products. In a jar testing procedure, PAC can be weighed and dosed directly to the jars. In some cases, a slurry of PAC can be prepared and used as a stock solution for dosing the jars. The following paragraphs include instructions for preparation of carbon slurry:

1. Add 200–500 mL of distilled water to a clean 1-L volumetric flask.
2. Weigh out 2.0 g of the PAC onto an aluminum pan or plastic weighing dish.
3. Using a squeeze bottle of distilled water, rinse all the PAC into the volumetric flask.
4. Fill the volumetric flask to the 1-L mark with distilled water.
5. Cap and shake the volumetric flask.
6. The strength of the stock solution will be 2,000 mg/L. Therefore, 1.0 mL of the stock solution added to a 2-L jar will be equivalent to a dose of 1.0 mg/L.
7. Store the PAC stock solution in the refrigerator. Most PAC stock solutions of this concentration can be used within a week.

The PAC solution must be stirred well, preferably continuously with a magnetic stir bar, to keep the solution homogenous before use. This will prevent PAC from stratifying (i.e., settling) in the stock solution. Furthermore, depending on the particle size of distribution of PAC, it can be challenging to transfer with pipettors. PAC can potentially clog the tip of the pipettor, which can result in loss of accuracy. For that reason, it is recommended that the tip of the pipettor be cut with a pair of scissors to allow proper transfer of PAC solution and dosing.

Sample Collection and Analysis

At the end of the jar tests, samples need to be collected and analyzed for a range of water quality parameters. These water quality parameters typically include temperature, pH, alkalinity, turbidity, and UV254. In some cases, taste and odor, TOC, and DBPs and DBP precursor can be analyzed as well. For measurements of floc charge taken with zeta potential or streaming current analyzers, samples may need to be taken during the rapid-mix step before floc particles become too large. With some of the analysis (such as UV254), it is important to filter the samples before analysis. In preparation for jar testing, it is important to ensure that sufficient supplies are available (chemicals, filters, etc.) for water quality testing.

If testing includes laboratory analyses that cannot be conducted immediately, samples must be preserved according to standard analytical practices. Preparations for a jar test must ensure that the appropriate preservatives are on hand. Commonly used preservatives may include nitric acid for metal analyses, sodium thiosulfate for quenching DBP samples, or ethylenediamine for preserving byproducts of chlorine dioxide disinfection. Refer to documentation for the appropriate method or consult the laboratory conducting the analyses.

Data Collection and Documentation

It is critical that the data from the jar test be recorded so that the results can be used in the future when a similar set of water quality conditions arise. During jar testing, several events may occur simultaneously, and the actual evaluation of the test results may not be made until several days after the test are performed because of the analysis time associated with some of the water quality parameters (such as TOC). Significant observations may be difficult to recall if not written down. This documentation requires preparation of data sheets and test protocols that not only are for data collection but are also reminders for the steps to take during a jar test. The data sheets should therefore be designed so that the form holds all relevant information for the jar test in the sequence that the data are collected.

Figure 7-11 shows an example of a data sheet. In preparing for the jar test, use of such a data sheet allows entry of source water quality parameters, concentrations of the chemicals, and the hydraulic characteristics of the full-scale plant and corresponding jar test parameters. Do not underestimate the importance of visual observations for a trained operator; for example, the relative speed with which flocs form the jars (e.g., jar 3 formed visible floc before jar 1) and the appearance of the jars (e.g., milky, cloudy, good phase separation, large flocs). Space must be available on the form for noting such visual observations.

It is best to prepare these data sheets using computer programs. These tools allow both the rapid calculation of removal efficiencies and the graphical representation of results. They also allow for filing the documents that can be searched, analyzed for trends, etc., when needed.

CONDUCTING THE JAR TEST

After the preparation tasks described in the previous section have been performed, the actual jar test can be conducted. At this point, the operator should have defined the study goals (e.g., optimizing coagulant or polymer dosage) and the testing parameters (hydraulic characteristics of the plant, points of application, etc.). Testing and analytical equipment should be ready (jars, stirrer, properly labeled sample containers, turbidity meter, pipettes, etc.). All reagent solutions should be prepared (coagulants, polymers, oxidants, etc.), and a data sheet should be available in a convenient spot.

The operator should ensure that all chemicals are properly labeled and that the reagent solutions are thoroughly mixed. The reagent containers should be placed near the jar test equipment in the order that they are used. Pipettes and syringes should be labeled and placed in or next to the corresponding reagent containers. Automatic pipettes should be set to the correct volumes.

Date:		Concentration (mg/L)
Time:		
Coagulant:		
Oxidant		
Polymer:		

Source Water				
pH	Turbidity (ntu)	Alkalinity (mg/L as $CaCO_3$)	UV_{254} (cm^{-1})	TOC (mg/L)

Jar number			1	2	3	4	5	6
Rapid mix		*G* (s^{-1})						
		rpm						
		Duration (s)						
Flocculation		*G* (s^{-1})						
		rpm						
		Duration (s)						
Coagulant dose (mg/L)								
Volume of coagulant added (mg/L)								
Oxidant dose (mg/L)								
Volume of oxidant added (mL)								
Polymer dose								
Volume of polymer added (mL)								
Coagulation pH								
Settling Velocity (cm/min)	Depth of Sampling (cm)	Time of Settling (min)	Turbidity (ntu)	Turbidity (ntu)	Turbidity (ntu)	Turbidity (ntu)	Turbidity (ntu)	Turbidity (ntu)
TOC (mg/L)								
UV_{254} (cm^{-1})								

rpm—revolutions per minute, TOC—total organic carbon, UV254—ultraviolet absorbance at 254 nm

Figure 7-11 Example of a jar testing data sheet

The jars and the paddles of the stirring mechanism should be cleaned by wiping with a damp cloth and rinsing with warm tap water to remove any residue from previous jar tests.

The data sheet (Figure 7-11) should be a good guide for conducting the jar test:

1. Treatment performance data are often expressed in terms of percentage removal. Therefore, an important first step is to determine the quality of the source water to be tested. Such data may be obtained from treatment plant records or determined during the jar test. Also, pH and alkalinity data may help determine necessary additions of acid or base.

2. Enter the names and concentrations of the chemicals to be added on the data sheet. This information is necessary to determine the volumes of chemicals to be added during the jar test.

3. Enter the *G*-values for the rapid mix and flocculation stages of the full-scale plant on the data sheet. If the effect of varying mixing intensities is to be evaluated, use the appropriate range of *G*-values. Convert these values to the appropriate rpm in the jar test. Depending on the jar test equipment used, refer to Figure 7-8 or Appendix B or jar test equipment manual to determine the correct rpm value.
4. Enter the detention times for the rapid mix and the flocculation stages of the full-scale plant in the datasheet. If the effects of detention times are to be evaluated, use an appropriate range of durations.
5. Enter the coagulant dosages on the data sheet. If the test will determine an optimal coagulant dosage, it is useful to select dosages in increments of 5–10 mg/L for alum, or equivalent dosages if other coagulants are to be tested. Smaller increments may be used for fine-tuning the optimal coagulant dosage. Then, calculate the volume of coagulant to be dispensed into the jar to obtain the desired coagulant dosage.
6. If applicable, proceed with all the other chemicals similarly.
7. Fill the jars with the water to be tested and position them under the stirring apparatus so they are centered with respect to the impeller shafts.
8. Lower the impellers or paddles so that they are about one-third from the bottoms of the jars.
9. Begin the flash-mix period on the basis of the previously determined values. Do not forget to record the starting time. Dispense the desired quantities of chemicals as rapidly as possible into the jars. Dispense the chemicals in the same sequence as at the full-scale plant unless the effect of moving the point of application is to be evaluated.
10. After the rapid-mix period, decrease the mixing speed to the predetermined value for the flocculation period. If the full-scale treatment process includes tapered flocculation, reduce the mixing speed at the end of each flocculation stage.
11. After the flocculation period, stop the mixer. Collect samples at the end of the sedimentation time, which is previously calculated to simulate the full-scale plant. Sample withdrawal may be accomplished using a syringe, a fixed sampling port, or a pipette. The first portion of the sample taken from a fixed port should be discarded. When a syringe is used, samples should be taken from the same depth as the fixed port.
12. After sampling, conduct the laboratory analyses. As with any analytical work, repetition is important to assess the amount of variation and enhance quality of data. Whenever possible, tests should be run more than once and data compared. If an analysis cannot be conducted immediately, samples shall be preserved according to specific analytical applications. Refer to documentation for the appropriate method or consult the laboratory conducting the analyses.
13. Enter laboratory results on the data sheet.

Standard Operating Procedures

As with other routine tasks, it is a good idea to develop a detailed SOP for the jar test that is specific to a particular plant. The SOP helps ensure that all operators conduct the test in the same manner and serves as a training tool. Examples of SOPs used by water agencies for conducting jar tests are in the Appendix C of this manual.

INTERPRETING AND PRESENTING THE RESULTS

Successful interpretation of jar test results is possible only when adequate data have been collected and recorded. Comparisons of jar test data require key information such as all water quality data, hydraulic data, and the types and dosages of the chemicals used. Once all the data are collected, the easiest way to evaluate them is to prepare charts and graphs using computer spreadsheet programs. Well-prepared charts and graphs show how well the jar tests simulate the full-scale plant as well as the effects of alternative treatment options.

Guidelines for Charts

Percentage removal versus absolute values. Jar test data can be easily converted into percentage removal values using the following formula:

$$\text{Percentage Removal} = \left(1 - \frac{\text{Final Value}}{\text{Initial Value}}\right) \times 100 \qquad \text{(Eq 7-22)}$$

This method may be chosen to assess treatment performance when initial water quality parameters vary or no absolute concentration is specified as the treatment goal.

When there is a specific treatment goal in place, use of percentage removal data may obscure actual differences, so the initial values or ranges of values should be included in the data report. For example, consider the following data in Table 7-4 for a system that has established a treatment goal to maintain settled water TOC below 2.5 mg/L:

Although the percentage removal is higher in jar test 1, the hypothetical treatment goal of less than 2.5 mg/L in the settled water is reached only in jar test 2.

Types of Charts

Several methods are available for reporting or reviewing the results of jar testing. The graph type selected depends on the preference of the operator and the intended use. A bar graph is suitable only for comparing a few values in different categories in a single plot. Figure 7-12 represents a comparison of jar testing results with the results from a full-scale plant. The percentage removal values for turbidity, UV absorbance, DOC, and trihalomethane (THM) formation potential (THMFP) were comparable between the full-scale plant and the jar tests. In addition, similar amounts of THMs were formed as a result of prechlorination. These results indicate that the jar testing protocol successfully simulated the full-scale treatment plant. The bar graph in Figure 7-13 demonstrates how jar test results may aid in selecting a suitable synthetic organic polymer for enhanced solids–liquid separation. The chart indicates that nonionic polyacrylamide performed best in the jar tests at all three sampling times, i.e., overflow rates.

An x–y (scatterplot) graph is suitable for displaying a series of data points. Multiple lines can be used to express results under various conditions. For example, in Figure 7-14, UV absorbance data were collected from six jars that received increasing alum dosages.

Table 7-4 Example of a specific treatment goal

Description	Jar Test 1	Jar Test 2
Source water TOC (mg/L)	8.0	3.0
Settled water TOC (mg/L)	4.0	2.0
Percentage removal	50.0	33.3

TOC—total organic carbon

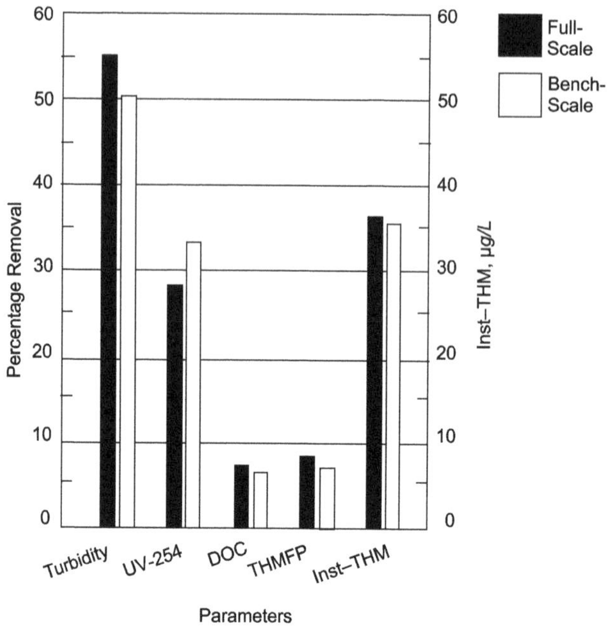

Source: City of Phoenix, Ariz. 1989

DOC—dissolved organic carbon, THM—trihalomethane, THMFP—THM formation potential, UV-254—ultraviolet absorbance at 254 nm

Figure 7-12 Example of a correlation between jar test results and full-scale plant performance

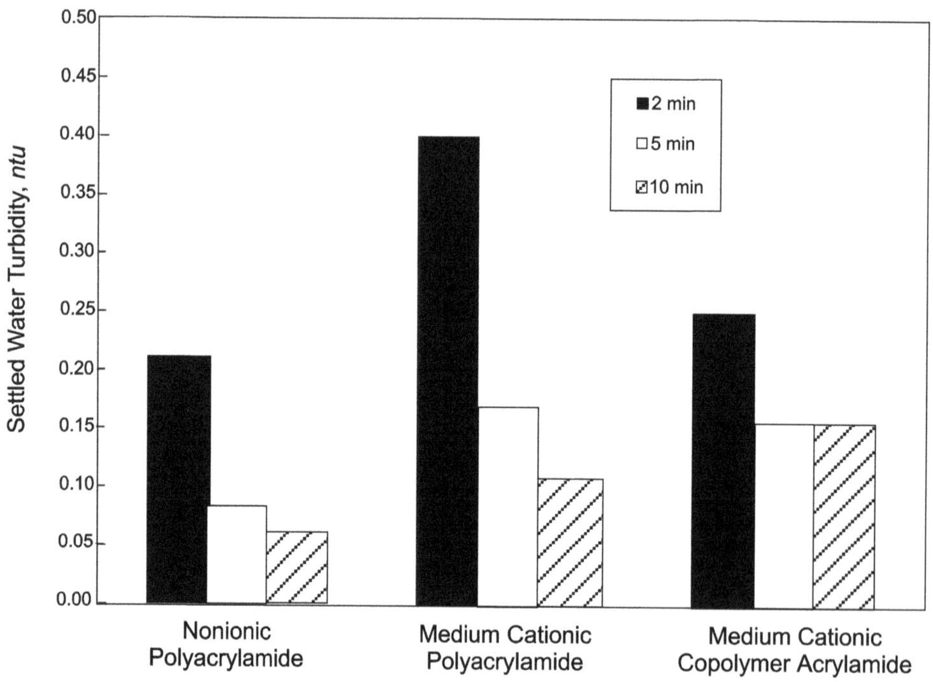

Source: AH Environmental Consultants 1997

Figure 7-13 Use of the jar test to select coagulant aids: turbidity versus settling time

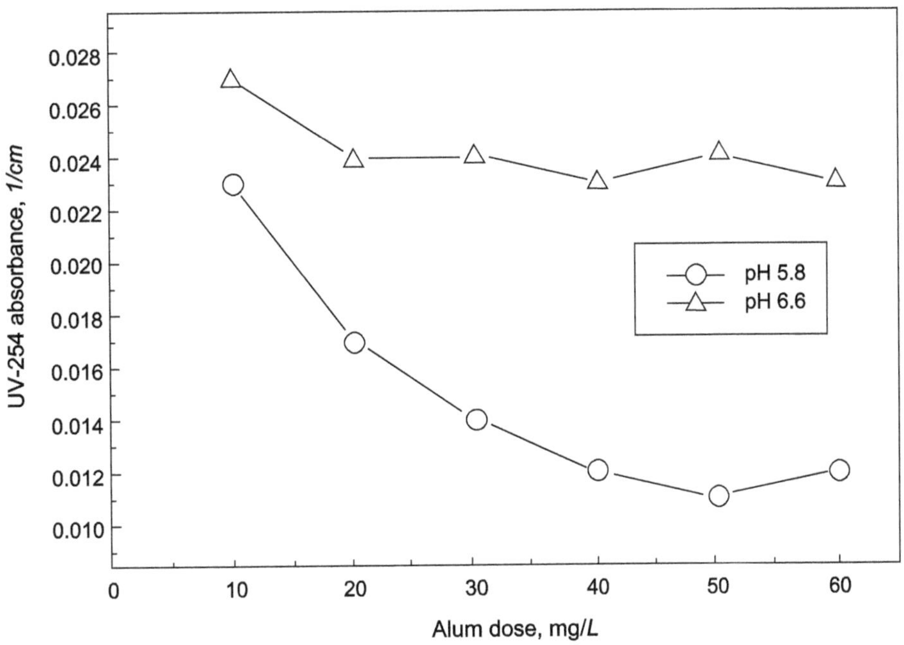

Source: AH Environmental Consultants 1997

UV254—ultraviolet absorbance at 254 nm

Figure 7-14 Use of the jar test to optimize the coagulation pH: UV254 versus alum dose

The test was then repeated at a different pH. This graph shows UV254 absorbance versus alum dosage, and each line represents the data series for one pH level. The plot indicates that coagulation at pH 5.8 yielded lower UV254 absorbance than coagulation at pH 6.6. Another example of an x–y graph, shown in Figure 7-15, illustrates the effect of flocculation time on settled water turbidity at two different settling velocities, i.e., overflow rates. The graph in Figure 7-16 was used to determine the optimal polymer dose for a treatment plant on the basis of jar tests. The logarithmic scale on the y-axis helps illustrate the differences in turbidity.

Topographs can be used to simulate a three-dimensional view of the collected data. They are commonly used to represent large data sets, and they often require special software packages. Figure 7-17 shows a topograph of THMFP as a function of applied ozone and alum dosage.

SPECIAL APPLICATIONS

Bench-Scale Evaluation of Non-Coagulation- or Non-Filtration-Based Processes

In conventional water treatment processes, additional chemical pretreatment (oxidation, softening, adsorption, etc.) other than coagulation may be needed to achieve sufficient treatment. To validate and optimize a chemical pretreatment scheme, a jar test may be used. Additionally, jar tests may also be used to evaluate the effects of lag time between the addition of two or more chemicals and to compare performance of alternative chemicals (coagulants, oxidant, adsorbents, etc.). This section describes selected applications for which jar tests can be used in conjunction with the desired pretreatment methods.

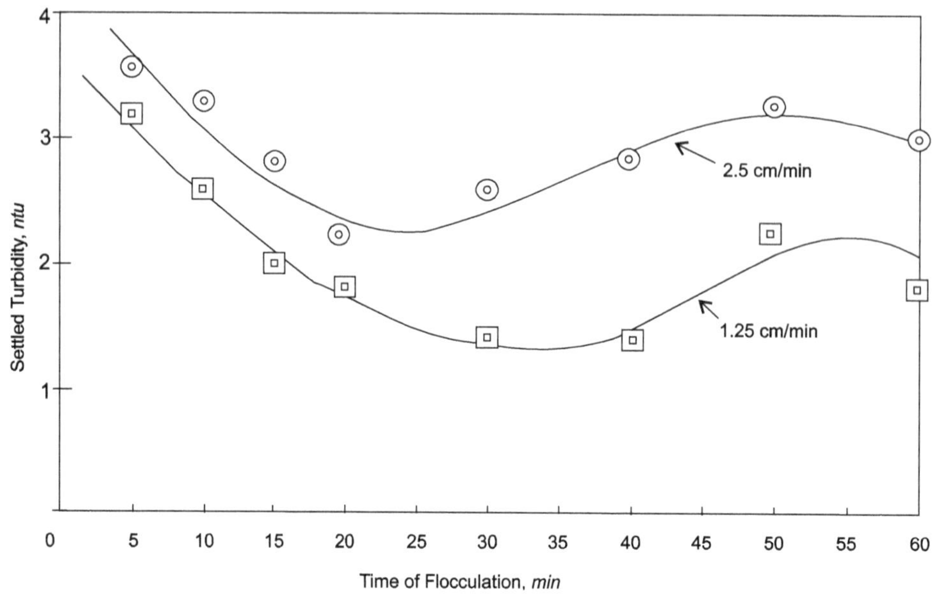

Source: J. Edward Singley and Tim Brodeur

Figure 7-15 Example of the use of the jar test: flocculation time versus settled turbidity

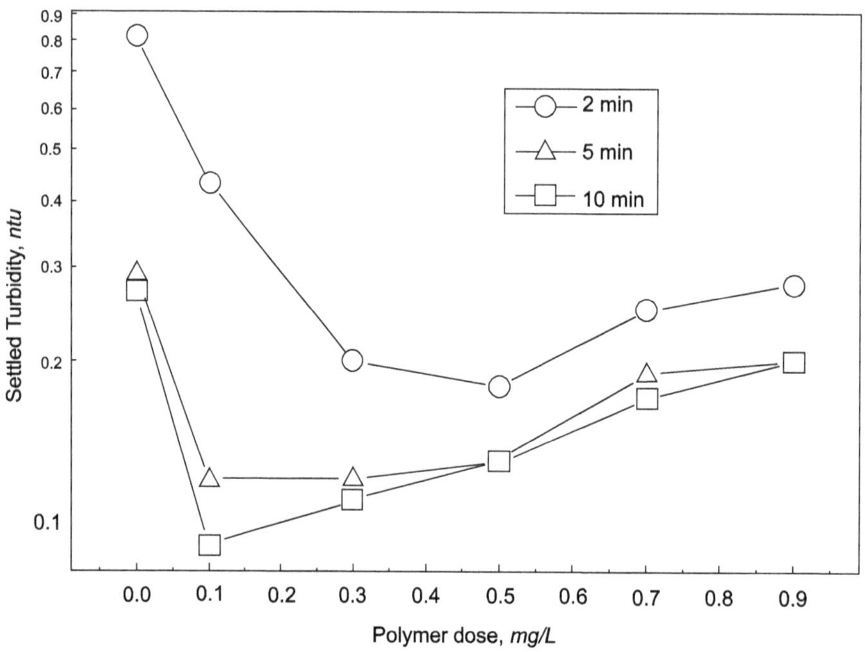

Source: Environmental Engineering & Technology 1987

Figure 7-16 Use of the jar test to determine the optimal polymer dose: turbidity versus dose and settling time

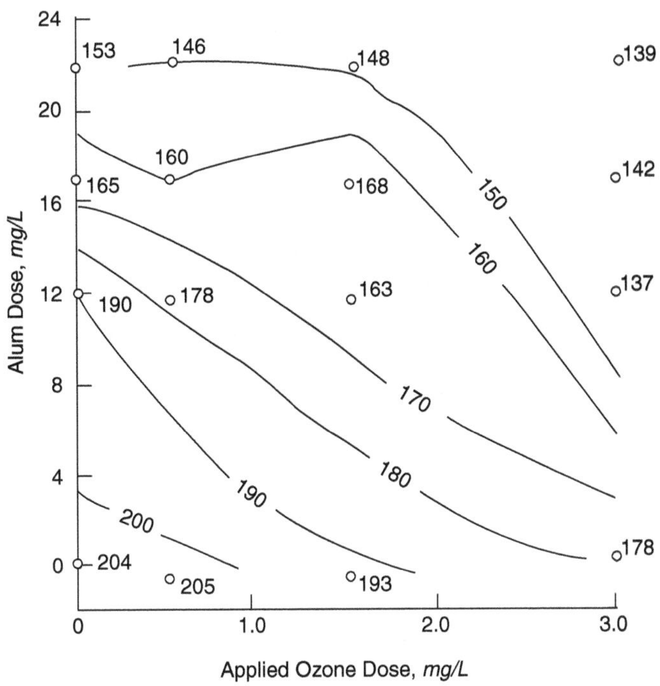

Source: AH Environmental Consultants 1997

Figure 7-17 Alum dose, ozone dose, THMFP topograph

Jar Tests for Ion Exchange Adsorption

The ion exchange (IX) process is commonly used in a fixed-bed configuration using natural zeolites or ion-selective resins to remove specific constituents in water. This section does not cover IX resins used in a fixed-bed configuration. Magnetic anion exchange resins may be applied in a non-fixed-bed configuration to reduce DOC content before coagulation. Jar tests can be used to assess the required resin treatment for removing target contaminants such as DOC. It is recommended that resin vendors be involved in conducting customized jar tests using magnetic IX resins.

Jar Tests for Softening Plants

Jar tests at lime-softening plants may be undertaken for a variety of reasons: verifying theoretical calculations of chemical dosage needed, evaluating settling efficiency after softening, evaluating changes in process operation, evaluating alternative coagulants for improved treatment, and evaluating treatment when source water quality has changed.

Treatment trains used in lime softening include single-stage treatment, two-stage treatment for excess softening, and split treatment.

Various treatment sequences are used at lime-softening plants; thus, a single procedure for performing jar tests cannot be prescribed. Instead, the jar test approach must be tailored to the plant in a way that mimics the full-scale process to the extent of which is feasible. The subsequent list provides examples that help ensure that a jar test mimics the lime treatment conditions at a WTP.

- If softening sludge is recycled back into the rapid mix in the treatment train, consider adding an appropriate percentage of settled lime-softening sludge to the jar test in the rapid mixing step.

- For split treatment, reserve some raw or pretreated water to add into the second stage of lime softening.
- For two-stage treatment, settle and decant clarified water from the jars treated in the first stage and perform the second stage of treatment with the clarified water.
- If clarification is done before lime softening, consider the use of clarified water from the treatment plant (lime-softening basin influent water) as the source water for jar tests.
- If the clarification step follows lime softening in the plant, use softened water from the plant for jar tests involving clarification of the softened water.
- To account for the impurities in commercial lime being used at the plant, either obtain lime slurry of a known strength from the plant or prepare a fresh slurry with the lime used for full-scale treatment.
- To dose lime slurry in tests with multiple jars, remove measured amounts of stirred lime slurry from the stock slurry using a pipette or syringe with a tip large enough that slurry particles will not block it. Dose the measured amounts into small beakers (one for each jar). To add the correct volume of slurry to a beaker, draw the exact volume needed into the pipette. After draining the slurry from the pipette into the beaker, rinse the pipette with laboratory-grade water, adding the rinsed-out water into the beaker. After all the beakers have the appropriate slurry quantities, add lime slurry from the beakers to all the jars simultaneously, and then rinse remaining slurry from each beaker into its respective jar. Lime residue left in the pipette or in a beaker would result in incorrect calculations of lime dosages used in the jar test.
- When chemicals other than lime are added in lime softening at the plant, add them in the same sequence in the jar test.
- To evaluate recarbonation for pH adjustment, acid could be added to lower the pH of the lime-treated (settled and decanted) water to the level attained by adding carbon dioxide in the plant.

REFERENCES

Allgeier, S.C., and R.S. Summers. 1995. "Evaluating NF for DBP Control with the RBSMT." *Journal AWWA*. 87(3):87-99.

Amburgey, J., A. Amirtharajah, B. Brouckaert, and N. Spivey. 2004. "An Enhanced Backwashing Technique for Improved Filter Ripening." *Journal AWWA*. 95(12):81-94.

Anderson, W.B., C.I. Mayfield, and P.M. Huck. 2008. "Endotoxin Release From Drinking Water Biofilters." *Journal of Water Supply: Research and Technology–AQUA*. 57(8):585-597.

AWWA. 2014 (6th ed.). M12, *Simplified Procedures for Water Examination*, Sixth Edition. Denver, CO: AWWA.

Biscardi, P.G., and S.J. Duranceau. 2017. "Preozonation Effects on Organic Foulants in a Coagulation-Ultrafiltration Membrane Process." *Journal AWWA*. 109(2):15-24.

Bundy, M.M., W.J. Doucette, L. McNeill, and J.F. Ericson. 2007. "Removal of Pharmaceuticals and Related Compounds by a Bench-Scale Drinking Water Treatment System." *Journal of Water Supply: Research and Technology–AQUA*. 56(2):105-115.

Bunker, D.Q., J.K. Edzwald, J. Dahlquist, and L. Gillberg. 1995. "Pretreatment Considerations for Dissolved Air Flotation: Water Type, Coagulants, and Flocculation." *Water Science & Technology*. 31(3-4):63-71.

Choi, K.Y., and B.A. Dempsey. 2005. "Bench-Scale Evaluation of Critical Flux and TMP in Low-Pressure Membrane Filtration." *Journal AWWA*. 97(7):134-143.

Cornwell, D.A., and M.M. Bishop. 1983. "Determining Velocity Gradients in Laboratory and Full-Scale Systems." *Journal AWWA*. 75(9):470-475.

Desjardins, C., B. Koudjonou, and R. Desjardins. 2002. "Laboratory Study of Ballasted Flocculation." *Water Research*. 36(3):744-754.

DiGiano, F.A., S. Arweiler, and J.A. Riddick Jr. 2000. "Alternative Tests for Evaluating NF Fouling." *Journal AWWA*. 92(2):103-115.

Elhadi, S.L.N., P.M. Huck, and R.M. Slawson. 2006. "Factors Affecting the Removal of Geosmin and MIB in Drinking Water Biofilters." *Journal AWWA*. 98(8):108-119.

Gonzalez Galvis, J.P. "Development of a Large Batch Bench-Scale Dissolved Air Flotation System for Drinking Water Treatability Tests." PhD diss., University of Ottawa, 2019.

Graham, N.J.D., C.C.S. Brandao, and P.F. Luckham. 1992. "Evaluating the Removal of Color From Water Using Direct Filtration and Dual Coagulants." *Journal AWWA*. 84(5):105-113.

Hering, J.G., P.Y. Chen, J.A. Wilkie, and M. Elimelech. 1997. "Arsenic Removal from Drinking Water During Coagulation." *Journal of Environmental Engineering*. 123(8):800-807.

Hudson, H.E., and E.G. Wagner. 1981. "Conduct and Uses of Jar Tests." *Journal AWWA*. 73(4):218-223.

Jang, N.Y., Y. Watanabe, and S. Minegishi. 2005. "Performance of Ultrafiltration Membrane Process Combined with Coagulation/Sedimentation." *Water Science & Technology*. 51(6-7):209-219.

Jarvis, P., P. Buckingham, B. Holden, and B. Jefferson. 2009. "Low Energy Ballasted Flotation." *Water Research*. 43(13):3427-3434.

NSF International. *Drinking Water Treatment Chemicals—Health Effects*. NSF/ANSI/CAN 60. Ann Arbor, MI: NSF International, 2020.

Park, P.K., C.H. Lee, S.J. Choi, K.H. Choo, S.H. Kim, and C.H. Yoon. 2002. "Effect of the Removal of DOMs on the Performance of a Coagulation-UF Membrane System for Drinking Water Production." *Desalination*. 145:237-245.

Pernitsky, D.J., R.E. Cantwell, E. Murphy, N. Paradis, J. Boutilier, and G. Bache. 2011. "Use Zeta Potential to Improve Direct Filtration Operation." *Opflow*. 37(2):20-23.

Scheuch, L.E., and J.K. Edzwald. 1981. "Removing Color and Chloroform Precursors From Low Turbidity Waters by Direct Filtration." *Journal AWWA*. 73(9):497-502.

Smith, S.D., and M. Edwards. 2002. "Bench-Scale Evaluation of Innovative Arsenic-Removal Processes." *Journal AWWA*. 94(9):78-87.

Tanaka, T.S., and M. Pirbazari. 1986. "Effects of Cationic Polyelectrolytes on the Removal of Suspended Particulates During Direct Filtration." *Journal AWWA*. 78(12):57-65.

Teixeira, M.R., V. Sousa, and M.J. Rosa. 2010. "Investigating Dissolved Air Floatation Performance with Cyanobacterial Cells and Filaments." *Water Research*. 44:3337-3344.

Wagner, E.G., and H.E. Hudson. 1982. "Low-Dosage High-Rate Direct Filtration." *Journal AWWA*. 74(5):256-261.

Weiss, W.J., E.J. Bouwer, W.P. Ball, C.R. O'Melia, C. Arora, and T.F. Speth. 2003. "Comparing RBF with Bench-Scale Conventional Treatment for Precursor Reduction." *Journal AWWA*. 95(12):67-80.

Chapter 8

Testing for Process Evaluation and Control

Chapter Summary

- Testing is more than a design tool. It provides value in ongoing process optimization.
- Testing does not have to be a continuous effort or full-time job.
- Developing a list of questions to answer helps focus testing.
- The chapter provides examples of nontraditional uses of pilot equipment for process monitoring.

Process evaluation and control testing is a critical tool for utilities to continuously improve the quality and cost of water provided to their customers. Depending on the goals of the testing, bench-, pilot-, or full-scale testing, or some combination, may be most appropriate to address the desired outcome. Bench-scale testing is a critical tool in determining the required chemical doses for the treatment process, which is covered in detail in the previous chapter. This chapter focuses on the larger decision-making framework around testing, focusing on the approach to developing goals and getting valuable answers. Parallel with this 4th edition of Manual M37, AWWA is completing a new pilot-testing manual (M76). With that in mind, this chapter is focused on testing for process evaluation and control at existing water treatment plants (WTPs).

There are a range of reasons utilities and other water professionals conduct testing studies, including

- optimization of an existing treatment process,
- regulatory needs associated with new or unproved–unregulated technologies,
- process monitoring for early confirmation of operating conditions,
- process evaluation or validation for a new plant or major capital improvements, and
- establishing design criteria for new or upgraded processes.

This manual focuses on the first three of the bullets in this list. Additional information can be found in AWWA Manual M76 for those who are focused on process evaluation or validation for a new plant or major capital improvements or who are starting a pilot-testing program from scratch.

ESTABLISHING A TESTING FRAMEWORK

Establishing Goals

Well-defined goals will highly increase the chances of a successful pilot or full-scale study. It is important that testing goals are discussed and identified at the very beginning of the project and focused on answering key questions for the facility or the utility. The purpose of testing typically focuses on improving treatability and performance and developing greater insight into the existing chemical and mechanical treatment processes to improve reliability and lower operations and maintenance (O&M). Usually, testing goals can be determined by identifying the information gaps that exist. These two categories are not mutually exclusive, and information for both categories can exist within the same project.

The following is a brief list of common pilot tests.

- Coagulation
  - Coagulant testing for optimization of turbidity, organics, disinfection byproduct (DBP) formation, corrosion control, etc.
  - Testing of alternate coagulation, coagulant-aid, and floc-aid polymer combinations.
  - Minimizing fluctuation of seasonal or source water variation effects on full-scale facility using different oxidants.
- Filtration
  - Addition of filter aid to improve filter ripening.
  - Evaluating various media profiles.
  - Evaluating various loading rates.
  - Improving iron and manganese (Mn) removal.

Determining Key Performance Indicators

Once the type of study has been determined, it is important to define what parameters need to be monitored and what the target goals will be for those identified parameters. A few examples would be

- target turbidity (ntu) for settled and or filtered waters;
- target levels of ultraviolet (UV) absorbance at 254 nm (UV254) or total organic carbon (TOC) removal from source water levels;
- target DBP values to be achieved;
- target corrosion indicator values;
- potential chemical cost reductions, extended service life goal, or reduced O&M efforts;
- target unit filter run volume (UFRV); and
- target level of trace contaminants removal.

Determining Necessary Testing Resources

Duration

The duration of a study should be defined as well as possible before the testing begins. Not all studies need to be 6–12 months long with gigabytes of data. Depending on the situation, a six-week study is capable of answering questions regarding coagulation adjustments for turbidity, organics removal, and/or DBP reduction. In instances in which a facility already has an operational pilot system, operators can obtain useful data in one to two days to determine treatment changes for variations in source water turbidity or organics loading. The same can be said for operating filters. A pilot-scale granular media filter skid can provide head loss development, effluent turbidity response versus loading rate, filter-to-waste rates and duration, and run times within 2 or 3 filter cycles with stable influent conditions.

Some study durations are determined not by the utility or a supporting consultant but by the governing regulatory agency. For instance, if the proposed testing is to be used for regulatory approval to change primary coagulant or adjust filter loading rates, then conversations should take place to determine what the governing agency requires for the type, quantity, and duration of the test and the required corresponding data.

The duration of the pilot should be balanced with the variability of water quality. If water quality changes drastically from season to season, then an ideal pilot will capture data in multiple seasons. If time or budget is limited, then choosing the season with the most difficult water quality is advised.

Staffing

Dedicated pilot or full-scale testing operators are not necessary for all studies; however, if a utility has dedicated staff for research and optimization studies, that is a benefit. Items to consider include the following:

- How long will the proposed study be?
- Can the additional samples required be collected with the full-scale plants' daily rounds?
- How much on-line monitoring and analytical instrumentation will be required to minimize staffing requirements for sampling and increase testing equipment operation time?
- What process automation and control will be included to minimize effects on staffing? Examples can be variable-frequency drive (VFD) pumps, modulating control valves, and automated backwashes.
- Is automatic data logging or data analytics included to minimize data transcription efforts?
- What are the maintenance requirements to keep the test going?

Site Requirements

Several site requirements need to be addressed well before the arrival of the pilot equipment. Information on selection and sizing of pilot plants is discussed in greater detail in Chapter 6 of AWWA M76. Basic considerations include

- indoor or outdoor installation;
- footprint;
- electrical and communications connections;
- availability of process water streams (source water, clarified water, etc.);

- full-scale finished water for testing, filter backwashes, washdown, and safety (fire, chemical spill, etc.);
- effluent discharge locations for treated water and process wastes;
- laboratory and operator facilities; and
- safety assessment (access–egress, safety showers, eye wash stations).

SPECIAL CONSIDERATIONS FOR COAGULATION, FLOCCULATION, AND FILTRATION PILOT-TESTING

The following information is focused on piloting equipment; however, this information can also be applied to testing of larger demonstration-scale or full-scale equipment in many cases.

Coagulation

Coagulation pilot equipment typically comes as a component of a coagulation, flocculation, and clarification skid, especially if the skid is rented or leased as opposed to purchased as a custom unit. Leased units tend to include more processes to appeal to more customer utilities, thereby reducing cost for the pilot-plant manufacturer. Generally, the coagulation process in a pilot plant should have the following components and controls:

- Source water supply—In some cases, source water can be delivered directly into the pilot skid through a pressurized feed line from the full-scale plant (e.g., a tap from the source water feed line). Changes in the operation of the full-scale system (e.g., a reduction in flow) can change the pressure being delivered to the pilot plant, which can affect flow to the test skid. To maintain constant flow to the pilot skid, the skid must have an accurate flow control valve and flowmeter to regulate the source water flow, and this can be difficult at pilot scale. In other cases, source water is delivered into a source water storage tank or overflow box. Water is then pumped or flows by gravity to the downstream process equipment. Inflow is set slightly higher than outflow, and the excess water is sent to an overflow. This maintains a constant flow and constant head to the pilot skid. Flows, tank level or pressure, and temperature monitoring should be included because storage of source water indoors can have a significant effect on source water temperature and pilot system performance.
- Coagulant feed and rapid mix—Chemical feed and mixing equipment are installed downstream of the pressurized direct source water connection or the source water tank.
  - o Rapid mix tanks or inline mixers should be installed that represent the intended full-scale process. Several mixers may be used to allow the pilot plant to introduce polymers or preoxidants before the addition of coagulants.
  - o Depending on the pilot-testing goals, it may be advantageous to have more than one chamber with chemical addition, adjustable mixing energy mixers, drains, isolation valves, and the ability to run the rapid mix chambers in series or as just one chamber.

It is highly recommended that all chemical feeds, flows, mixing intensity, and sensors be controlled and/or the data are sent to a data collection system to ensure consistency of treatment and preservation of all produced data.

- o Ensure that there is a representative flow measurement or calibrated metering pump available for each chemical feed.
- o Because of the low chemical feed rates of pilot plants, low-flow chemical feed pumps or dilution of the stock chemical may be required. Consult the chemical vendor to verify how much a given chemical can be diluted and how long the chemical will remain active at that dilution. Some chemicals, such as polyaluminum chlorides and polymers, have a reduced shelf life once they are diluted.

Flocculation

Flocculation studies in a pilot plant are typically part of a coagulation study that will require the use of a coagulation–flocculation–clarification skid. Flocculation is somewhat of a unique process to evaluate and optimize because common practice in determination of performance is based on measurements for downstream processes. As an example, flocculation efficiency is often determined by qualitative key performance indicators (KPIs) such as settled turbidity or filter UFRV, as well as by qualitative visual observations regarding floc size and "fluffiness." As technology improves, there will be more analytical equipment available to quantify floc formation, efficiencies, and size through direct measurement of the floc itself. There are currently some instruments that use cameras and a proprietary algorithm to quantify floc shape, size, quantity, and density in an attempt to add quantifiable data to the direct measurement of the floc. Despite the limited analytical instrumentation for direct flocculation evaluation, there are still recommended controls and equipment for a successful piloting of the flocculation stage. Typical evaluation of flocculation formation is based on downstream process performance. As an example, granular media WTP, overall filter UFRVs (as described in Chapter 1) may be a better KPI for flocculation than either settled water turbidity or floc appearance. Clarifier turbidity is a function of particle number, size, and optical properties, and lower clarifier turbidity does not always correspond to longer filter runs or lower filter effluent turbidity. UFRV takes ripening, breakthrough, and head loss rate into consideration.

- Multiple basins—Having multiple basins in series with the capability of isolating to the use of just one will allow for a more accurate full-scale flocculation matching at the pilot scale by allowing for larger or staged flocculation that may be present in the full-scale facility.
- Variable-speed mixers—Variable-speed mixers are recommended in each of the flocculation basins to properly mimic the full-scale mixing intensity that is present in the full-scale plant. It is not recommended to specify a single-speed flocculation mixer on the pilot skid on the basis of a mixing intensity calculation of the full-scale flocculators. Often, there are more variables affecting what is the actual mixing intensity or mixing energy present in the full-scale flocculators than are accounted for in the mixing intensity calculation. Adjustment of the mixing intensity on the pilot-skid flocculators allows for better plant matching. Once baseline plant matching is complete, additional testing can be conducted to determine whether increasing or decreasing flocculator mixing speeds could improve floc formation and settling rates at full scale. There is also a relatively low cost of having controllable variable-speed flocculators in the pilot skid.
- Illuminated viewing windows—Although the "human-eye" observation of the size and density of floc formation is not necessarily a quantifiable variable, it is a useful tool in assessing potential causes of poor settling performance. It gives the pilot-plant operator the capability of assessing for several conditions such as, but not limited to, pin-floc or floc-shear. A powerful flashlight shone from the top can also be helpful.

- Chemical addition ports—The presence of chemical addition ports in each of the flocculation basins on the pilot skid allows for testing the effects of flocculant aid polymers, oxidizer addition after coagulation, and pH adjustment after coagulation. This is a relatively low-cost addition that can add significant flexibility to optimization testing.

Clarification–Sedimentation Processes

Pilot-scale testing rarely involves the use of conventional sedimentation basins because of the large footprint required to achieve sufficient clarification and because the design of these conventional units tends to be based on well-established design guidelines. Instead, high-rate clarification processes such as tube or plate settlers are typically used in pilot-scale clarification.

The surface loading rate (SLR) in pilot-scale clarifiers may be modified in one of two ways. First, the flow rate through the clarification process can be increased or reduced, which will proportionally increase or decrease the SLR. However, flow requirements or limitations from either upstream or downstream processes may limit how much the SLR can be varied through flow modification. Alternatively, if the clarifier is designed to allow for its settling characteristics to be modified (e.g., by adding or removing plates or increasing or decreasing the number of tubes), then the SLR can be modified while maintaining similar flow rates through the process.

Other technology-based clarification processes, such as dissolved air flotation, solids-contact clarifiers, sludge blanket clarifiers, and ballasted flocculation, may also be available in smaller units designed for pilot-scale testing. Such processes are typically installed in portable skids or trailers and are typically procured from the equipment manufacturers for temporary demonstrations or leases. The mechanics behind these processes are described in Chapter 3. Because these processes are technology-based, specific installation requirements and operations will be recommended by the equipment manufacturer.

See Figure 8-1 for a 10 GPM Flocculation and Sedimentation Pilot module.

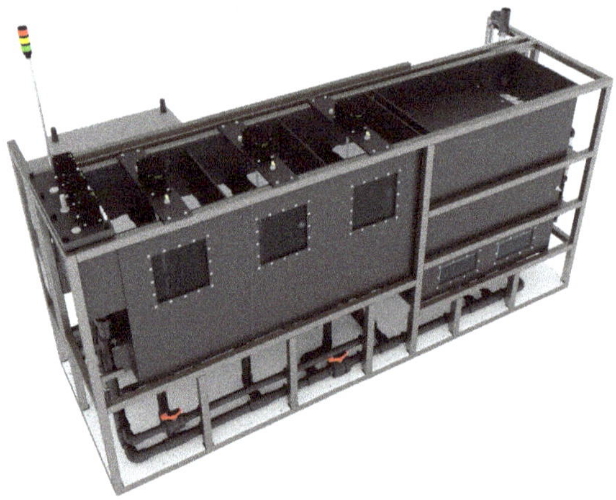

Source: Intuitech

Figure 8-1 10 GPM Flocculation and Sedimentation Pilot module with three-stage rapid mix, three-stage paddle flocculation, and lamella plate clarifier

Filtration

This section focuses on granular media filtration, although many of the required components and data collected are also recommended in membrane filtration pilot studies. A significant amount of information can be collected and calculated about filter performance and efficiencies from filtration pilots.

A typical nonautomated pilot-scale filter column will have

- manual valves to control when the filter is filtering, backwashing, or filter-to-waste;
- a VFD feed pump and a flowmeter to help maintain a constant flow during filtration and backwash stages;
- an inline or flow-through turbidity meter for monitoring particulate removal; and
- sample ports at various depths in the column for analysis of media properties or development of head loss through the bed during testing.

An automated pilot-scale filter skid may contain

- Programmable logic controller (PLC) controlled automated valves to control filter stages of filtering, backwash, and filter-to-waste;
- a VFD feed pump and flowmeters to control flow through the filter, provide the option of multiple backwash rates (staged low, medium, and high backwashes), and control filter-to-waste rates;
- head loss sensors;
- inline or flow-through turbidimeters; and
- sample ports at various depths in the column for analysis of media properties during testing.

The immediate difference between automated and nonautomated filter columns is the extra data. A manual filter will require more attention and can only backwash when someone is present. An automated filter column can backwash whenever the defined backwash parameters are met because the PLC controls the valving. Head loss, effluent turbidity, or time can be defined in the PLC and then used as backwash initiators. With the addition of head loss, flow, time, automated backwashes initiated by set parameters, and the significant increase in data points, other information can be calculated:

- Head loss slope or head loss development
- Unit filter rate volume calculations
- Particulate removal efficiencies
- Evaluation of improvements in filter-to-waste time based on varying time and flows for staged backwashes
- Media degradation studies by extracting media from pilot columns and sending to a laboratory

Some items are different in pilot-scale filters versus setup and operation of a full-scale granular media filter:

- Most pilot filters will not have a gravel bottom or a commercial-style underdrain similar to a full-scale granular media filter. Pilot filters will have a stainless-steel or plastic screen in the bottom of the column for media retention. This does not significantly affect pilot performance or operations.

- Because of the wall effects present in narrow pilot filter columns, it is generally recommended to use columns of at least 6 in. in diameter for dual- and tri-media filters. For specialty filters at high loading rates, even larger columns may be needed, depending on testing goals.
- Also, most pilot-scale filter columns are pump fed instead of gravity fed. Particle removal and characteristics have demonstrated the same performance in a pump-driven pilot-scale filter as a gravity-fed filter if care is taken to minimize hydraulic disturbances. Careful hydraulic design of pilot filter columns is necessary to ensure constant flow rates with minimal hydraulic disturbance. It is critical to confirm that the pump that is used for supplying water to the pilot filter columns has a proven recorded history of being able to provide constant flow to avoid effects in filter performance based on flow and pressure variations. Peristaltic or progressive cavity pumps can minimize this issue but can cause pulsation problems that can result in unsteady effluent turbidity or particle counts if not designed properly. Additionally, it can be difficult to source high-quality filter effluent flow control valves for the low flow rates of pilot filter columns.
- Water flows more easily at the interface between media and the surface of the filter wall. A higher ratio of wall contact to the surface area of the media can result in marginally higher effluent turbidity values. This is because some of the particles are not trapped by the media next to the wall. However, a well-designed pilot filter operating with proper coagulation will have no trouble meeting a typical filter effluent turbidity goal.
- See Figure 8-2 for a Granular Media Filtration Pilot module.

Source: Intuitech

Figure 8-2 Granular Media Filtration Pilot module with four 6-in. columns and automated valves for automatic backwashing and autonomous operations

EXAMPLES AND SPECIAL TESTING

Coagulation Monitoring Downstream of Rapid Mix

Some WTPs, especially conventional WTPs, install permanent pilot filters immediately downstream of their rapid-mix process. These pilot filters are controlled and monitored directly by the WTP supervisory control and data acquisition system and used as a critical process control to provide early warning of chemistry upsets in the full-scale system.

In these systems, the coagulated water is passed directly through the pilot filter before flocculation or clarification. Although the pilot filter effluent turbidity rarely matches the full-scale filter effluent turbidity, the pilot effluent turbidity under optimized chemistry conditions provides a baseline for what should be expected following effective coagulation. When a significant increase in pilot filter effluent turbidity is observed, it alerts the operators that the chemistry is no longer optimized and allows them to make coagulation changes before the flocculation and clarification basins are full of poorly coagulated water.

This approach decreases the response time for chemistry upsets from several hours to several minutes by removing the retention time in the flocculation and clarification basins from the amount of time required to observe the effect of changes in chemistry.

Spiking for Various Contaminants

Depending on the pilot-testing driver (new technology, process optimization, or regulatory), the test plan may need to include objectives for spiking specific contaminants into the pilot feed to augment the concentrations that are naturally occurring in the actual source water. Artificial spiking events examine more challenging or extreme water quality conditions that would normally be expected to occur naturally during the pilot test. This could represent the effect of a wildfire, flood, or severe algae bloom. The opportunity to collect water for testing during these events may not exist, and spiking can be used to simulate extreme conditions. When possible, bench-scale testing can be used as proof of concept for treatability of the artificial spiking water or to narrow the conditions to be explored at the pilot scale. Bench-scale testing can be conducted in the laboratory to promote conditions for better control and repeatability. When pilot-testing is needed, considerations for spiking include the following:

> Contaminant spiking can be included in the pilot test plan to evaluate
>
> ✓ Changes in water quality that deviate from the original treatment process design criteria
>
> ✓ Seasonal effects, such as lake turnover, algal blooms, low alkalinity runoff, or high turbidity from rain events
>
> ✓ Infrequent changes to source water as a result of natural disasters (forest fire, drought, flooding, etc.)
>
> ✓ Manmade effects to the watershed (agricultural runoff, chemical spills, dredging, drilling, etc.)

- Planning—It is important to define the objectives for the artificial spiking phase. The intent is to simulate worst-case events to understand how full-scale treatment could respond. To do so, measurable criteria for determining success must be defined. For example, the test results must determine the operational limits in which the treatment process can maintain the system's water quality goals. During the planning phase, historical data are reviewed to characterize the challenge water. Once the appropriate maximum constituent concentrations and background matrix are identified, the sampling parameters, frequency, and locations can be defined for monitoring. Because pilot effluent is not sent to the distribution system, regulatory involvement is not required for planning but can be beneficial if regulator buy-in is required for process changes.

- Preparation—Whenever possible, seize the opportunity to collect water during extreme water quality events and use it for conducting the testing. If it is not possible, representative test water(s) must be prepared. When appropriate, laboratory-grade chemicals can be used to spike chemical constituents to the water. Implement safety protocols for the handling of hazardous chemicals and be sure to plan for the management of excess chemical and concentrated residual waste disposal. For natural events, sediment or burned material from a recent runoff or wildfire event can be used to mimic the effects of an extreme weather event. Use source water as close to the background matrix as possible to represent the anticipated seasonal water quality for the artificial spiking condition. A caveat, however, is that there is significant difficulty in simulating the complexities of actual watershed behaviors from an artificial adjustment. This type of spiking is more in line with understanding trends than in accurately predicting full-scale performance.
- Execution—Artificial spiking phases should be conducted later in the test plan so that baseline performance conditions are determined first. Allow for transition time between baseline and challenge test phases so performance can stabilize. Purposeful testing of over- or under-responses (e.g., chemical feed rates, hydraulic loading rates) can also be explored.

Other Special Conditions

Pilot-scale and full-scale testing can also be used to test processes that are dependent on coagulation. These tests could include

- pipe loop testing from pilot plant-treated water evaluating effects from coagulant changes between sulfate- and chloride-based coagulants because the ratio between the two can affect galvanic corrosion in the distribution system, and
- evaluating treatability of residuals produced by different coagulants at the pilot scale to determine potential effects and solids production, dewatering chemical needs, and dewatering technologies for a full-scale implementation.

In addition to the small list of potential optimization studies that can be conducted, Table 8-1 demonstrates a format that can be used to keep track of proposed studies, the infrastructure or process affected, the variables to control, and the ever-important reason for testing. Creating a table like this can be beneficial in the early stages of planning a pilot or full-scale study. Outside of creating a list of proposed studies, it also helps to organize several of the key drivers and basic study requirements that can be used for determining study size, equipment, duration, and more.

Table 8-1 Pilot study types and benefits

Study to Conduct	Location	Variables	Reason for Testing	Downstream Processes Affected
Various coagulants	Rapid mix	Coagulant types	Improve removal of organics, turbidity, DBP precursors	Flocculation, clarification, intermediate ozone, filtration, membrane, UV
Coagulant dose	Rapid mix	Dose and concentration	Improve removal of organics, turbidity, DBP precursors	Flocculation, clarification, intermediate ozone, filtration, membrane, UV

(continued)

Table 8-1 Pilot study types and benefits (*Continued*)

Study to Conduct	Location	Variables	Reason for Testing	Downstream Processes Affected
Coagulation pH	Pre-rapid mix or rapid mix	Dose and concentration of coagulant, addition of pH adjusters (sulfuric acid or lime [caustic])	Improve removal of organics, turbidity, DBP precursors	Flocculation, clarification, intermediate ozone, filtration, membrane, UVA
Oxidizer addition	Pre-rapid mix or rapid mix	Oxidizer types, dose, concentration	Improve removal of organics, DBP precursors	Flocculation, filtration, intermediate ozone, membrane, UV
Coagulant aid	Pre-rapid mix or rapid mix	Type and dose of various polymers	Improve floc formation and particulate removal	—
PAC	Pre-rapid mix	PAC dose	Improve removal of organics, DBP precursors	Flocculation, intermediate ozone, filtration, membrane, UV

DBP—disinfection byproduct, PAC—powdered activated carbon, UV—ultraviolet, UVA—ultraviolet absorbance

UNDERSTANDING HOW TO ENSURE SUCCESSFUL PILOT-TESTING

Baseline or Plant Matching

Baseline or plant matching may be important, depending on the testing objectives. During the plant matching phase, the injection points, mixer speeds, flocculator rpm, etc., are set to achieve similar performance between the full-scale facility and the pilot plant. This ensures that the pilot-plant results can be relied on to accurately represent the expected performance of the full-scale plant.

Coagulation chemical doses in the pilot plant should be set to be the same as the full-scale facility, and this should result in similar pH, particle charge, and TOC or UV254 removal values. Differences greater than 10 percent could indicate a physical problem in either the pilot or full-scale plant. For example, stock chemical concentration or dilution values could be inconsistent, chemical feed pumps could require calibration, or equipment of hydraulic differences in mixing may be affecting the results. Controlled jar tests using high-precision laboratory equipment such as calibrated micropipettes for chemical dosing may be valuable to troubleshoot differences. Removal of TOC or UV254 can be especially valuable in this regard, as the removal of these parameters should be very close for similar chemical dosing conditions.

Pilot-scale clarifier turbidity may differ from pilot values because of subtle hydraulic differences in the flocculation or clarification equipment. Again, clarified turbidity values within ±15 percent of each other can be considered sufficient. Well-built pilot filters should be able to produce a filtered-water turbidity within ±10 percent of that achieved by the full-scale filters. Regardless of the magnitude of the differences in results, it is important that the pilot and full-scale plants respond in the same way and relative magnitude to a process change. For example, if the coagulant dose is reduced slightly, both plants should show the same response, such as a decrease in TOC removal, a slight increase in filter effluent turbidity, and a slight reduction in UFRV. A larger coagulant dose reduction should result in a larger change in the aforementioned parameters and could result in poor filter ripening or early turbidity breakthrough.

A detailed example is shown in the sample data in Table 8-2 and Figure 8-3 in which the pilot plant is dosing 43 mg/L of coagulant, which is exactly the same as the full-scale facility. The coagulation pH compared with that of the full-scale plant is essentially the same. This then allows for the evaluation of the pilot plant's ability to match resulting parameters such as settled turbidity and other downstream parameters and processes. As a side note, the 11-h sample set shown previously was used for description and should not be considered as a representation of the adequate amount of time to determine a successful plant matching phase.

Table 8-2 Plant matching conditions for coagulation stage

	Coagulant Dose *ppm*			Coagulation pH		
Time	Full Scale	Pilot Train 1	Pilot Train 2	Full Scale	Pilot Train 1	Pilot Train 2
8:30	43	43	43	6.63	6.57	6.61
9:30	43	43	43	6.66	6.63	6.62
10:30	43	43	43	6.57	6.57	6.59
11:30	43	43	43	6.55	6.61	6.57
12:30	43	43	43	6.61	6.59	6.63
13:30	43	43	43	6.59	6.64	6.65
14:30	43	43	43	6.63	6.67	6.66
15:30	43	43	43	6.62	6.59	6.63
16:30	43	43	43	6.58	6.52	6.59
17:30	43	43	43	6.52	6.64	6.58
18:30	43	43	43	6.56	6.61	6.59
19:30	43	43	43	6.60	6.58	6.62

Figure 8-3 Plant matching coagulation pH results

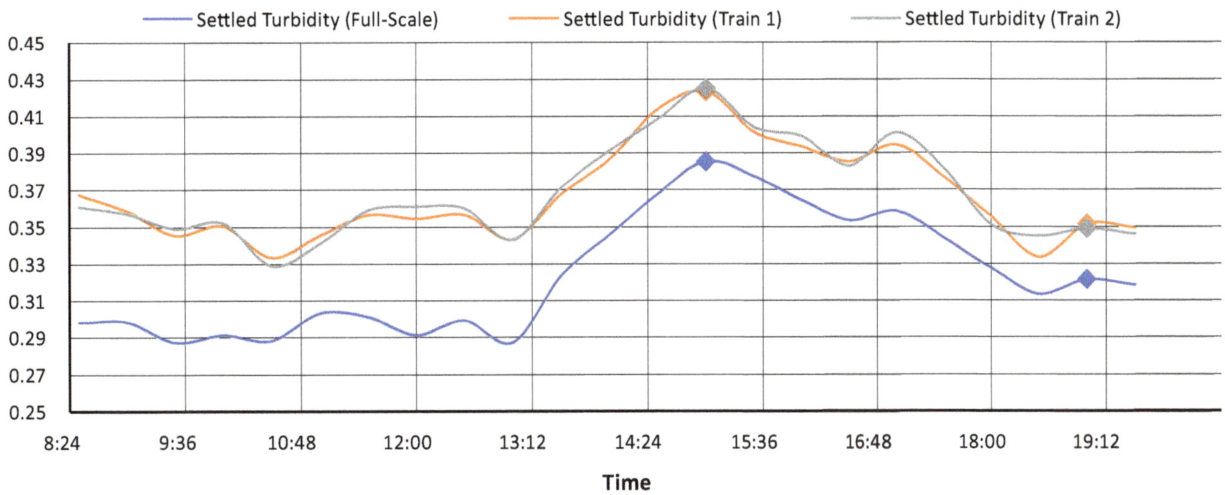

Figure 8-4 Plant matching settled turbidity results

In Figure 8-4, the comparison of the full-scale settled turbidity values versus the settled turbidity values of a two-train pilot plant shows a difference in values between full and pilot scale. There is almost no difference between settled turbidity rates between the two settled pilot trains, but the offset in values is somewhat consistent between the full-scale and pilot-scale values. This does not indicate an inability to plant match or mimic the full-scale facility. It merely demonstrates there is some difference in the hydraulic characteristics at the pilot scale. An important point in Figure 8-4 is that the offset between the full-scale and pilot trains stays relatively constant throughout both phases of the event, which begins around 13:12. As an added note, if a two-train pilot system is used, then one train can be used to plant match while the other train is used to test optimized conditions. As long as the plant matching train maintains values similar to those of the full scale (even with the determined offset from the early plant matching phases), then any variation in results between the two pilot trains should be indicative of a full-scale implementation of the optimized conditions on the other pilot train.

Expected Instrumentation Values for Pilot-Testing

There are several factors that contribute to results on instrumentation in a pilot plant. Different types of instrumentation can report values differently when comparing pilot scale with full scale. This does not mean the values are incorrect, inaccurate, or nonrepresentative of the full-scale comparison. It simply indicates a need to understand how the instrumentation results compare in a smaller and different hydraulic environment from the full scale. Table 8-3 is a list of various instrumentation equipment and an explanation as to how the results may vary as well as a brief discussion of how to have certainty despite the variation. This table does not indicate that the equipment does not work correctly at the pilot scale, but it does show that sometimes the size of the infrastructure treating the water can have an effect on the produced value recorded by the analytical instrumentation. The information is still quite usable, especially if baseline and plant match testing was conducted at the beginning of the study, as discussed in the section titled "Baseline or Plant Matching."

Table 8-3 Expected values from pilot instrumentation

Analytic	Treatment Locations	Range of Variation From Full-Scale Application	Typical Cause of Variability From Full Scale if Variability Exists Between Pilot and Full-Scale Tests
pH	Multiple	Should be the same as variation between same probes in full-scale treatment and within manufacturers stated range of accuracy.	Different type of probe than full scale, electrical interference, unique hydraulic mixing at either scale of treatment.
Turbidity	Raw	Should not vary more than manufacturers' specifications if compared with same equipment at full scale.	Inlet for pilot-plant source water is not in the correct location, or clogs on pilot intake are causing filtering of particulate.
Turbidity	Settled	Can vary by more than manufacturers' stated range of accuracy when compared with full scale. Variation from full scale usually trends higher than full scale.	Hydraulics are often quite different in a pilot versus full scale, which can lead to variations in settled turbidity. Adjusting flow or settling rates as well as baselining will offset variations to provide significance to the testing.
Turbidity	Filtered	Can vary more than manufacturers' stated range of accuracy when compared with full scale. If so, this is usually the result of a hydraulic constraint in the pilot equipment. Developing a baseline will account for the variance and allow for significant and usable data to still be collected.	Column filters typically provide a smaller surface area and a greater ratio of filter wall to media surface than a full-scale granular media filter. Values may not match exactly, but baselining to develop a known offset is an accepted way of validating testing performance despite the variation.
Particle counts	All	Typical steady state readings should be within orders of magnitude of full-scale operation. However, pilot system samples are much more prone to significant spikes than full-scale systems.	High filter loading rates and significant sidewall effects can be present in filtered-water particle samples. Additionally, the potential for vibration affecting particle counts in pilot filters is much higher than in full scale.
Chlorine (free and total)	Multiple	Should not vary more than manufacturers' specifications if compared with same equipment at full scale.	Different types of instruments; hydraulics at either scale is such that representative samples cannot be collected (incomplete mixing). Contact time may be less in a pilot than in full scale.
Total organic carbon	Multiple	Should not vary more than manufacturers' specifications if compared with same equipment at full scale.	Different types of instruments; hydraulics at either scale is such that representative samples cannot be collected (incomplete mixing).
Total trihalomethane	Multiple	Should not vary more than manufacturers' specifications if compared with same equipment at full scale.	Different types of instruments; hydraulics at either scale is such that representative samples cannot be collected (incomplete mixing). Contact time with precursors often is less in a pilot than in full scale.

Chapter 9

Instrumentation for Monitoring and Control

Analytical instrumentation is important for the monitoring and control of coagulation and filtration. On-line sensors are commonly used for monitoring changes in source water quality, monitoring and control of coagulant feed, clarification and filtration processes, and for regulatory compliance. This chapter provides a description of the most commonly used instruments in a water treatment plant (WTP) and focuses on details about the instrument such as the measurement principal, data interpretation, and troubleshooting. Additional technical details on on-line sensors as well as further information on the selection, specification, and integration of on-line monitors into water treatment operations can be found in *Online Monitoring for Drinking Water Utilities* (Hargesheimer et al. 2002).

PROCESS APPLICATION

On-line monitoring applications are found throughout most WTPs. Table 9-1 summarizes the instrumentation discussed in this chapter and indicates typical use for various stages of the process. Detailed application of these instruments to the process can be found in the relevant process chapters.

COMMON CONSIDERATIONS

Several common considerations should be given to the selection, installation, and maintenance of instruments:

- Select instruments with the proper range. An instrument will function best and with the least maintenance if the instrument's range and design are suitable for the anticipated range of conditions.

Table 9-1 Typical on-line sensors used in coagulation and filtration applications

On-line Parameter Measured	Source Water Monitoring	Coagulation	Clarification	Filtration
Turbidity	●	○	●	●
Particle count	◐	○	◐	●
Alkalinity	●	●	◐	Not used
Streaming current	◐	●	○	○
Zeta potential	◐	●	○	○
Total organic carbon	◐	●	◐	◐
Ultraviolet absorbance	●	●	◐	◐
pH	●	●	◐	◐
Head loss (differential pressure)	Not applicable	Not applicable	○*	●
Sludge level	Not applicable	Not applicable	◐	Not applicable

Source: David Pernitsky

● = on-line monitoring very useful for process monitoring and control
◐ = on-line monitoring moderately useful for process monitoring and control
○ = on-line monitoring not commonly used for this process
○* = used for contact adsorption (roughing filter) clarifiers

- Understand the environment under which on-line instruments are to be operated. Environments that have temperature extremes, high humidity, or are corrosive require appropriate instrument designs that are tolerant to these conditions.
- Before selection of the instrumentation, ensure that the data collection and transmission method is appropriate for the utility's supervisory control and data acquisition (SCADA) system.
- If the instrument is used for regulatory compliance, ensure that the instrument has proper calibration and that its design meets the specified criteria that are outlined by the regulatory compliance method.
- Locate the instrument appropriately. An instrument located close to the sample provides little time for sample line interference and provides the most "real-time" data. However, instruments also need to be placed at a location where proper calibration and maintenance can take place.
- Sidestream instruments are complete flow-through instruments in which the sample is taken out of the main process stream and continuously passes through the instrument or sample cell.
- In situ instruments have the sensor portion of the instrument placed directly into the main process stream. In situ instruments do not require a sample to be transported to and from the instrument.
- Set up sample lines to sample in the middle of a process streamflow to avoid air bubbles at the top and sediment at the bottom, as well as any wall effects. If sampling in a tank, sample far enough into the tank to avoid a dead spot, away from any chemical addition points, and away from any mixing blades.
- If a sample pump is needed, it is desirable to locate it on the downstream side of the sensor if possible; this allows it to draw the sample through the analyzer first so that pump interferences are minimized (Hargesheimer and Lewis 1995).
- Sample tubing should be composed of inert materials so that particulate sticking and particulate shedding are minimized. Ideally, the best material for sample lines is inert to process fluids such as high-density polyethylene or

polytetrafluoroethylene (PTFE). Special consideration should be given when using PTFE because it typically contains per- and polyfluoroalkyl substances (PFAS), so the effect of PFAS on the sample and waste stream should be considered. Also, tubing that is opaque (black is best) will keep organisms that require light from growing on the inside surfaces of sample lines. Avoid use of low-density or soft plastics.

- Understand the requirements for calibration verification and maintenance. It is important to budget adequate resources for operation and maintenance of these instruments. To understand better what sort of budget and resources will be required, ask the manufacturer for their calibration procedure and calibration kit cost (if they provide one).
- Perform calibrations as instructed by the manufacturer. The instructions were developed by people who have thorough knowledge of their instruments and how to optimize instrument performance. The operational procedures that are provided by the respective manufacturers should be followed.
- Standard operating procedures (SOPs) should be developed to ensure that maintenance is conducted at timely intervals.

TURBIDIMETERS

Turbidity has long been an important indicator of water quality, and today, turbidity measurements are the standard tool among drinking water utilities to determine the performance of particulate removal processes. Thus, turbidity is a regulatory parameter that is used to gauge filter effectiveness and to help maintain consistency in the quality of treated drinking water.

Measurement Principle

Turbidity measurements use light scattering as an indirect measurement of the amount of particulate material in a water sample. All turbidimeters consist of a light source, a sample cell, and a light detector (photodetector). Light is scattered by the particles in the sample, and the scattered light is measured by the detector. Less light reaches the detector when the particle concentrations are higher.

The light-scattering behavior of particles in different water samples depends on the size, shape, and other properties of the individual particles, the overall particle concentration, the wavelengths of the light source, and the optical geometry used to measure the scattered light. Therefore, different water samples that have identical mass concentrations of suspended solids may have very different turbidity measurements. In general, it is difficult to detect particles that are much smaller than the wavelength of the incident light. Particles must be similar in size or larger to significantly affect the path of the light and result in scatter.

Commercially available instruments are designed with different types of light sources, as described in Table 9-2, and with different numbers and orientations of photodetectors, as shown in Figure 9-1. Figure 9-2 shows the response of each of these detectors in one instrument as turbidity increases. Historically, a single detector located at 90° to the light source was used to maximize sensitivity in low turbidity applications. This design was referred to as a nephelometer. To improve detection at higher turbidities, instruments with multiple detectors were introduced, and the responses from the multiple detectors were combined in a mathematical algorithm (ratio designs). Instruments may combine the design features of several different categories, and new designs are always

Table 9-2 Types of light sources available for turbidimeters

Light Source	Spectral Bandwidth	Interferences	Comments
Tungsten filament incandescent lamp	400–1,600 nm (peak at 950 nm).	NOM PAC	Can use optical filters to increase sensitivity to small particles Lamp is very sensitive to applied voltage and needs a stable power supply.
LEDs	Usually 860 nm ± 60 nm to comply with ISO 7027-1 specs	Wavelength minimizes interference with NOM.	More stable over time than tungsten
Laser	660 nm is commonly used. Very small bandwidth; availability in a variety of wavelengths	Depends on the wavelength	Can provide more accurate detection of small particles

LED—light-emitting diode, NOM—natural organic matter, PAC—powdered activated carbon

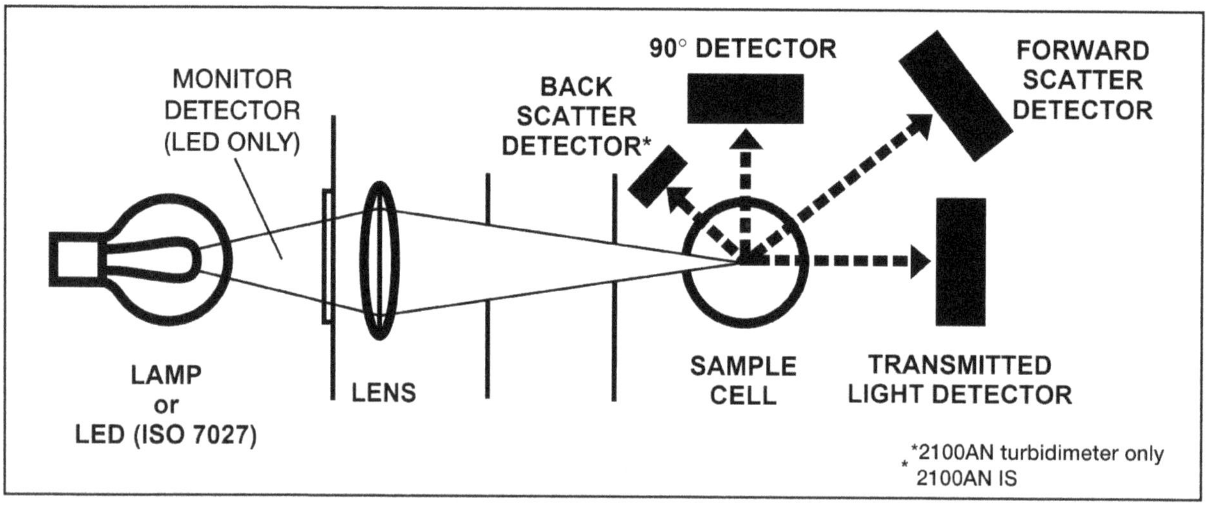

Source: Hach Company

Figure 9-1 Typical turbidimeter optical geometry

being developed. More details regarding the science of light scatter, its relationship to WTP turbidity measurements, and descriptions of instrument design can be found elsewhere (*Standard Methods* 2005, Sadar 1998, USEPA 1999).

Generally, turbidimeters are selected on the basis of providing the required accuracy over the expected turbidity range. Certain instrument designs are specified by regulation or standard on the basis of application. For example:

- US Environmental Protection Agency (USEPA)—The USEPA publishes a list of approved analytical methods for compliance with their series of surface water treatment rules, which includes turbidity (USEPA 2017). This list is periodically updated to reflect advances in analytical technology and the introduction of new instruments to the marketplace.

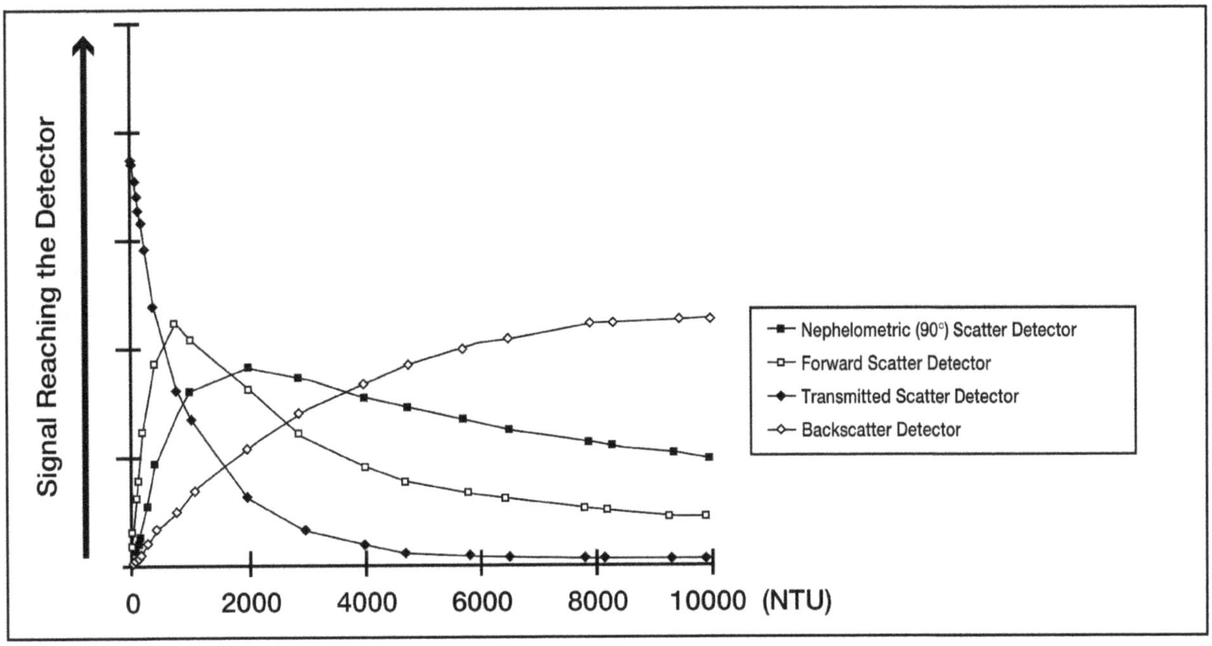

Source: Hach

Figure 9-2 Variation in photodetector response with increasing sample turbidity

- Organization for Standardization (ISO) Method 7027-1—European utilities typically comply with this standard, which requires the use of an 860-nm incident light source with one or more detectors, of which the 90° detector is the primary detector in the ratio measurement.

TYPES OF INSTRUMENTS

Turbidimeters are available in on-line, portable, and benchtop designs.

On-line Turbidimeters

On-line turbidimeters can be either sidestream or in situ.

Sidestream turbidimeters are designed to continuously measure turbidity as a sample passes its optical elements (i.e., light source and detector). Because the sensor is not located in the process stream, it can be larger and more complex, which allows more flexibility in the optical design. Sidestream turbidimeters are therefore typically better for low-range turbidity measurements such as filter effluent, and most low-range turbidimeters are of this type. The passage of samples into the instrument helps to condition the sample by removing bubble interference, as well as preventing the penetration of stray light from outside sources. These instruments are sensitive to flow, pressure, settling of particles, and fouling of optical surfaces.

In situ turbidimeters are commonly placed in open channels or tanks, although they are sometimes mounted inside pressurized pipes. In situ turbidimeters typically have the light source and the light detector(s) on the same surface but are positioned to form the required optical angles. This optical geometry can provide a wide range of measurement but limits the sensitivity of the measurement. Thus, in situ turbidimeters are typically used for higher-turbidity samples, such as source waters, filter backwash or other waste streams, and sample points upstream of the filters. Generally, these instruments should not be used for monitoring of very clean waters, such as those with turbidities less than 0.5 ntu.

Portable and Benchtop Turbidimeters

Depending on the instrument design, portable turbidimeters may not meet regulatory requirements if reporting turbidity data for compliance with regulations, and this should be checked before purchase.

For most benchtop turbidimeters, samples are manually obtained from the process flow, transferred into a measuring cell, and prepared for measurement (capping, cell polishing, and degassing). When measuring samples, it is important to ensure that the proper sample preparation procedures are followed to obtain an accurate measurement. Manufacturers of benchtop turbidimeters typically provide detailed information on the preparation of sample cells for measurement.

Data and Interpretation

On-line turbidity data is essential for ensuring regulatory compliance and for optimizing filter backwash, filter ripening, and for terminating filter runs if turbidity breakthrough occurs. Figure 9-3 depicts results of on-line turbidity versus filter run time for filter optimization.

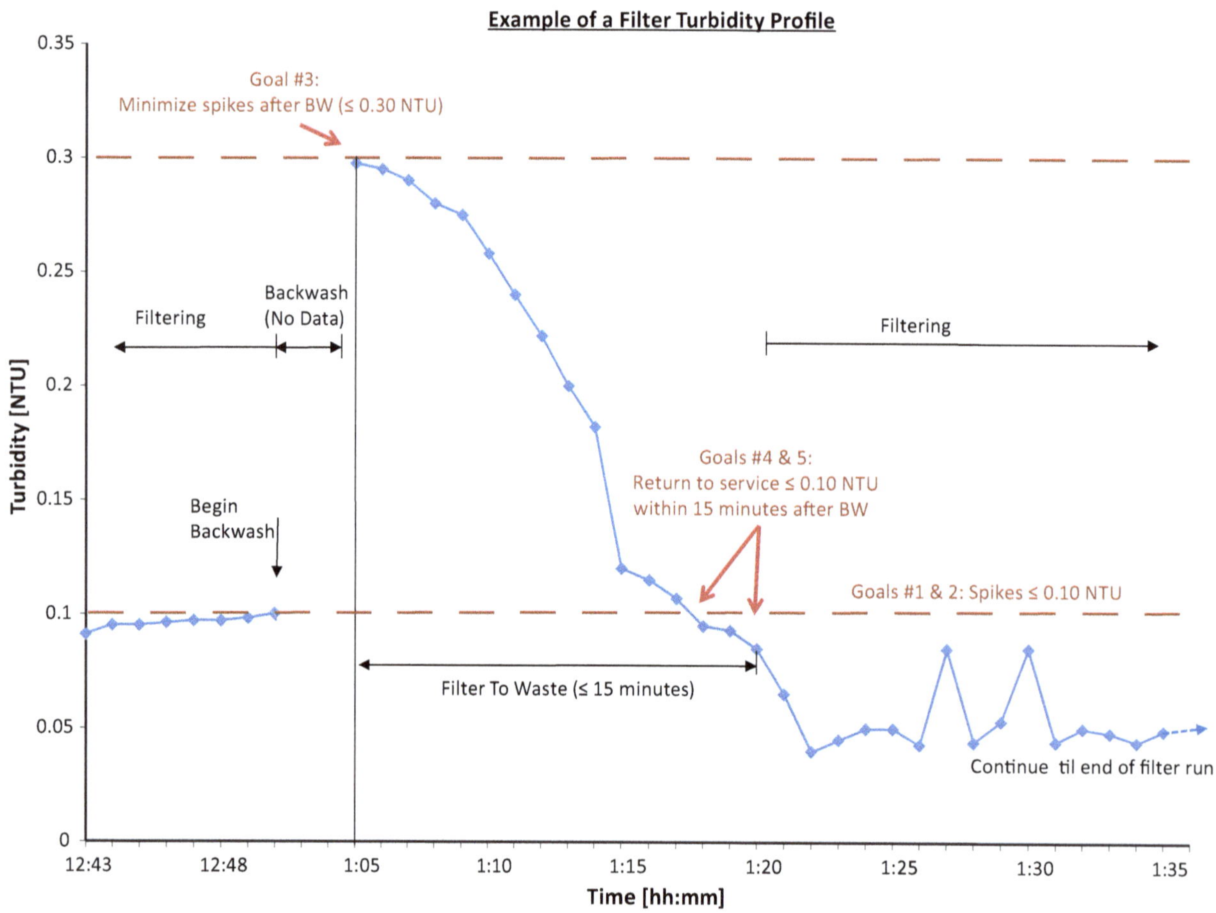

Figure 9-3 Monitoring on-line turbidity versus filter run time for filter optimization

Setup, Standards, and Calibration

Benchtop turbidimeters should be set up and operated in a location free of dust and debris, away from direct sunlight or sources of heat, and in an area where the temperature is stable. The location will ensure that the instruments maintain stability and accuracy at the critical low end of the measurement range—the area where regulatory compliance is a critical goal of the WTP.

Turbidimeters should be calibrated according to the manufacturer's instructions, which for most instruments include specific calibration points and calibration standards. In older literature, procedures were provided that allow for an operator to calibrate an on-line turbidimeter through the use of a benchtop turbidimeter (*Standard Methods* 2005). However, this is no longer recommended (USEPA 1999). Most benchtop turbidimeters will have a higher level of stray light, and this can contribute positive errors to exceptionally clean water samples, especially for those in the range of 0.05 ntu and lower. Typically, this will generate more error than if calibration is performed with defined turbidity standards.

Although on-line turbidimeters should not be calibrated on the basis of benchtop turbidimeter readings, if the effluent turbidity from one filter in a bank of filters is substantially higher than the turbidity of water from the other filters, a comparison of on-line and benchtop turbidimeter data for the filters can reveal whether both sets of data are similar. If benchtop turbidimeter data show low turbidity for all filters, including the one with the high on-line reading, the high on-line reading should be considered suspect.

Best Practices and Troubleshooting

All instruments require maintenance, and turbidimeters are no exception. Experience may dictate that instrument sample chambers be cleaned or flushed more frequently than the manufacturer's minimum recommendation. The higher the turbidity of the sample, the more frequent the maintenance. When cleaning instruments, also clean or replace all sample lines, valves, bubble removal devices and flow control devices, and any surface over which the sample flows.

Beyond regular maintenance, to prevent measurement error, the following best practices are recommended:

- Operate the instrument within the manufacturer's recommended flow range. These flow ranges are designed to optimize the removal of bubble interferences and prevent particle fallout. If samples possess a high degree of dissolved gases, a sample flow at the lower end of the specified range is recommended. If particle sedimentation is of concern, operate at the higher end of the flow range to keep heavier and larger particles suspended.
- Most instruments have a recommended measurement strategy for minimizing noise so that representative data can be consistently generated. These settings include bubble rejection algorithms, signal averaging, and data logging frequency. Refer to the instrument manufacturers and regulatory authorities for recommended or required protocols for setting up these instruments and the logging of measurement data.

Users of in situ turbidimeters should be aware of several interferences. In addition to the same interferences that affect sidestream turbidimeters, in situ turbidimeters also have potential interferences from ambient light and from reflections within the sample process. Consult the specific manufacturer for best practices when using these instruments.

Some of the most common turbidimeter problems and how to best troubleshoot them are listed in Table 9-3.

Table 9-3 Common turbidimeter problems and troubleshooting approaches

Symptom	Troubleshooting Activities
On-line turbidimeter reads a verification standard higher than expected.	Clean the turbidimeter to ensure all settled particles have been removed.
Turbidity is higher than expected as compared with the secondary measurement device.	Clean or replace sample cells or optical equipment that have scratches or other imperfections. Install a bubble trap or correct cell pressure for flow-through instruments or degas sample cells to eliminate entrained air. Reduce condensation on the outside of the sample cell in humid environments with desiccant or dry instrument air.
Particles continually settle in the turbidimeter.	Increase sample flow rates.
Sample has a high degree of noise.	Reduce the flow or increase the pressure of the sample chamber (if applicable) because bubbles are likely not being removed. Verify that sample flow rate to the instrument is constant. Changes in sample flow rate can have significant effects on measurement stability.
The instrument will not verify after calibration.	Recalibrate: Ensure standards are prepared accurately, the instrument is properly cleaned and flushed, and that the calibration instructions are being followed.
Field instrument does not match SCADA value.	Correct scaling of field and SCADA signals and filtering of signal in SCADA system.
A new turbidimeter does not read exactly the same as the old instrument.	May be a result of advances in technology or different optical geometry. Consult the manufacturer for the explanation.
Readings continually decrease over time.	The lamp may need replacement. Check the optical components for fouling or deterioration. If identified, clean or replace the component as necessary.
Calibration standard in the turbidimeter will not become stable.	If the standard was poured into the measurement chamber, it is likely that the chamber was dirty, and particulates are causing the poor stability. Flush everything and recalibrate.
Significant bubbles form on the internal surfaces of the instrument chamber.	The surfaces are not wetted. Soak the surfaces with an oil-cutting detergent (dish soap or laboratory detergent) for several hours to several days.

Source: Adapted from Michael Sadar

SCADA—supervisory control and data acquisition

PARTICLE COUNTERS

Particle counters are used to count particles and measure their size. Many types of particle counter technologies exist, but light obscuration particle counters are the type most commonly used to monitor the effectiveness of sedimentation and filtration performance.

Measurement Principle

Light obscuration particle counters, as shown in Figure 9-4, are based on the ability of a particle to obscure or scatter light as it passes between an optimally controlled light source and a detector, which produces a current. As a particle passes through the "detection zone," it results in a net decrease in the amount of light that would otherwise reach the detector if the particle were not present. Theoretically, a larger decrease in detector current correlates to a larger particle. The number of particles that pass over the detector in a given volume of fluid (detector events) also correlates directly to the number of particles present at that particular time.

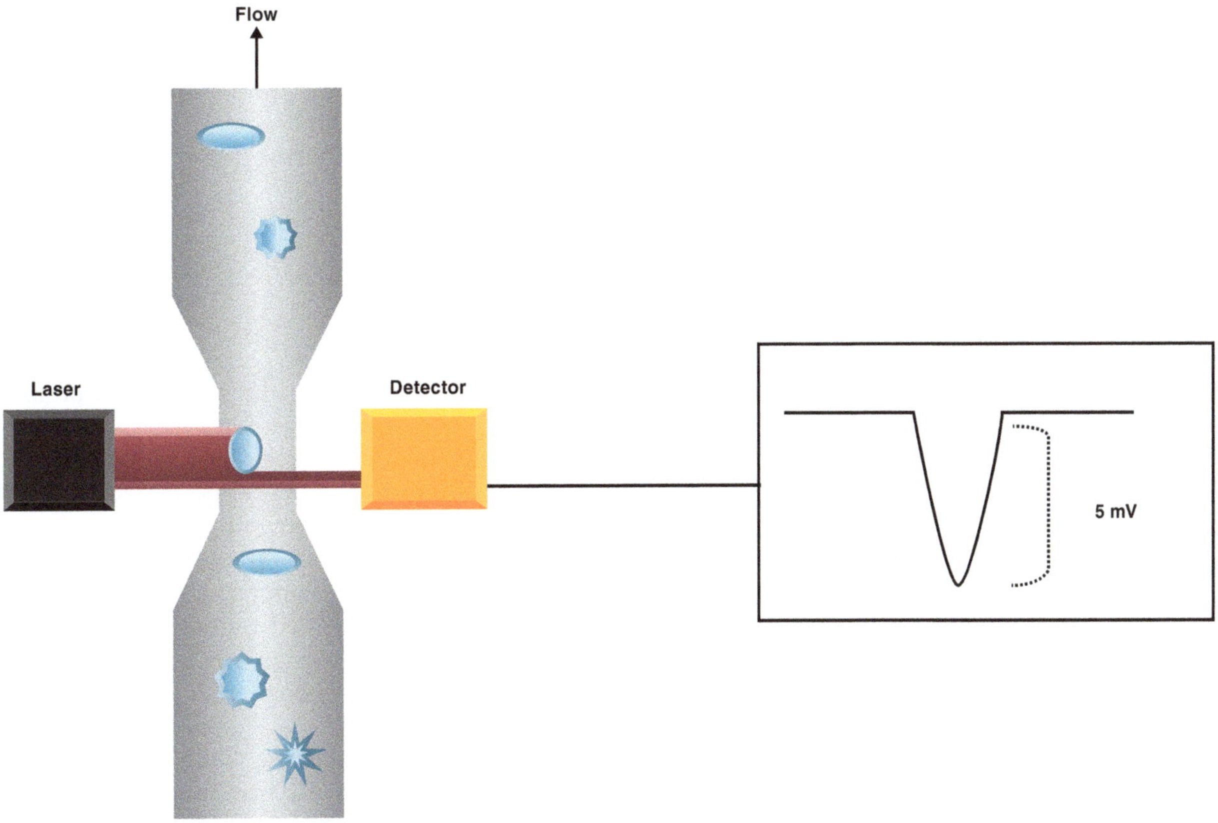

Source: Chemtrac Systems

Figure 9-4 Theory of operation for a light obscuration particle counter

Types of Instruments

The most common types of particle counters used in the drinking water industry use light blocking technology. Although unable to measure less than 1 or 2 μm, these devices are much more affordable and potentially easier to use and maintain as compared with submicron particle counters, which use alternative detection methods such as light scatter or dynamic light scattering.

For the light blocking type of particle counter, there are both on-line and benchtop versions. The on-line version is most often used to measure filter effluent samples on a continuous basis. The benchtop particle counter can be used to verify the data from on-line particle counters; however, one must be careful when collecting filter effluent samples because it is very difficult to capture a truly representative sample. One should always expect the benchtop test results to be slightly higher than the on-line results as a result of contamination. Some portable models do have an option for working in both the benchtop mode (i.e., grab sample) or on-line mode. The on-line mode allows the customer to take the portable unit out into the field to do on-line testing or to run in series with an on-line particle counter to verify the results.

Data and Interpretation

The typical standard used to calibrate the sensor is polystyrene latex (PSL) spheres, which have a very precise diameter. Real-world particles differ in their respective shape, refractive indexes, and absorptive characteristics. Thus, when samples are analyzed by a particle counter, the sizes are determined by analyzing how much light the particle blocks or scatters away from the detector and comparing that with a calibration curve that was developed using the PSL spheres. If a particle happens to block as much as a 10-μm sphere, then the particle is classified as a 10-μm sphere even though the actual shape and size may be significantly different.

See Chapter 6 for further discussion of the interpretation and application of particle counter and turbidity data to filter effluent.

SETUP, STANDARDS, AND CALIBRATION

Setup

Location of sampling and sample flow rates are significant in the setup of an on-line particle counter.

Particle counting is most valuable for filter effluent monitoring. However, particle counting can be used upstream of filtration. In this application, it is important to consider that the sensor will require more frequent cleaning and that sizing of particulate may be grossly inaccurate because of floc breakage as the sample passes through the sensor. Floc breakage during counting could also be a problem if particle counters are used to assess clarifier performance by counting flocculated water and clarified water in plants where sweep floc coagulation is practiced. Breakage of floc in the counter sensor would appear as a larger number of small particles rather than smaller numbers of large particles and would produce incorrect particle count data.

Air bubbles can also create significant error by scattering light that generates false-positive results and causes counts to appear in the above-20-μm range. Air bubbles can arise from degassing caused by a pressure drop or increasing water temperature or entrained air resulting from backwashes. These interferences can be reduced or even eliminated if adequate bubble removal devices are used.

Calibration

Cleaning. Particle counters have an indicator (e.g., alarm light-emitting diode [LED], 0–100 percent readout) to inform the user when the sensor optics need to be cleaned. Some particle counter technologies have specialized flow cell construction and can be cleaned only by using certain procedures and chemicals. Cleaning of the flow control apparatus and bubble traps is necessary to ensure consistency and a high level of particle counter performance.

Verification by comparison. A simple way to verify particle count results is to run the sample in series between two units. If two particle counters are mounted within 20 ft of one another, it is usually possible to run a long piece of tubing from one sensor's outlet to the inlet of the other sensor, as illustrated schematically in Figure 9-5. Once the two units are connected in this manner and the flow is verified to be in correct range, the units can be left to run in series for hours or days to verify that the measurements from the two

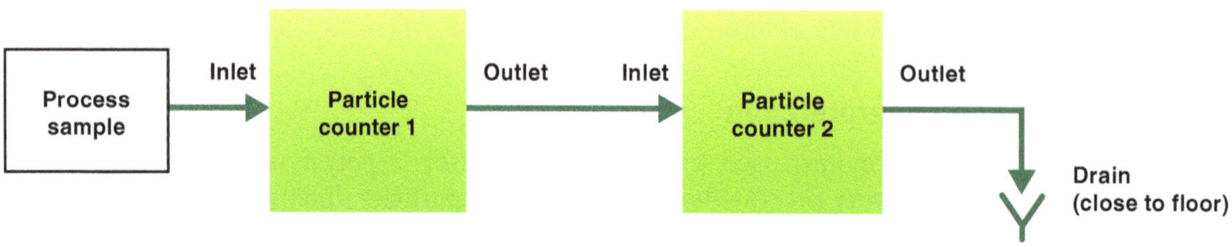

Figure 9-5 Verification of reading by sampling in series

units are in suitable agreement. This can be an easy way to verify whether there is an issue with a given filter that may be showing higher or less stable counts than neighboring filters.

Instrument calibration and verification. Calibration of a particle counter is required to correlate the detector's response to a given particle size and to verify the instrument is counting accurately. Most manufacturers recommend recalibration on an annual basis. However, if the sensors continue to perform as expected and pass verification, a different schedule can be considered.

Particle counter standards. The standards used to calibrate particle counters are PSL Spheres that can be purchased from various manufacturers and a wide range of options. However, calibration is typically done by manufacturers or a third party.

Laser replacement. The types of lasers used in light blocking particle counters generally have a life expectancy of seven to nine years.

Best Practices and Troubleshooting

Before purchasing a particle counter, the following key factors should be considered:

- Number of size channels measured and available reporting formats (e.g., particles/mL, counts/100 mL, raw counts, volumetric reporting)
- Lower and upper limits of particle sizing range and upper limit of particle detection range (i.e., largest particle that can be detected but not sized)
- Cleaning procedure
- Cost of calibration (both onsite and at factory) and recommended frequency
- Availability of calibration procedure (documents and videos) as well as any proprietary software or required accessories
- Expected frequency and cost of laser replacement

Common particle counter problems and troubleshooting guidance are provided in Table 9-4.

ALKALINITY ANALYZERS

Alkalinity is important for coagulation control purposes and for ensuring that treated water is stable and noncorrosive.

Measurement Principle

Alkalinity is a bulk measurement of the acid-neutralizing capacity of the chemical constituents present in a water sample. Alkalinity is measured by titrating the water sample with acid. The volume of acid is then mathematically converted to an alkalinity concentration, typically expressed as mg/L as calcium carbonate ($CaCO_3$). A single titration end point may be used for determining total alkalinity, or multiple end points may be used to determine detailed alkalinity speciation. Any compound that can accept a proton (H^+) is included in the definition of alkalinity. In fresh waters, carbonates (CO_3^{2-}, HCO_3^-) are the predominant alkalinity species, with lesser amounts of hydroxide (OH^-), silicates (SiO_2^-), phosphates (PO_4^{3-}), and borates.

Because defined laboratory methods exist for performing a titration to determine alkalinity, benchtop analyzers are not typically used in the water industry.

On-line alkalinity analyzers typically measure alkalinity by titration in a semibatch mode. A water sample is taken into the measurement cell, and the amount of acid required

Table 9-4 Common particle counter problems and troubleshooting approaches

Symptom	Troubleshooting Activities
Instrument readout indicates fouled sensor.	• Ensure sample is flowing through sensor at the proper flow rate. • Clean sensor with brush or other cleaning device that is supplied or recommended by the manufacturer. • If regular cleaning does not restore clean optics, try cleaning with manufacturer-approved cleaning chemicals. • If the aforementioned steps do not restore clean optics, the issue may be condensation on the optics. This is only likely to be a potential cause if condensation is also noticed on other surfaces such as metal piping or sample tubing. If condensation is suspected, disconnect sample from sensor for 24–48 h and install fresh desiccants inside the unit and ensure all cable fittings and the enclosure are tightly sealed using only dry materials (avoid wet or putty type sealants because these will release). • A faulty laser or damaged flow cell/optics can also be cause for a fouled sensor indication. Repair action would be required.
Particle counts are unstable and/or high.	• Clean sensor with brush or other cleaning device supplied or recommended by the manufacturer. • Install clean, clear tubing coming off top of sensor. Start flow and tap on tubing to get rid of any air bubbles. Check back 1 h later to see if air bubbles are collecting in tubing. If bubbles have reappeared, the elevated/unstable counts may be the result of air bubbles passing through the sensor. • Install another particle counter in series or swap out the particle counter with another instrument to see if the measurement results appear the same or different. If results are the same, the indicated particle counts are more likely to be accurate. Otherwise, a calibration or repair action is required. Lasers that are not set up properly or failing are often a cause for unstable particle counts, especially when changes in particle counts are very drastic and sudden.
Particle counts are always very low or zero.	• Ensure sensor optics are adequately clean and that the sensor has proper flow rate. • Install another particle counter in series or swap out the particle counter with another instrument to see if the measurement results appear the same or different. If results are the same, the indicated particle counts are more likely to be accurate. • Perform verification in accordance with manufacture recommendations. Verification should be done at the lowest size detectable (e.g., 2 μm).

to reach either a pH (most common) or a colorimetric end point is recorded. A typical instrument is shown in Figure 9-6.

Because many natural waters do not have significant concentrations of noncarbonate alkalinity species, total inorganic carbon (TIC) measurements are often used as an estimate of the alkalinity. Most total organic carbon (TOC) analyzers measure TIC.

Data and Interpretation

Alkalinity measurements are typically made on the source water to detect changes in quality and potential effects on coagulation. Source water alkalinity can change quickly in some water sources during rain events. Because metal-salt coagulants consume alkalinity and drop the pH, if source water alkalinity concentrations are low, the addition of acidic coagulants such as alum or ferric coagulants can drop the pH to values that are too low for good floc formation.

Setup, Standards, and Calibration

Calibration is accomplished by analyzing buffered solutions of known alkalinity or by comparing the on-line instrument with a laboratory titration. For instruments that rely on a pH end point, the internal pH probe may also require periodic calibration.

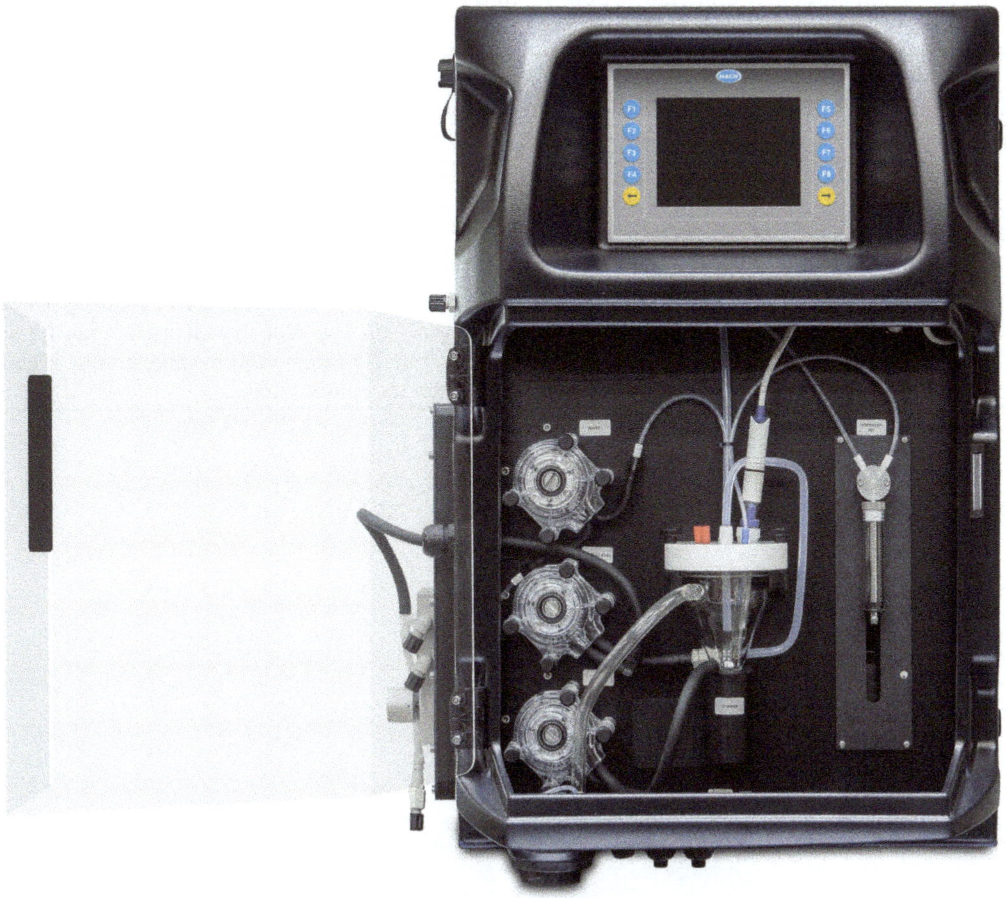

Source: Hach

Figure 9-6 Typical example of a titration-based alkalinity analyzer

Best Practices and Troubleshooting

Some general guidelines can be followed to ensure that alkalinity measurements are representative of the sample and that these measurements provide the operator with information to be proactive regarding their processes.

- Select a measurement technology that is best suited to the water chemistry. TIC–TOC analyzers are only appropriate to provide alkalinity measurement in samples where only carbonate alkalinity species are present. Similarly, colored waters may not be suitable for titration instruments that use a colorimetric end point analyzer.
- Prefiltration may be necessary to ensure that suspended solids do not interfere with the alkalinity measurement or foul or damage the instrument.

Common problems and troubleshooting approaches for alkalinity analyzers are shown in Table 9-5.

STREAMING CURRENT MONITORS

Streaming current monitors (SCMs) measure the electrical charge characteristics of particles in water and can be used to control or monitor coagulation and sludge conditioning

Table 9-5 Common alkalinity analyzer problems and troubleshooting approaches

Symptom	Troubleshooting Activities
Sluggish response	• pH electrode may become fouled by soaps, oil and grease, or inorganic precipitates. Electrode may require more frequent cleaning, or titration time can be extended to allow electrode to come to equilibrium in the presence of these interferences.
Inconsistent readings	• Check reagent levels. • Look for leaks and slipping of tubing on reagent pumps. • Look for air in reagent lines.
Plugging with solids	• Source water monitoring applications with moderate-to-high suspended solids concentrations, tubing, valves, and the reaction chamber may be subject to plugging by accumulated solids. This may require more frequent cleaning, or the installation of a filter on the instrument inlet.

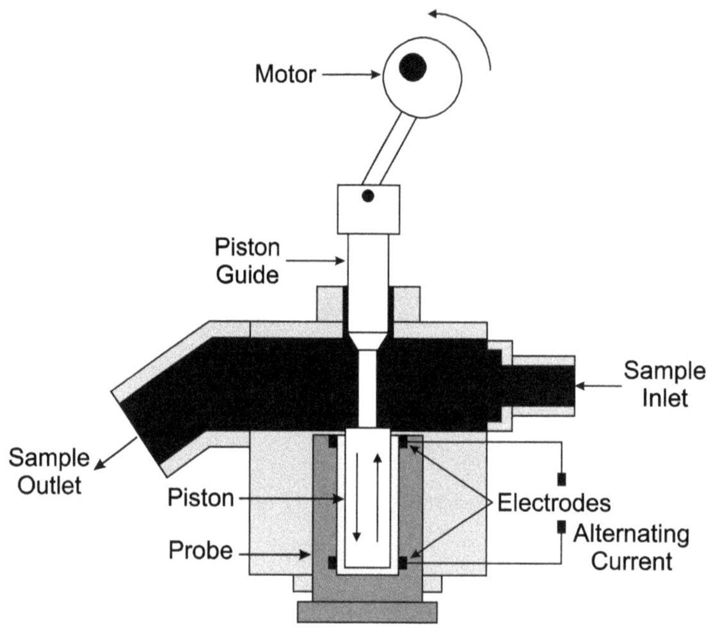

Source: Chemtrac Systems

Figure 9-7 Schematic diagram of an SCM

processes. These instruments are typically installed on-line, downstream of coagulant or polymer addition. The postcoagulation particle charge determined by the instrument is used to inform chemical feed rate. A change in the streaming current reading away from a predetermined set point indicates that the coagulant dose is either lower or higher than the optimal dosage to create neutrally charged particles.

Measurement Principle

A simplified cross section of a SCM is presented in Figure 9-7. A sample of water flows through the chamber where a small piston generates voltage and current as electrically charged particles attached to the surfaces of the piston and the inner boot move relative to each other. Electrodes in the cylinder sense the voltage/current, and electronic processing generates an output signal.

There is no standardized methodology for measuring streaming current, and the magnitude of the current generated in any given instrument depends on its design, the water chemistry, and the sampling conditions. Therefore, different responses will be seen when comparing different instruments. More information on sensor design and the relationships between streaming current and particle charge can be found elsewhere (Dentel and Kingery 1988).

Types of Instruments

There are two types of streaming current devices used in the water treatment industry: on-line and benchtop.

On-line streaming current. On-line streaming current devices are commonly referred to as SCMs or streaming current detectors. These devices require a sidestream sample that is taken at the point or shortly after coagulant addition and mixing. There are multiple suppliers of on-line streaming current instruments and various options offered.

Benchtop streaming current. Benchtop versions of streaming current are referred to by many different names, including laboratory charge analyzer, coagulant charge analyzer, particle charge detector, and particle charge analyzer. The testing methods associated with Benchtop streaming current devices vary widely.

Data and Interpretation

The performance of a streaming current instrument must be optimized and monitored until robust operational and maintenance protocols have been developed for each site. Several of the more common factors affecting instrument readings are described in Table 9-6.

Table 9-6 Common factors affecting instrument readings

Parameter	Effect
Water quality	• Changes in turbidity, TOC, and coagulant dose affect streaming current. • Increases in turbidity and/or TOC (both negatively charged) result in more negative readings, indicating a need to increase coagulant.
pH	• pH affects the surface charge of particles, the charge of the functional groups on NOM molecules, and the charge associated with dissolved species of aluminum (Al)– and iron (Fe)–based coagulants (Pernitsky and Edzwald 2003). • More variable or higher pH result in less reliable charge readings. • pH changes >0.3 units may require adjustment of streaming current set-point values.
Coagulant type	• Different coagulants are less effective at different pHs; outside their effective pH range, they will show little charge response.
Lag time	• Lag time is the time it takes for water to travel from point of chemical addition to sensor; decreases in plant flow may increase lag time. • A long lag time makes the response less timely.
Upstream chemicals	• Upstream chemicals affect streaming current.
Foul and probe wear	• Iron, manganese (Mn), and other contaminants can foul readings. • Moving parts of streaming current analyzer can wear over time.

NOM—natural organic matter, TOC—total organic carbon

SETUP, STANDARDS, AND CALIBRATION

On-line Streaming Current

Installation. The sample point must be located after thorough mixing to ensure uniform dispersion and reaction of coagulant. Sampling between 30 and 60 s has been found to work well for rapid response in coagulation monitoring or control. In many cases, the most ideal sample point will be at the rapid mix chamber or immediately after an inline or static mixer. Consult the manufacturer for guidelines based on experience at similar facilities. It is also advisable to evaluate different sampling points to ensure maximum instrument response with a minimum of signal fluctuation. If possible, perform this evaluation under a variety of operating conditions. Samples should generally be taken before flocculant aid addition because these high molecular weight polymers can cause readings to be less stable or even cause sensor fouling if not well mixed.

During initial setup of an on-line streaming current analyzer, WTP operators may want to evaluate the response time for coagulation changes to be observed on the SCM readings as well as conduct a test to determine the observed change in SCM reading on the basis of a defined coagulation dose change. Operators may also want to take an SCM reading of the source water to establish a baseline value for what the SCM would read if there is a loss of coagulant.

Normal operation for coagulation control or monitoring with an SCM is based on establishing "optimized" operating conditions for the WTP coagulation process and then either recording the SCM value at that point or "zeroing" the SCM at that value (most common) to allow the operator or control system to maintain the SCM reading at the target or zero value, as long as there are no measured changes in source water quality or coagulated water pH.

Maintenance. The maintenance for streaming current sensors can be broken down into "short-term" and "long-term" categories. Short-term maintenance involves cleaning the probe and piston per manufacturer's instructions. Depending on water quality, types of contaminants, and upstream chemicals being used, the required cleaning frequency may be weekly to monthly. Long-term maintenance is required to address wear and other mechanical issues that develop over time. Streaming current sensors have moving parts that wear down and will need to be eventually replaced. Replacement interval of probe and piston parts depends on the amount of abrasive solids in the sample and can range from every six months to every six years.

Calibration. No industry standard calibration procedure exists for streaming current instruments. Because the most likely cause of inaccurate streaming current readings is a deteriorated probe and/or piston, a more common verification method is to swap out the existing probe and piston with a new set of parts and perform comparative testing between the old and new set. If the new set shows a notable improvement in response, then the old set should be discarded. Review manufacturer recommendations on performance testing procedures.

Best Practices and Troubleshooting

The following is a summary of general best practices guidelines and troubleshooting tips.

- Qualification—It is important to qualify an application for streaming current technology. Discuss onsite and/or laboratory testing services with streaming current supplier.
- Make occasional adjustments to coagulant dosage to check response of streaming current reading. Maintenance may be required if response seems smaller, sluggish, or if readings appear unstable. Once or twice a year consider installing a new probe and piston and compare the responses. This is an efficient way to identify when an older probe and/or piston needs to be replaced.

Table 9-7 Common SCM problems and troubleshooting approaches

Symptom	Troubleshooting Activities
Streaming current reading showing signs of instability and oscillation in values	(1) Clean the sensor according to supplier instructions. (2) Ensure good mixing of coagulant. (3) Inspect sensor and look for loose motion in motor and drive linkage. (4) Check sensor wiring. (5) If using automatic coagulant control based on streaming current reading, put control to "manual" and see if oscillation stops. If so, PID settings may need to be adjusted.
Streaming current reading suddenly very negative and/or no longer responding to coagulant feed changes	(1) Check flow of coagulant. (2) Check flow to sensor. (3) Ensure piston is moving up and down. (4) Clean probe and piston to remove any staining or scale. (5) Check pH because a large increase in pH can reduce response to coagulant. (6) Check sensor wiring. (7) Replace probe and piston with a new set to compare response.

PID—proportional-integral-derivative, SCM—streaming current monitor

- When a streaming current reading goes more positive or negative, look for logical explanations like changes in turbidity, organics, or flow (if not flow pacing) that would explain the change in streaming current reading and that would support a dosage adjustment being made in the direction and magnitude that streaming current readings indicate.

Common streaming current analyzer problems and troubleshooting guidance are provided in Table 9-7.

ZETA POTENTIAL METERS

Zeta potential provides a measurement of net floc charge that is reported in millivolts (mV). As discussed in Chapter 1, coagulant conditions need to be adjusted to produce floc particles that are near-neutral charge to grow floc properly for optimal removal in downstream processes. For each specific WTP, there is generally an upper and lower range of zeta potential values that results in good plant performance.

Measurement Principle

Zeta potential is not directly measured. Instead, electrophoretic mobility is measured, and zeta potential is then calculated from the electrophoretic mobility.

To measure electrophoretic mobility, a sample is placed in a chamber called a measurement cell. Then, an electric field is applied across the cell. Particles move toward the electrode of opposite charge. The particles move with a velocity that is proportional to their zeta potential and in a direction that indicates whether their charge is positive or negative. The sign of the charge (positive or negative) and velocity is measured and expressed per unit field strength as their mobility.

Types

Most laboratory zeta potential/electrophoresis instruments are laser Doppler velocimetry (LDV) instruments. Microscope based electrophoresis instruments are no longer commonly used. The key difference between the two categories is that with a microscope based system the particle movement is manually observed in a microscope. An LDV instrument uses a laser based interferometric technique to determine the electrophoretic mobility from a Doppler shift of the light. This is automated and operator independent.

On-line Instrumentation. An on-line commercial instrument is shown in Figure 9-8. This instrument continuously measures zeta potential and provides data resolution that is enough for the control of a rapidly changing processes.

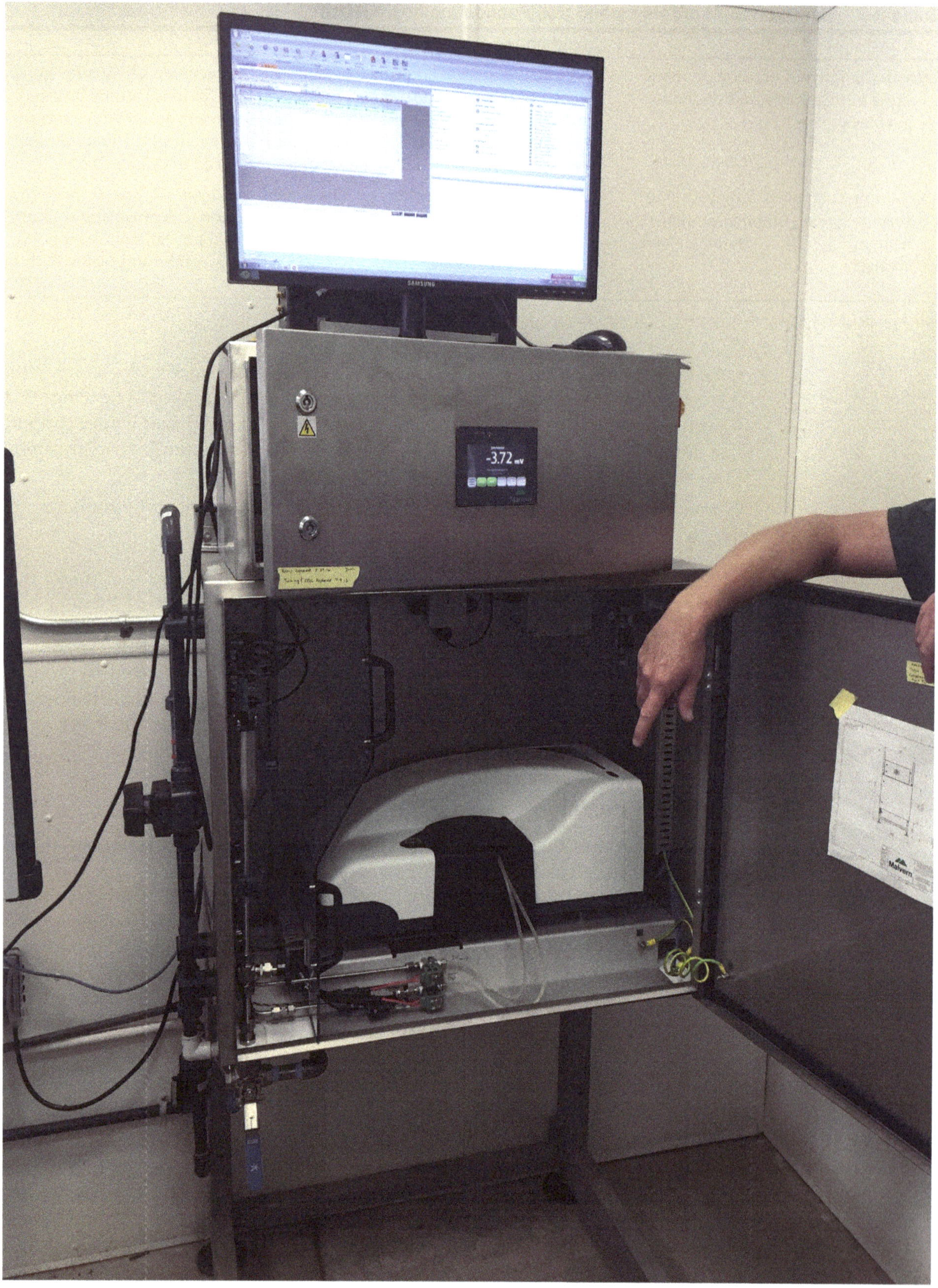

Source: Ali Leeds

Figure 9-8 On-line zeta potential measuring system

Source: Vinnie Hart

Figure 9-9 Optimal zeta potential differs between WTPs and depends on the coagulant and process

Data and Interpretation

Although a zeta potential of zero suggests that particles are neutral and would be optimal for coagulation, in practice, the optimal zeta potential depends on the plant process, as shown in Figure 9-9.

Testing should be conducted at implementation of the use of zeta potential to determine the particular zeta potential target that works most effectively for a particular WTP.

Setup, Standards, and Calibration

Zeta potential instruments are validated by using a standard. If the results do not meet specification, then it should be assumed that the instrument requires repair. Each manufacturer of zeta potential instrumentation typically also has a transfer standard that can be used to validate the performance of the instrument.

Best Practices and Troubleshooting

Always run multiple measurements. The better the quality of the data, the more repeatable the answers obtained will be. When comparing measurements,

- check the results to ensure that count rate remains constant,
- check the results to see that the mean zeta potential and standard deviation from measurement to measurement are similar, and
- ensure that these raw data are similar between measurements.

Table 9-8 Common zeta potential analyzer problems and troubleshooting approaches

Symptom	Remedy
Standard is not reading correctly.	• Confirm the standard is fresh and prepared to manufacturer's specification. • Contact manufacturer for repair.
Reading is inconsistent.	• Ensure the sample is well mixed and representative of the process. • Sample is aggregating.

When introducing the sample into the cell, ensure to not introduce air bubbles. Air bubbles interfere with particle motility and detection.

Zeta potential is a function of temperature. In the case of samples being measured that may be colder than room temperature or warmer, Laser Doppler electrophoresis instruments can be set to the same temperature as the sample so that changing temperatures during the measurement are not an issue.

If the instrument experiences condensation buildup during periods of humid atmospheric conditions, check the back of the instrument for a port that allows the system to be purged with dry air to avoid condensation buildup.

Common zeta analyzer problems and troubleshooting guidance are provided in Table 9-8.

TOC ANALYZERS

On-line measurements of TOC became popular after the USEPA and other regulators stipulated that WTPs achieve specific TOC removals through coagulation. Chemical properties of TOC are important not only for coagulation optimization but also for understanding and minimizing the formation of disinfection byproducts (DBPs).

Previously, TOC analysis was limited to laboratory instruments; however, robust on-line analyzers can now be used to monitor and determine TOC removals quickly and efficiently.

Measurement Principle

TOC is an aggregate measurement of the carbon content of dissolved, suspended, and colloidal organic matter present in water and is the most commonly used measure of natural organic matter (NOM) in drinking water practice (Potter and Wimsatt 2009). TOC measurements do not provide any information regarding the structure, size, or chemical properties of the NOM being measured but offer a direct, simple, and nonspecific measurement of all organic compounds in a sample. Some structural insight can be provided via ultraviolet (UV) absorbance at 254 nm, which is shown to correlate well to aromatic compounds typical in NOM. In terms of size, TOC measurement does not generally include settleable solids, and dissolved organic carbon (DOC) includes only the portion of sample that will pass through a 0.45 μm filter.

All TOC analyzers do the following three steps: (1) remove and/or measure excess TIC, (2) oxidize TOC to carbon dioxide (CO_2), and (3) measure or detect that CO_2. Steps 2 and 3 are depicted in Figure 9-10.

In the first step, acid (typically phosphoric acid, but sometimes hydrochloric or sulfuric) is added to the sample to reduce the pH and convert TIC (HCO_3^-, CO_3^{2-}, etc.) to CO_2. The TIC is removed by either diffusing through a CO_2 permeable tubing or purged with a nonreactive gas. This TIC is removed to provide a more accurate TOC number.

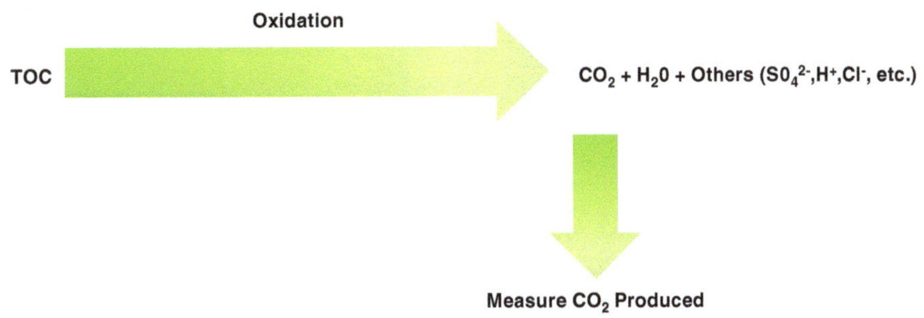

Figure 9-10 Measurement of organic carbon in TOC analysis

In the second step, one of three oxidation techniques is used in most commercial on-line TOC analyzers: high-temperature combustion, UV-persulfate oxidation (shown in Figure 9-10), and supercritical oxidation. UV-persulfate analyzers are most common for drinking water applications.

In the third step, the produced CO_2 is measured with a nondispersive infrared or conductivity detector after a CO_2 selective membrane and expressed as a mg/L or µg/L concentration of carbon.

Data and Interpretation

To quantify TOC removal percentages, TOC measurements can be taken at the influent and effluent of the clarification and filtration processes. Most TOC analyzers can be configured with at least two streams to provide real-time inlet, outlet, and removal percentages. Typical data analysis time ranges from 2 to 10 min per repetition. Measurements taken downstream of clarification will provide the quickest notification of changes in coagulation performance and serve as feedback for dosing optimization. For rapidly changing water sources, such as during a storm event, early warning of increased TOC levels may be best obtained by continuous source water analysis.

Care should be taken on source water samples that contain high levels of chlorides. Persulfate units can handle chloride content up to 300–500 mg/L. Chloride concentrations greater than 0.05 percent can inhibit oxidation of organic matter because of the slowing of the persulfate oxidation in the presence of excess chlorides. For seawater applications, it is recommended to use a technology capable of handling at least 3 percent salt solutions. A high salts kit can be used with combustion units, and newer technology such as two-staged water oxidation and supercritical water oxidation is able to handle salt solutions.

Advanced water treatment systems, whether for potable reuse or industrial pure water, can achieve TOC levels less than 0.1 mg/L. In these cases, on-line units tend to be more accurate than grab samples in the laboratory because of the increased risk of sample error or contamination.

Setup, Standards, and Calibration

The location of sampling and sample characteristics like salts, particles, and level of organics will help determine the best technology. The instrument should be in a clean, dry, temperature-controlled environment to facilitate maintenance and to preserve the integrity of the reagents. A method of disposing of the used reagents, either a drain connection or a holding tank, will be required. Depending on the analyzer, this might involve neutralization or compliance with local waste disposal regulations.

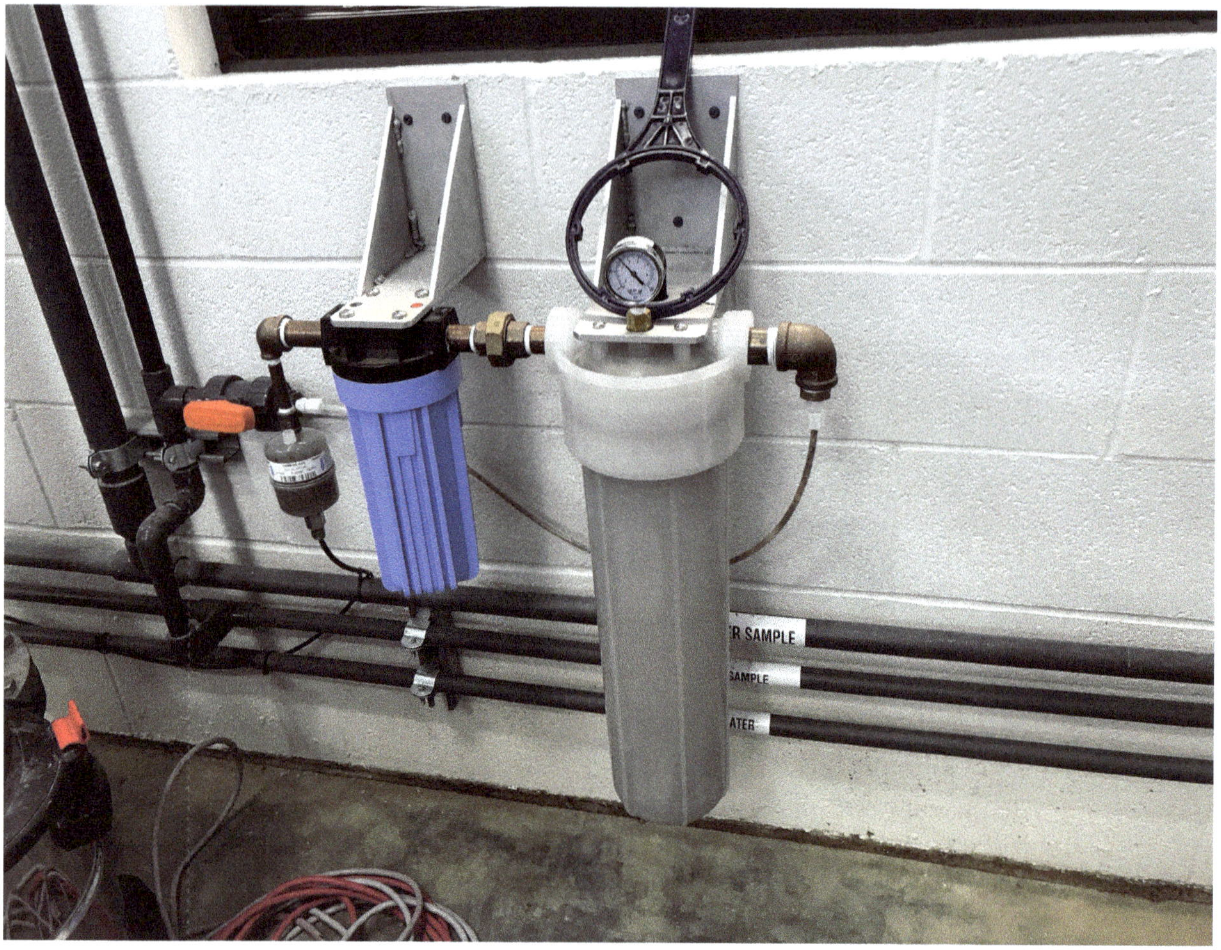

Source: Amanda Scott

Figure 9-11 Typical TOC prefiltration apparatus

The location of the sample point for the instrument will depend on the application. Source water TOC samples should be obtained before the addition of water treatment chemicals. On-line TOC analyzers are designed to measure suspended, colloidal, and dissolved organic matter, but tubing sizes inside the analyzers are not built for large sediments and particulates. Depending on the manufacturer's recommendations, some applications measuring source waters will require the installation of a prefilter upstream of the on-line analyzer, as shown in Figure 9-11. This is especially common on high-turbidity source waters. Prefiltration may remove suspended particulates as well as inorganic silts and sediments, which are not part of the TOC measurement. For high-turbidity waters, a multistage filtration might be used to minimize filter changeout, but for normal or treated waters usually only a small filter is needed.

When compared with other sensors, TOC analyzers tend to have fewer interferences and less drift but might require more frequent maintenance. This maintenance might include a weekly visual inspection with actual active maintenance, such as changing out a reagent, on a monthly or quarterly basis. Calibration requirements vary by manufacturer, but most have improved to 6- or 12-month frequency and can be either single point or multipoint.

Table 9-9 Common TOC analyzer problems and troubleshooting approaches

Symptom	Troubleshooting Activities
High variability	• Check surrounding environment for vibrations, high VOCs, and extreme ambient temperature. • Check flow-through analyzer. If tubing blocked or pump not working, flush system (consider H_2O_2, acid, or cleaning solution) and replace any necessary parts. • Check the reagent delivery and amount, whether it is dispensed by a vial, valve, syringe, cartridge, or carboy. Ensure fittings are tight to avoid bubbles. • Low TOC samples in the low ppb range will inherently have higher variation. • Extreme high concentration inorganic carbon samples might cause variability in TOC readings.
Poor accuracy	• Drift as a result of calibration instability. Consider recalibrating. Check for chloride content in persulfate oxidation.
Low TOC	• Check the reagent delivery, especially for the persulfate oxidizer, and check that the UV lamp is on and that heating is to the desired temperature. Check oxidizer level and age because it will degrade over time. • Check catalyst for fouling and consider replacing. • Contamination in carrier gas or CO_2 scrubber for combustion units. • Check gas flow causing low recovery. • Check acid delivery and inorganic carbon removal by examining pH of waste stream. If less inorganic carbon is removed, variability in TOC (from TC–IC) increases.
High TOC	• Check acid delivery and sparge time for inorganic carbon removal by examining pH of waste stream and gas flow. If less inorganic carbon is removed, in NPOC mode the TOC will be high. • High background as a result of contaminated check standard. Run diagnostics.

CO_2—carbon dioxide, H_2O_2—hydrogen peroxide, NPOC—nonpurgeable organic carbon, TC–IC—total carbon less inorganic carbon, TOC—total organic carbon, UV—ultraviolet, VOC—volatile organic carbon

Best Practices and Troubleshooting

To ensure proper function of an on-line TOC analyzer, there is a need for proper sample flow, TIC removal, oxidation, and detection. If one of those is missing, results could be compromised. Most on-line TOC analyzers have onboard diagnostics as well as defined errors and warnings. Table 9-9 highlights troubleshooting tips across manufacturers and methods to maintain uptime and reliable results.

UV ANALYZERS

Measurement of UV absorbance has been shown to be a useful alternative to TOC measurements for characterizing NOM in drinking water treatment (Edzwald et al. 1985). Most organic compounds found in natural waters absorb UV radiation, with carbon–carbon double bonds and aromatic ring structures strongly absorbing light at 254 nm.

UV absorbance is therefore useful as an indirect measure of organic matter present in the water for coagulation process control. The exact relationship between UV absorbance and TOC concentration is unique for each source water source. However, for a given source water source, increases in UV absorbance indicate increasing NOM.

Measurement Principle

Absorption is a process in which a chemical decreases the ability of electromagnetic radiation (i.e., a light beam) to transmit through media that are highly transparent to light, such as water. In this way, absorbance can be used to measure impurities present.

UV analyzers measure the light that is transmitted through a sample of water of known thickness and compare that measurement with a measure of light that is transmitted through a sample of pure water (often deionized or otherwise purified water). The resulting ratio is reflected as an absorbance reading calculated mathematically from the raw data.

Two terms are commonly used to describe how the intensity of a light beam going through a water column is affected by the impurities in the water. These are transmittance and absorbance, as illustrated in Figure 9-12.

Absorbance (A) of a water sample is defined by the following equation with parameters, as illustrated in Figure 9-12:

$$A = \log \frac{I_0}{I}$$

UV transmittance can be calculated from the absorbance on the basis of the following equation:

$$T(\%) = 10^{-A} \times 100$$

Although absorbance and transmittance are unitless quantities (absorbance is the logarithm of a ratio), they change depending on the path length ("b" as per Figure 9-12) traveled by light through the sample water. Therefore, it is very important to express absorbance values per path length used for the measurement. Most frequently, absorbance and transmittance are normalized to a 1 cm path length, which results in units of cm^{-1} and percent (%)/cm, respectively.

Types

UV analyzers are available as benchtop, portable, and on-line units. In addition, the analyzer might be a single wavelength (typically 254 nm), multiple wavelengths, or

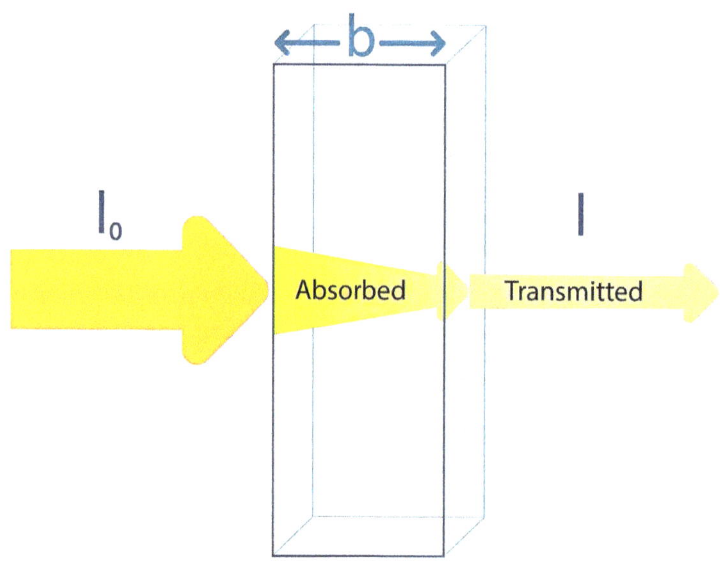

Source: ABB

Figure 9-12 Absorbed and transmitted light over a water column of path length "b"

a full-scan UV-Vis unit. Different types of lamps may be used by the manufacturers to provide absorbance readings at different wavelengths.

Benchtop Analyzers

Benchtop units are usually laboratory-grade full-spectrum (UV or UV + Visible) spectrophotometers. They are sensitive and powerful analysis tools and can be used for grab sample analysis and as reference units for other absorbance measuring instruments. These units tend to be used for the analysis of small volumes of samples for spot checks rather than continuous monitoring purposes.

Portable Analyzers

Portable units, as shown in Figure 9-13, can be single or multiple wavelength analyzers. They are best suited for sampling in the field for quick analysis. They may also be used as reference units for real-time analyzers. Portable units are much less expensive compared with benchtop full-spectrum analyzers.

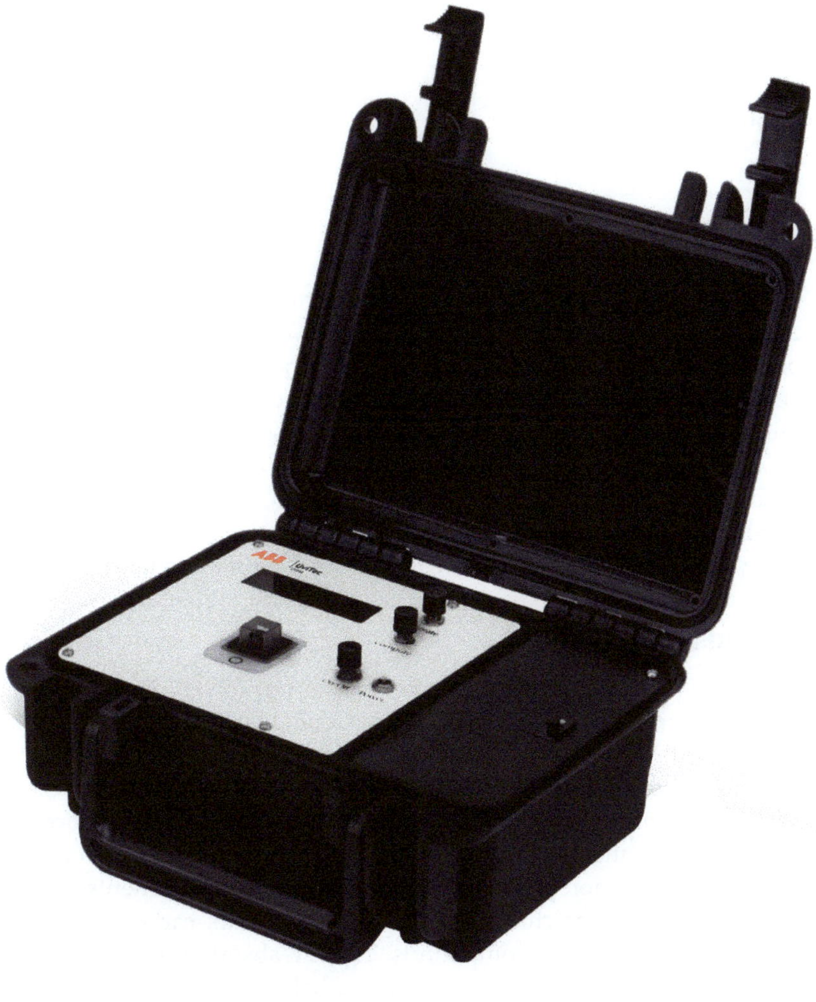

Source: ABB

Figure 9-13 Portable UV254 meter

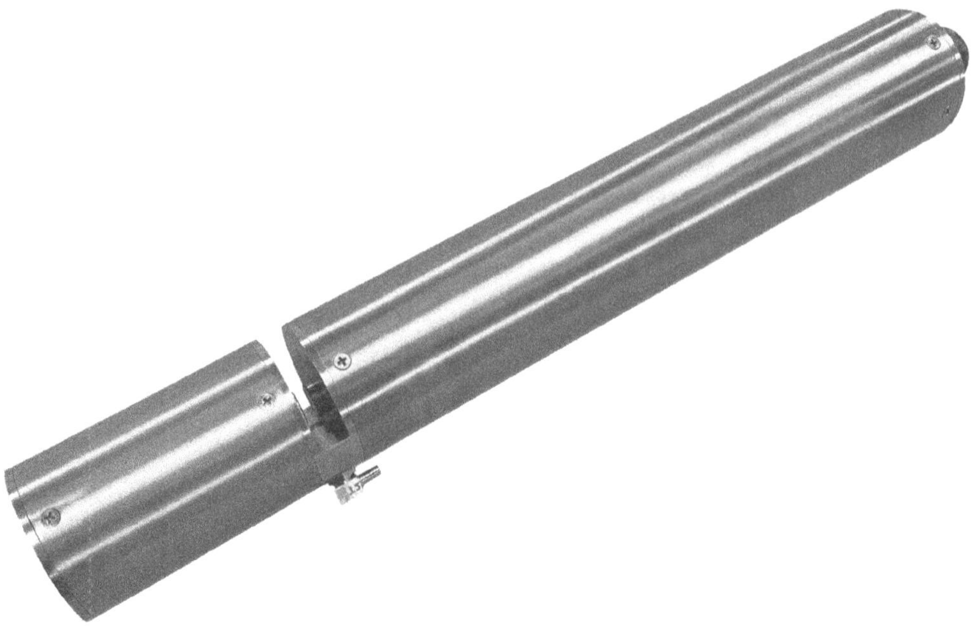

Source: ABB

Figure 9-14 Probe-style absorbance sensor

On-line Analyzers

On-line analyzers are probes that go directly in the water and measure in situ. They are cleaned by pressurized air or wipers. These two methods of cleaning are effective against biofouling but may not work well against mineral deposition. Calibration of inline analyzers is more time and labor intensive than calibration of sidestream-style analyzers. Figure 9-14 shows a probe-style absorbance sensor.

Sidestream analyzers bring the water to the sensor compartment from either a pressurized source or by using a pumping system, as shown in Figure 9-15. These analyzers typically use chemical cleaning, which is very effective against mineral deposition type fouling. Sidestream analyzers are easy to access for maintenance and calibration purposes.

Data and Interpretation

Measurements of UV absorbance do not account for all organic matter present because organic matter without aromatic rings or double bonds (e.g., sugars, alcohols, and simple aliphatic acids) will not be measured. In addition, the analytical procedure may require filtration of the sample, so results tend to be related more to the concentration of DOC than the concentration of TOC. However, the UV-absorbing fraction of NOM represents organics that are most readily removed by coagulation, and therefore is potentially more relevant to coagulation process optimization than measurements of TOC alone.

Oxidants such as chlorine and ozone can break carbon–carbon double bonds and ring structures in organic compounds. The UV absorbance of oxidized water can thus be lower than the UV absorbance of the water before oxidation, even though no organic carbon was removed by oxidation.

UV absorbance is used along with DOC to calculate the specific UV absorbance (SUVA) of the water, which is valuable for assessing NOM removal as described in Chapter 1.

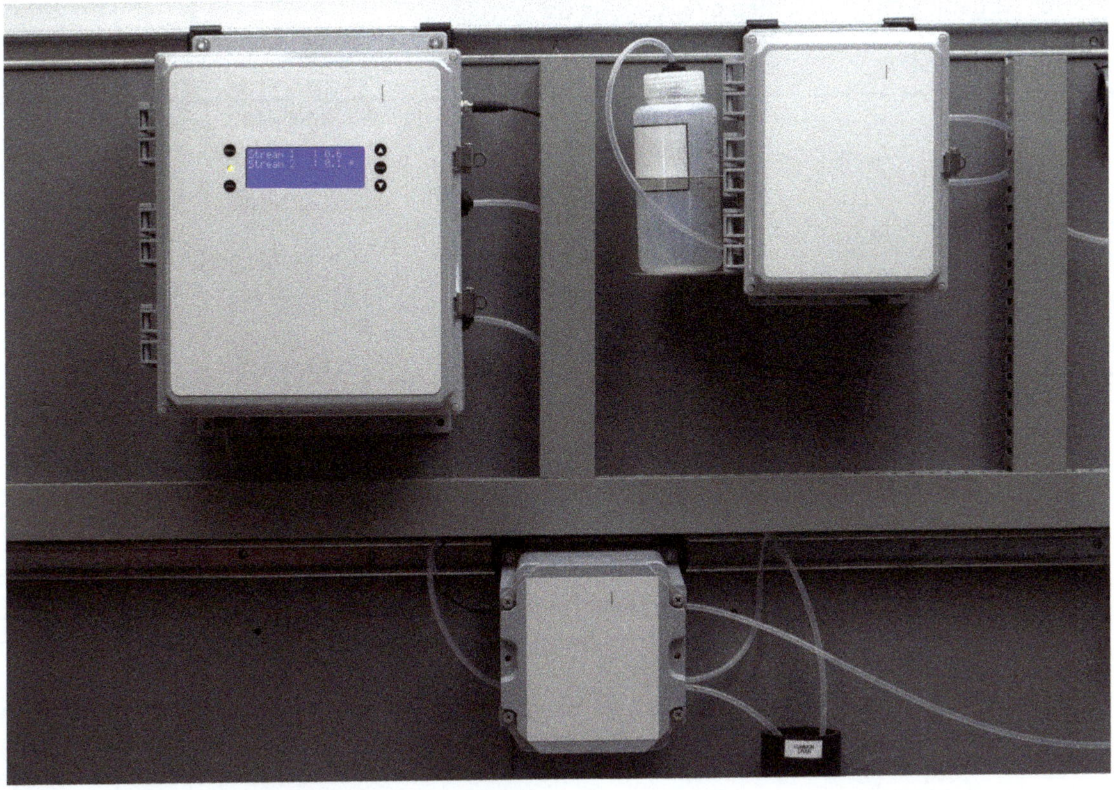

Source: ABB

Figure 9-15 UV analyzer, automatic cleaning system, and dual feed system enabling time-shared analysis of two separate streams in the plant

In most cases, UV absorbance can be directly used as a parameter to monitor for relative removal through a treatment process rather than using it to correlate to DOC, TOC, or NOM. This may be especially true for the coagulation process in which the higher molecular weight and more hydrophobic fractions of NOM, which tend to absorb more UV radiation, are preferentially removed (White et al. 1997). Because the hydrophobic fraction of NOM typically forms more DBPs (Kavanaugh 1978), goals targeting DBP reduction via coagulation can be better achieved using UV absorbance monitoring.

Multiple wavelength and spectral absorbance measurements. Although the most common wavelength at which absorbance is measured is 254 nm in the water industry, there is no obligation to use this wavelength for correlating to DOC/TOC. Organics absorb in a wide band in the UV region, and many wavelengths from 200 nm to 360 nm may be used. More information pertaining to the characteristics of organics in the water can be obtained if absorbance is measured at multiple wavelengths, as illustrated in Table 9-10.

The composition of NOM has a significant effect on the coagulation process. For example, for equal concentrations of fulvic acids and humic acids, the required coagulant dose is twice as high for fulvic acids compared with humic acids (Rigobello et al. 2011, Edzwald 1993).

Using multiple wavelengths also allows for mathematical compensation for interfering substances. For example, suspended solids absorb light across the visible, near-infrared, and UV regions of the spectrum at a proportional ratio. Because soluble organics do not absorb visible light at 550 nm or longer wavelengths (e.g., near-infrared light at 850 nm), the effect of solids can be mathematically subtracted from the absorbance readings at 254 nm when absorbance is measured at 550 or 850 nm.

Table 9-10 Organics characteristics over various UV wavelengths

Wavelength and Notes *nm*	Organics Characteristics	Reference
260 with TOC	Evaluate treatability of organics by biological processes	Tambo and Kamei 1978
285	Correlates well with fulvic acid content of the water	Kalbitz et al. 2000
280	Aromatic content of fulvic acids	Chin et al. 1994
300	Characterization of humic substances	Artinger et al. 2000
320	PAHs	El Khorassani et al. 1998

PAH— polycyclic aromatic hydrocarbon, TOC— total organic carbon

Calibration

Absorbance instruments are typically calibrated at the factory, and primary calibration is rarely done in the field. Some manufacturers include an optical filter that can be used to verify the calibration of the instrument. If this calibration verification deviates too far from the manufacturer's specifications, it is usually recommended that the instrument be returned to the factory for service.

Typical quality assurance checks performed in the plant include checking the baseline absorbance of an organic-free blank solution, i.e., deionized or distilled water, and comparing absorbance readings with a well-maintained laboratory spectrophotometer or field UV absorbance photometer. Most calibration errors are the result of dirty sample cells or optics.

Maintenance requirements include replacing the lamp according to the manufacturer's recommendations, cleaning the optical surfaces of the sample cell, and replacing any upstream filters as necessary.

Best Practices and Troubleshooting

Suspended particles can scatter light and interfere with absorbance measurements, so prefiltration may be necessary upstream of UV absorbance analyzers for all but filtered water samples. A typical filter apparatus is shown in Figure 9-16. Alternatively, some analyzers measure absorbance at a second wavelength for turbidity compensation purposes. Independently collected turbidity readings can also be used for correcting the absorbance readings of a UV analyzer.

Fouling of the optical sample windows can also result in errors. Inorganic precipitation and biofilms can obscure the sample cell, leading to unexpectedly high absorbance measurements. Some instruments come with automated cleaning systems such as wiper mechanisms, pressurized air purging, and chemical cleaning to remove foulants. Similarly, the presence of other UV-absorbing substances such as dissolved iron, nitrate, and nitrite can interfere with measurements of NOM.

Standard Methods (2005) states that when UV absorbance of a water sample is to be measured, a cell path length providing an absorbance value between 0.005 and 0.95 per path should be selected for precise readings. Even when the manufacturer specs indicate that a higher absorbance reading is possible, for better accuracy, it is important to keep absorbance between these values and use dilution when shortest cell path length cannot provide readings less than 0.95 per path.

For benchtop analyzers, cuvette material must also be considered because it affects the data:

Source: ABB

Figure 9-16 Typical filter apparatus placed upstream of UV analyzer

- Fused quartz cell can be used to measure absorbance down to 215 nm.
- Synthetic fused silica cuvettes would be required to measure down to 180 nm.
- Regular glass is only usable down to 330 nm.
- Disposable cells made of plastic can be used only down to 350 nm.

For portable and benchtop units, cuvette orientation should also be consistent. If there is a marker on the cuvette, always insert the cuvette with the marker facing the same direction.

Common issues and troubleshooting options for UV analyzers are provided in Table 9-11.

If a saw-tooth pattern is observed, cleaning should be performed more frequently.

Table 9-11 Common UV analyzer problems and troubleshooting approaches

Symptom	Troubleshooting Activity
Erroneous reading	• Solids in the water may be interfering. Consider filtering or using an analyzer with additional wavelengths to compensate. • Compounds in the water, including treatment chemicals such as O_3, Fe^{3+}, MnO_4^-, $S_2O_3^{2-}$, ClO^-, H_2O_2, and Fe^{2+} may be interfering with readings; use analyzers with additional wavelengths to compensate.
Erratic data (see Figure 9-17)	• Check for bubbles or solids in samples or on cell windows. • Check tubing for clogging that creates bubbles.
Unusual readings	• Check for condensation of water drops on cell windows; desiccant dryers may be needed in humid environments. • Check lamp run time, which may be shortened by environmental factors; replace lamp if needed. • Fouling affecting UV absorbance may not be visible to the naked eye; test system with distilled or deionized water to confirm readings.
Absorbance data drifts or changes sharply with cleaning (see Figures 9-18 and 9-19)	• Increase the frequency of cleaning. • Change the cleaning fluid and/or method.

ClO^-—hypochlorite, Fe^{2+}—ferrous iron, Fe^{3+}—ferric iron, H_2O_2—hydrogen peroxide, MnO_4^-—permanganate, O_3—ozone, $S_2O_3^{2-}$—thiosulfate, UV—ultraviolet,

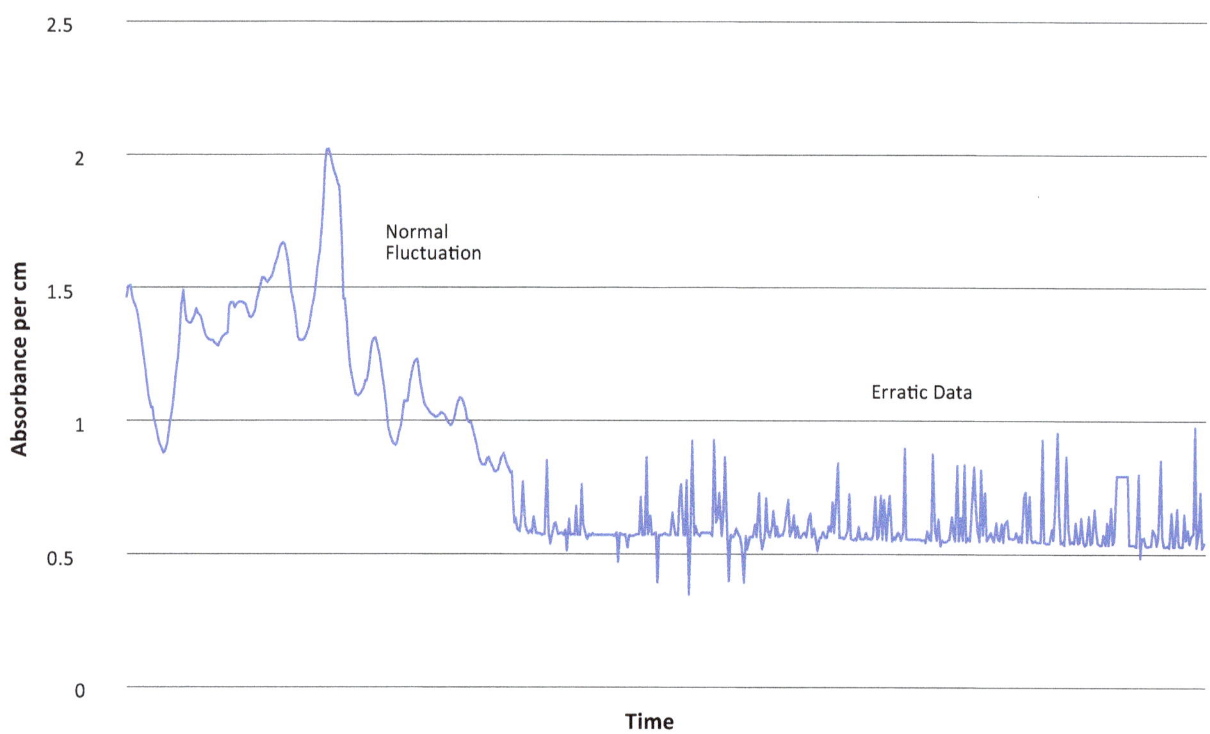

Source: ABB

Figure 9-17 Particles and bubbles can create erroneous readings

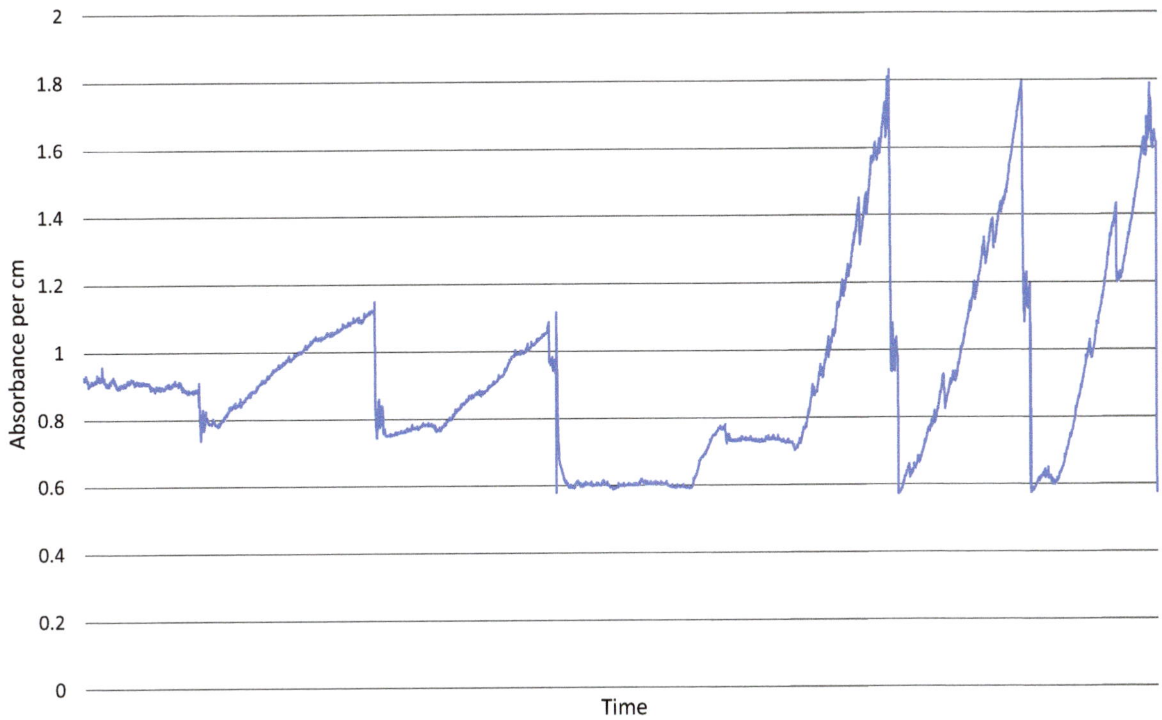

Source: ABB

Figure 9-18 Periodic sudden drops in data generally indicate automatic or manual cleaning

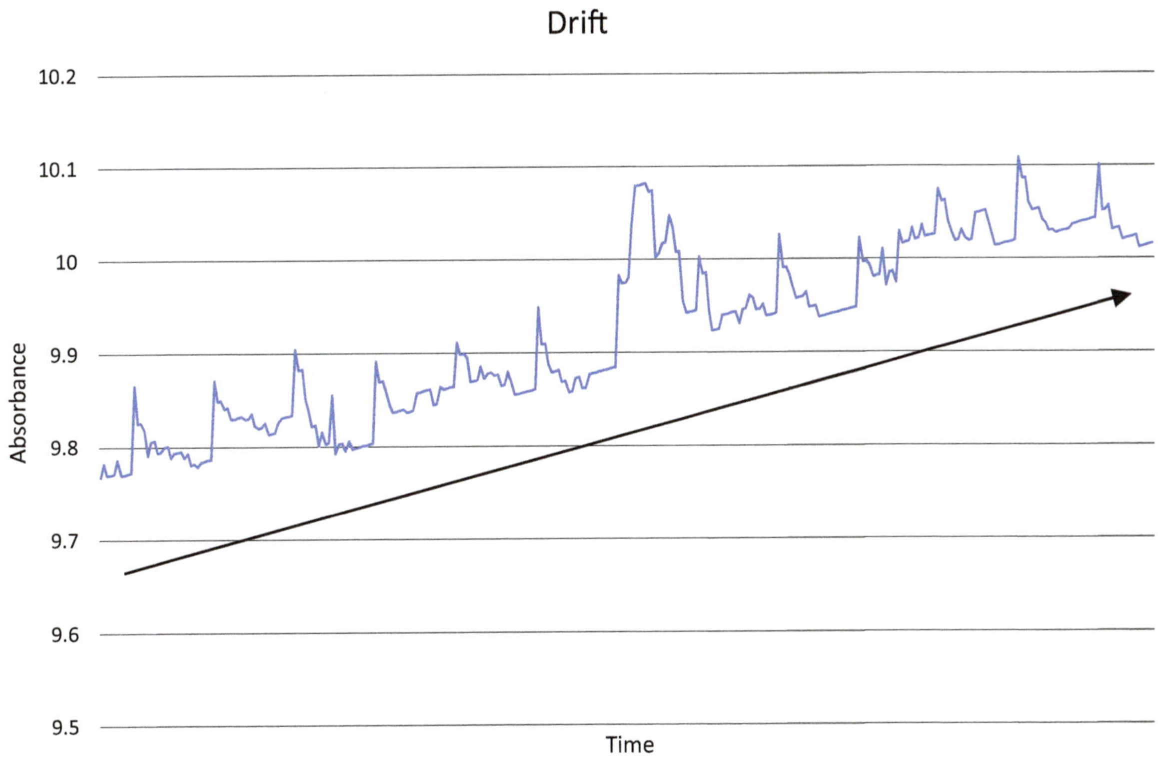

Source: ABB

Figure 9-19 Fouling of a cell may lead to an upward drift in the absorbance data

pH ANALYZERS

Measuring pH is fundamental to water treatment operations because it plays a role in performance across many processes.

Measurement Principle

pH analyzers measure the concentration of hydrogen ions (H^+) in a water sample. Pure water consists of molecules of H_2O. A very small fraction of these molecules in pure water disassociates forming H^+ and OH^- (hydroxyl) ions. At neutral pH (pH = 7), the concentrations of hydrogen and hydroxyl ions are equal. In acidic solutions (pH less than 7), the concentration of hydrogen ions in solution is greater than the concentration of hydroxyl ions. In basic solutions (pH greater than 7), the opposite is true.

In water treatment applications, hydrogen ion concentration is expressed in terms of pH, which is defined as the negative log of the hydrogen ion activity or concentration.

$$pH = -\log [H^+]$$

According to method 4500-H^+ (*Standard Methods* 2005), electrometric pH measurement involves determination of hydrogen ion activity by potentiometric measurement using a reference electrode and a glass electrode, in which the electromotive force produced in the glass electrode system varies in a linear fashion with pH. The pH electrode is made of a pH-sensitive ion conductive material. Electrons can be exchanged across the conductive glass by hydrogen ions, and this creates the mV potential.

The mV value does not provide enough information for the calculation of the pH value. The temperature measurement must also be known to correctly interpret the measured mV data, so temperature compensation is often provided to account for this. In practice, the reference point is always the value at 25°C, and this is the only way pH values in publications are comparable. Figure 9-20 shows the effect of temperature on mV reading at pH 2.00.

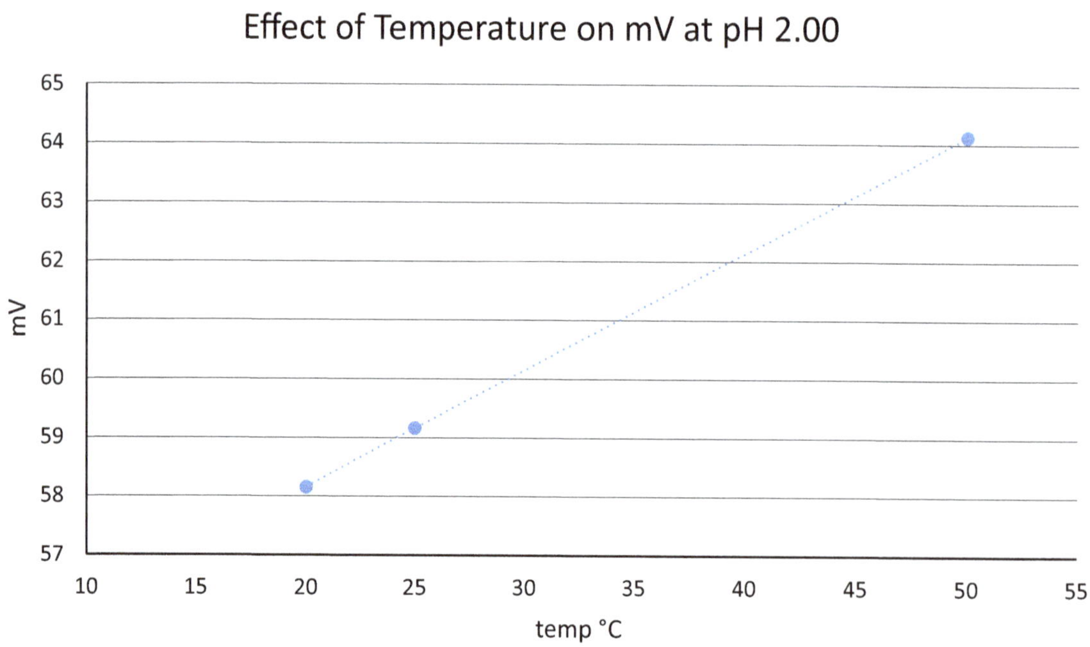

Source: Endress+Hauser

Figure 9-20 Effect of temperature on mV reading at pH 2.00

The baseline mV value for a change in pH of 1 unit at 25°C is 59.16 mV. Through age or wear, a pH electrode will have a slope that is no longer equal to 59.16 mV. As a rule, when the slope decreases to around 50 mV/pH unit, the electrode has reached the end of its life cycle. If the electrode zero-point shifts during use, this indicates that the reference solution, usually potassium chloride (KCl), is migrating from electrode into the process or the process media has penetrated the reference junction. Deviations greater than 30 mV (0.5 pH) indicate the life cycle end of the electrode.

Another important performance metric of a pH electrode can be the response time. Response time depends not only on temperature (the lower the temperature, the higher the impedance, the slower the electrode) but also on the performance of the reference junction (such as plugging) or the degradation of the pH-sensitive glass. Modern process pH measuring systems continuously measure and store sensor health data such as glass impedance and detect glass breakage.

At high pH values, sodium ions can interfere with pH measurement. *Standard Methods* (2005) recommends the use of a "low sodium error" electrode for measuring pH over 10 because of the possibility that standard glass electrodes can produce pH readings that are too low.

Types

The differences between a pH measurement in a laboratory environment using a benchtop pH meter and a continuous pH measurement in a process with an on-line pH meter are summarized in Table 9-12.

The pH electrodes for benchtop devices are usually very precise measuring electrodes because they do not have to be designed to accommodate high-pressure or high-temperature applications. As a result, the pH measurement in the process and in the laboratory may differ. Additionally, temperature, pressure, CO_2 in the air, etc., can affect the pH value.

Portable devices are best suited for field verification of process instruments.

Setup, Standards, and Calibration

pH sensors require calibration at commissioning and at regular intervals thereafter. The manual provided by the instrument manufacturer should be consulted for instructions on calibration procedures and the recommended frequency for calibration. To obtain valid pH data, plant staff must perform calibration and other maintenance procedures at least as frequently as specified in the instrument manual or more frequently as experience dictates. The typical life span of a pH electrode is six months to two years.

Table 9-12 Differences in benchtop and on-line measurements

Laboratory Measurements	On-line Process Measurements
Manual sampling: sample properties may change during transportation to laboratory	In process immersion, insertion, or in a retractable assembly
Laboratory conditions: constant, no overpressure, constant temperature, stirred, air contact	Process conditions: variable, process pressure, and process temperature
Short measurement time	Permanent measurement
Electrode: short exposure to stress	Electrode: constant exposure to stress
Cleaning daily under laboratory conditions	Cleaning during process shut down or with automatic cleaning systems
Calibration and adjustment daily under laboratory conditions	Calibration and adjustment during process shut down or with automatic calibration systems

Source: Endress+Hauser

Table 9-13 Contaminant and appropriate cleaning solutions

Contaminant	Cleaning Solutions
Light coating or scaling	Water spray from a faucet or a water bottle
Heavy coating or scaling	Hot water solution containing dishwashing detergent (without lanolin); carefully remove the coating with a brush or toothbrush
Alkaline or scale	5 percent hydrochloric acid or acetic acid (vinegar is acceptable)
Acidic coatings	Weak caustic (less than 4 percent sodium hydroxide)
Oil, grease, organic	Detergent, or if coating is tenacious, an organic solvent compatible with the electrode material

Source: Endress+Hauser

pH measurement between instruments is less influenced by instrument design than many other measurements because on-line and benchtop instruments share a fundamentally common design. Therefore, the use of benchtop pH meters to periodically check the validity of on-line pH data is encouraged as a means of having confidence in the on-line pH data. However, an on-line pH electrode is in continuous contact with the sample stream and is therefore conditioned to the temperature and ionic strength of the process stream. The pH probe of a benchtop instrument is typically stored at room temperature in a buffer solution. These differences can result in sluggish performance and/or measurement discrepancies, especially in cold waters.

A calibration check or a calibration is necessary depending on SOPs. A calibration or calibration check is performed using two buffers, which are selected such that the expected sample pH is between the two pH values of the buffers. If the electrode is reading within defined tolerances in these buffers, typically at ±0.1 pH, a calibration is not needed. Fresh buffer should be used for calibration because it can degrade over time.

Best Practices and Troubleshooting

pH probes require periodic cleaning. Table 9-13 provides a few cause-and-remedy scenarios for determining a cleaning method to use on the basis of the contaminant. After cleaning is accomplished, rinse the electrode and perform a calibration check. If the electrode falls within the tolerance band, it is ready for use; if not, consider another cleaning interval, use a more concentrated cleaner, or if it is as clean as it can get, perform a full calibration. After this is done, rinse the electrode in water, and put it back into the process or in its storage solution.

pH electrodes are fragile and need to be treated accordingly. For example, do not use knife blades or abrasive brushes for cleaning. Watch out for processes that contain abrasives (sand), thermal shock, high pressure, temperature, and vibration because these are detrimental to the electrode's life.

Most pH electrodes must be mounted so that the internal fill solutions in the active electrode and reference electrode are in the tip of the electrode and no air gap exists. Therefore, it is recommended that the electrode be mounted at a minimum of 15° from horizontal, as shown in Figure 9-21.

When mounting a pH electrode in a pipe run, avoid situations in which the process level fluctuates and causes the electrode to be exposed to air. Electrodes must always be kept wet, as shown in Figure 9-22, to make a measurement, and exposure to air for long periods may shorten the life of the electrode. If reagents are added inline, ensure that adequate reaction time and mixing energy have occurred before the point of pH measurement to ensure the mixture is homogeneous.

Keep flow velocities across the pH electrode to 10 ft/s or less.

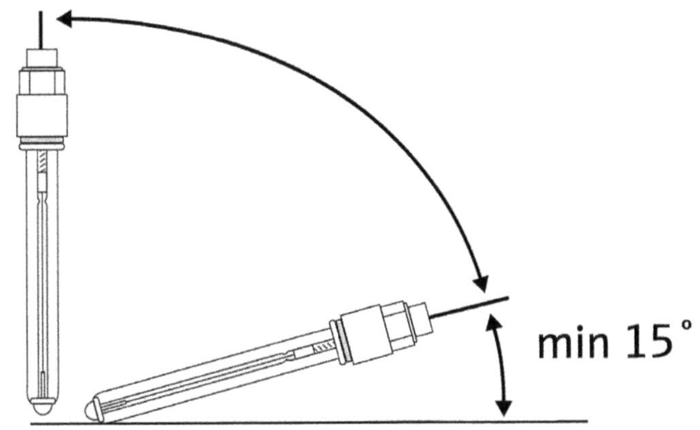

Source: Endress+Hauser

Figure 9-21 Proper pH electrode orientation

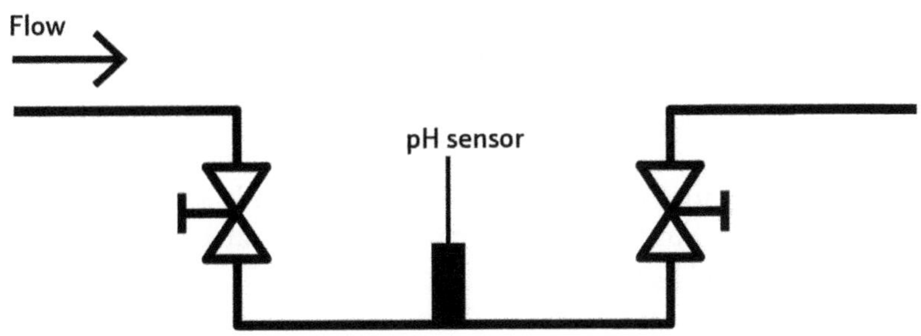

Source: Endress+Hauser

Figure 9-22 Correct installation of a pH electrode: slip stream mounting provides a full pipe even when the process is not flowing

Table 9-14 Common pH analyzer problems and troubleshooting approaches

Symptom	Troubleshooting Activity
Intermittent signal	• Confirm probe remains submerged under all process conditions; if not, adjust probe location
Inconsistent readings	• Check for adequate process mixing before measuring pH

When electrodes are not in use, they must be stored with the factory supplied protective caps in an appropriate wetting solution. An alternative is to store the electrodes in a clear beaker filled with 3.0 molar KCl or pH 4 or pH 7 buffer or tap water. The reason for using a clear beaker is that one can easily see the level of fill solution in use and refill it when the level starts to drop.

Common issues and troubleshooting options for pH analyzers are provided in Table 9-14.

DIFFERENTIAL PRESSURE

Differential pressure is a trusted form of measurement used in a wide variety of processes. The various uses include level indication in a closed environment, flow rates for fluids, monitoring of centrifugal pump performance, as well as identification of restriction in valves and filter systems. Differential pressure is frequently used in WTPs to monitor for filter underdrain clogging.

Measurement Principle

For differential pressure measurement, two relevant measurement points are identified in the piping system or vessel, as shown in Figure 9-23. The pressure at each location is transferred to a sensor/transmitter combination at which point the values are analyzed to identify any deviation present. The deviation, or differential pressure, between the two points can then be interpreted as the level in a vessel, flow of a process fluid, or potential head loss in a filter system.

Types

The common configurations for measuring differential pressure are described in Table 9-15.

Setup, Standards, and Calibration

Calibration frequency depends on the type of differential pressure system being used (impulse line or remote seal) and the location of the transmitter (indoor or outdoor). Systems with impulse lines mounted indoors typically should be calibrated every four to six years, whereas outdoor systems should be calibrated every one to four years. Remote seal systems should be calibrated twice as often.

Best Practices and Troubleshooting

Common differential pressure problems and troubleshooting approaches are provided in Table 9-16.

SLUDGE LEVEL MEASUREMENT

Efficient operation of sedimentation basins requires measurement of the level and density of settled solids in sedimentation basins. Therefore, measurement of the solids-liquid interface within sedimentation basins is critical for process control of sedimentation basin loading rate and sludge pumping rate. Solids-liquid interface detectors provide a measurement of the depth and density profile of solids within sedimentation basins. The focus of this section is on electronic sludge blanket level (SBL) analyzers that automatically detect the depth of liquid-solids interfaces.

Measurement Principle

The principle of measurement for SBL analyzers is either acoustical or optical. Acoustic sensor designs, shown in Figure 9-28, measure the travel time of ultrasonic echoes that are reflected by solids below a submerged transducer. In general, the more concentrated the solids particles, the more reflection, and the higher the signal at the receiver. Detection of interfaces is based on the change in signal and not the absolute value. An "echo" is a point at which signal intensity increases abruptly. Sludge level can be determined in samples

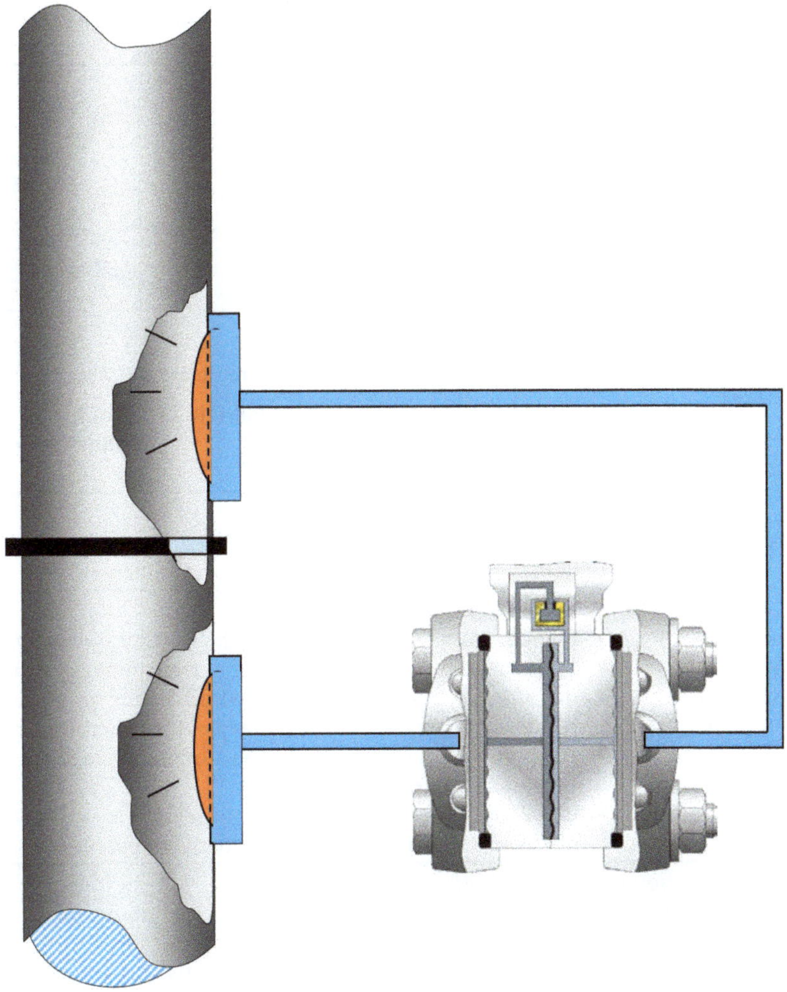

Source: Endress+Hauser

Figure 9-23 Example of differential pressure measurement

at times when one or more echoes are observed. The acoustic principle works best when attempting to locate the densest layer of thickened solids in relatively well-settling sludge. On the other hand, reliable determination of interfaces may not be possible in applications with nebulous layers of diffuse solids ("fluff").

The optical design measures the transmission of light across a gap. Solids interfere with transmission of light across the gap, which reduces the signal at the photodiode. The loss of light is directly proportional to the concentration of solids particles in the gap. One optical design, shown in Figure 9-29, consists of a sensor that travels through the water column at a programmed interval. The traveling design provides both a measurement of the depth of liquid-solids interface and total suspended solids (TSS) concentration. Interfaces are generally defined by default or user-defined static thresholds of TSS. Therefore, optical sensors will always return a signal if threshold values are reached but may not always locate the most-dense layer of sludge.

The sensor for a different optical design is a fixed vertical column of lights/detectors mounted under the water. Interface levels are defined by adjustable loss of signal (light) across the gap.

Table 9-15 Common configurations for differential pressure measurement

Traditional Differential Transmitter

- Uses single dP transmitter connected to the two measurement points in the system, as shown in Figure 9-24.
- One measurement point will be identified as the LP measurement point and the other as the HP measurement point.
- The LP and HP measurement points are then connected to the transmitter.
- Internally the transmitter uses a single sensor (dP = HP–LP).
- Pressure is transferred from the measurement points to the transmitter and sensor using impulse lines or remote diaphragm seal.

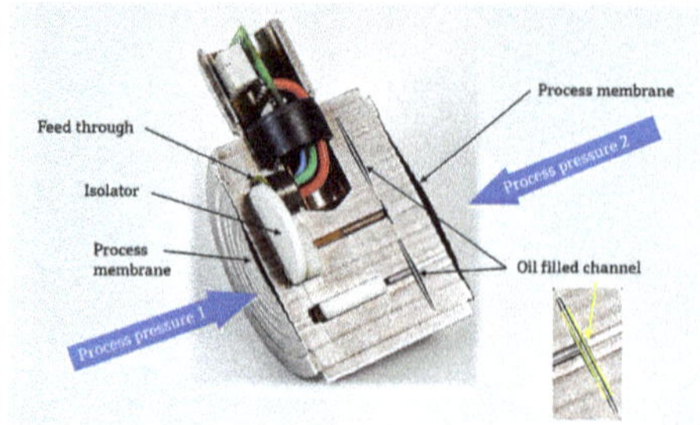

Source: Endress+Hauser

Figure 9-24 Traditional differential pressure transmitter

Impulse Lines:

- Tubing or pipe with direct access to the process.
- Process fluid is in direct contact with the impulse lines as well as the transmitters sensor.

Benefits:

- Direct contact provides the most accurate and fastest responding form of differential pressure measurement.

Potential Concerns:

- High process temperatures in contact with the sensitive sensor electronics.
- Wet/dry leg inconsistencies—The transmitter is calibrated on the basis of presence or lack of process fluid in the impulse lines. Evaporation or condensation changes the accuracy of the differential pressure measurement.
- Plugging—Plugged impulse lines no longer correctly transfer changes in the process to the transmitter.
- Chemical compatibility—The impulse lines as well as the process membranes of the sensor must be compatible with the process.

Remote Diaphragm Seals (shown in Figure 9-25):

- Closed hydraulic system connecting the process measurement points to the transmitter.
- Process contact is maintained using a chemically compatible diaphragm membrane.
- Behind that membrane is a capillary with specially formulated fill fluid used to transfer pressure information to one side of the transmitter's sensor.

Benefits:

- Provides temperature protection for the sensor.
- Chemical compatibility concerns are restricted to the process diaphragm and process connection.
- Elimination of any plugging concerns in slurries or dirty process conditions.
- No wet/dry leg issues.

Potential Concerns:

- The response time to changes in process pressure can be compromised because of the capillary effect.
- Potential thermally induced drift. As the remote diaphragm seal/capillary system is completely enclosed, process and/or ambient temperature changes will result in the expansion or contraction of the fill fluid. In turn, this will increase or decrease the force on the sensor diaphragm and differential pressure reading. Even when no actual change in process dP has occurred.

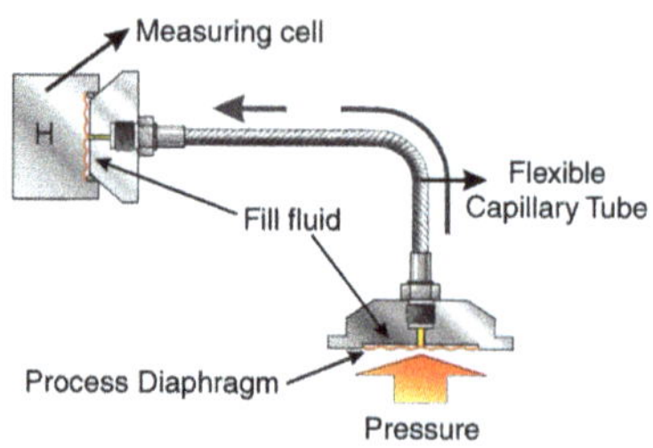

Source: Endress+Hauser

Figure 9-25 Remote diaphragm seal for pressure measurement

(continued)

Table 9-15 Common configurations for differential pressure measurement (*Continued*)

Differential Pressure Using Two Gauge-Style Transmitters (shown in Figure 9-26)

- Uses two measurement points in the system and compares the two values to arrive at a differential pressure.
- Measurement at the LP and HP locations is performed with individual/independent gauge-style transmitters.
- The pressure value measured by each transmitter is then relayed to a math module, PLC, or DCS that performs the differential pressure calculation.

Benefits:

- Modular design. Individual components can be easily replaced, if necessary, versus replacement of a complete capillary system.
- Elimination of capillaries and the possible associated issues.
- No wet/dry leg or plugging problems.
- Reduced installation/commissioning time.

Potential Concerns:

- The need for a second I/O should a PLC or DCS be used for the dP calculation.
- Added expense of the Math Module.

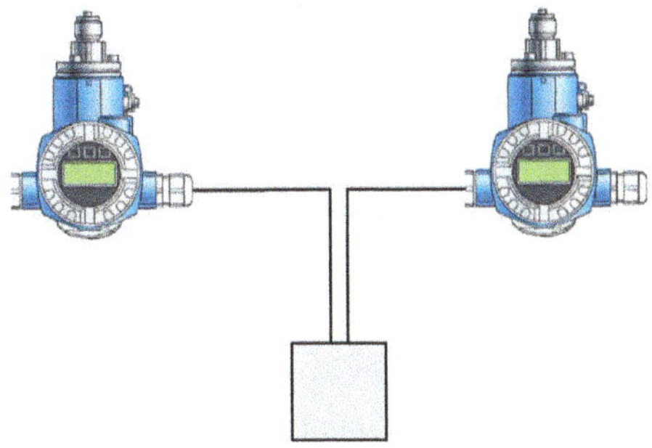

Source: Endress+Hauser

Figure 9-26 Differential pressure measurement using two gauge-style transmitters

Electronic Differential Pressure with a Single Transmitter (shown in Figure 9-27)

- Similar to using two independent transmitters.
- However, system performs the differential pressure calculation eliminating the need for a separate math module, a second I/O in your control system, or additional programing.

A "true" electronic dP solution consists of the following components.

- Two standard gauge sensors (just sensors, not transmitters).
- A single transmitter that receives the information from the LP/HP sensors, performs the dP calculations, and transmits the results to the control system.
- Standard 4 wire twisted pair shielded cable connecting the sensors to the transmitter.

Benefits:

- Individual components can be easily replaced, if necessary, versus replacement of a complete capillary system.
- Elimination of capillaries.
- No wet/dry leg or plugging problems.
- Reduced installation/commissioning time.
- Multiparameter information: dP, LP sensor, and HP sensor.
- Exceptional flexibility—placement of the transmitter and display for maximum visibility and accessibility.
- Single I/O point to controller.

Potential Challenges:

- Limited pressure/temperature ranges.
- Resolution errors when ratio of max head pressure to max level is too large.

Source: Endress+Hauser

Figure 9-27 Electronic differential pressure with a single transmitter

DCS—distributed control system, HP—high pressure, I/O—input/output, LP—low pressure, PLC—programmable logic controller

Table 9-16 Common differential pressure problems and troubleshooting approaches

Symptom	Troubleshooting Activity
Unresponsive to change	• Check for closed isolation valves that would separate the sensor from the process. • Check for plugs in impulse lines. • Examine diaphragm seal; confirm diaphragm volume is coordinated with sensor requirements.
Changes in readings without process changes	• Evaluate location; ambient temperature fluctuations may be responsible. • Examine impulse lines for evaporation or condensation.

Types

On-line and portable versions of SBL analyzers are available. On-line versions are mounted to the handrail above the desired monitoring location and operate on 120-VAC power. On-line SBL analyzers consist of a probe or sensor and a meter or controller that displays the relevant measurements, manages communications, and provides a user interface. The sensor component of on-line acoustic models is fixed in place below the water level and requires a deflector to lift the sensor out of the water on clarifiers with traveling skimmers. The sensor for the on-line version of traveling optical models is mounted at the end of a cable wound around a rotating drum. Limit switches activate drum operation and control sensor travel.

On-line SBL analyzers will be useful in applications in which precise control of interface level is desired to optimize process control of sedimentation. The continuous feedback can be used to enhance manual control or as input to control loops that automate sludge removal.

The most common portable sludge blanket measurement device is a core sampler, also referred to as a "sludge judge," which is manually lowered through the depth of a sedimentation tank and then raised for visual observation of the water column. An example of this type of device is shown in Figure 9-30. Most water treatment facilities own a core sampler and use it on a regular basis. Battery-operated portable versions are a substitute for a core sampler.

Portable electronic SBL analyzers consist of a gap-style optical probe on a cable attached to a handheld meter. Default threshold settings for the location of interfaces can be changed manually. The cable version has colored bands at regular intervals for visual depth determination. Another version has a pressure sensor to measure depth. The interface level function does not require calibration. A zero and span calibration is possible for the TSS concentration function.

Data and Interpretation

Depth measurements provided by SBL analyzers may include depth to the top of the solids-liquid interface and/or the depth of the thickened sludge layer. A profile of the solids concentration can also be displayed locally on some models. An echo profile from an acoustic-style sensor is shown in Figure 9-31. Depth measurements and profiles from acoustic SBL analyzers and fixed optical analyzers are updated continuously. Measurements from traveling optical sensors are updated at the interval that the sensor is lowered through the water column.

Interpretation of sludge level measurements is straightforward. The depth to the top layer provides an indication of how close solids are to the overflow weir. A high or rising top layer could indicate an overload of the clarifier or a process upset. Monitoring the

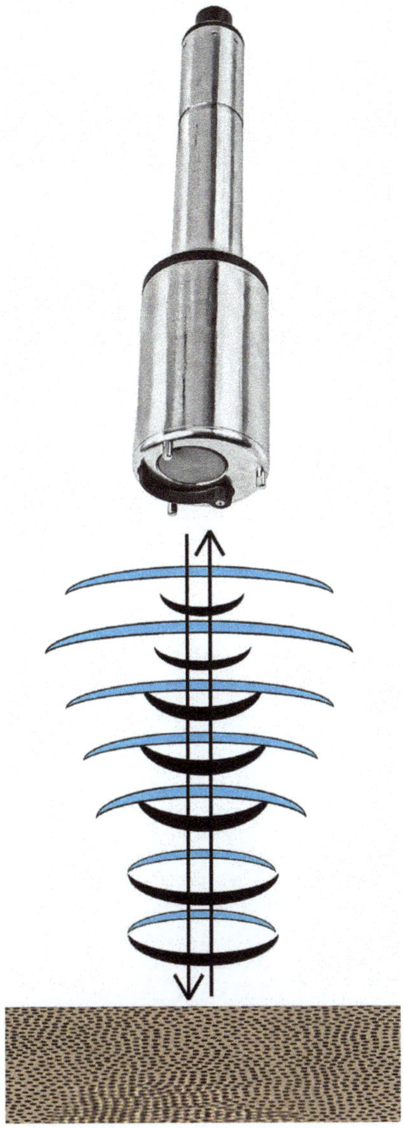

Source: YSI

Figure 9-28 Acoustic SBL sensor

thickened layer provides an indication of the depth of the relatively concentrated layer of solids. Sludge removal from the basin should be discontinued as the thickened layer approaches the bottom of the tank to minimize carryover of water to downstream solids handling processes. Sludge removal can be resumed when the dense layer reaches a desired threshold.

The solids profile provides an indication of how the solids concentration varies with depth. It is especially helpful during setup to identify interfaces and obstructions below water.

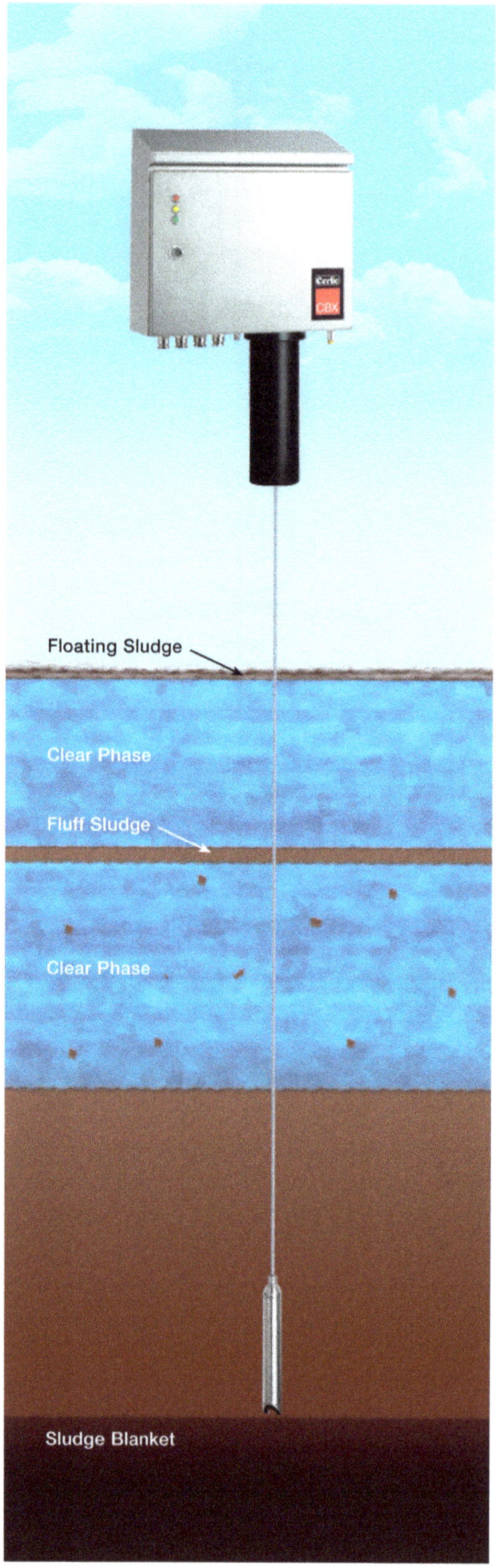

Source: Cerlic Environmental Controls

Figure 9-29 Optical SBL sensor with traveling photodiode

Setup, Standards, and Calibration

SBL analyzers are calibrated at the factory and do not require a field user calibration. A user calibration may be desirable to improve the accuracy of TSS measurement for the traveling optical design. A zero calibration is performed in a bucket of clean water. A sample (span) calibration is performed in a bucket of the sample and requires an accompanying laboratory reference measurement of the TSS concentration in the sample.

SBL analyzers require setup or configuration as part of installation and commissioning. Basic setup parameters include units of measurement, water depth, target interface(s), and location of blind zones to ignore. Acoustic models also have settings for temperature, depth

Source: Original photo by R. Smith. Face anonymized using AI.

Figure 9-30 Sludge blanket measurement with a portable core sampler

of sensor, and signal dampening to eliminate interference from traveling rakes. Acoustic models have many additional settings to adjust how the signal is processed to tune the measurement. Traveling optical models require input of the rake height and measurement interval.

Best Practices and Troubleshooting

Measure SBL at locations that are representative of the process and free from interferences. The most representative location for SBL measurement depends on the application. However, the following guidelines may be helpful:

- Circular clarifiers—The recommended SBL measurement location is from one-third to two-thirds the radial distance from the center.
- Rectangular clarifiers—The recommended SBL measurement location is typically the sludge hopper.
- Measure using a steel tape or otherwise physically determine water depth at the measurement location. Many clarifiers have sloped bottoms that complicate the ability to obtain depth at a specific location from a drawing. The tank bottom can be determined from acoustic sensors at the location where the signal intensity increases rapidly and to a relatively high level. For sensors on the end of a cable, the bottom can be determined at the location the cable goes slack as it is slowly lowered into the water.
- Select a sample location that is free from obstructions and interferences. For on-line acoustical sensors, be aware that the sensor generates a signal cone that expands with depth.

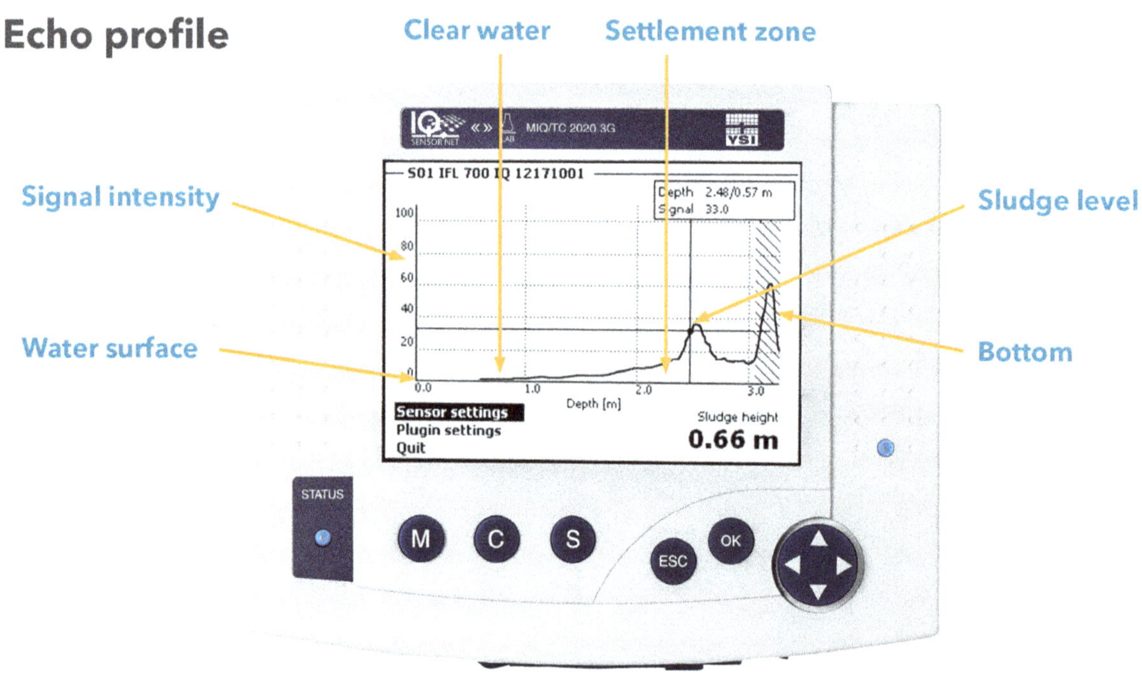

Source: YSI

Figure 9-31 Profile image from an acoustic SBL analyzer

- Choose a measurement location that is away from turbulence caused by influent, which can transport or resuspend solids and create a cloud around the sensor.
- Locate on-line acoustical sensors under water and above the SBL. This may seem trivial, but it may not be obvious particularly in covered basins.

Manufacturers' initial recommended cleaning interval varies from four weeks to three months, but longer intervals are sufficient in some applications. Moderate fouling will not interfere with sensor function. An overall decrease in signal may indicate excessive fouling. Light intensity can be increased to compensate for fouled LEDs or detectors on the fixed optical version.

Troubleshooting underperforming electronic SBL analyzers is challenging. Many facilities experiencing unsatisfactory performance of on-line versions of electronic SBL analyzers cite an inability to reconcile on-line measurements with the reference measurement, typically a sludge judge. Confirm the tank bottom and change settings on the basis of observation of the signal profile. It may be helpful to eliminate signal dampening during the troubleshooting process.

The reference measurement should also be examined when troubleshooting; the measurements obtained from a core sampler are low resolution (typically 0.5 ft.), very subjective to the technique and judgment of the operator, and sensitive to the design, integrity, and deployment of the sampler. Consider using a portable electronic sampler to provide better resolution and repeatability.

A summary of troubleshooting activities is provided in Table 9-17.

Table 9-17 Common SBL sensor problems and troubleshooting approaches

Symptom	Troubleshooting Activity
No echoes detected (acoustic models)	• Verify sensor is under water and above sludge • Adjust minimum intensity setting lower • Reduce signal dampening • Adjust process to improve sludge settleability • Remove accumulation from analyzer sensor
Extraneous echoes detected (acoustic model)	• Check for fixed interferences (walls, supports, rakes) around the probe that are reflecting the signal • Adjust settings to limit evaluation range • Adjust minimum intensity setting higher • Increase signal dampening
Readings fluctuate between values over time	• Examine basin flows to see if sludge is being disturbed by flow changes or movement of components (rakes, scrapers) • Increase signal dampening
Disagreement in core sampler and analyzer readings	• Confirm analyzer and reference samples are taken at the same location • Confirm check valve on core sampler is working properly (not slow or leaking) • Adjust SBL analyzer settings to adjust detection levels higher or lower to match core sampler • Remove accumulation or increase LED output (fixed optical models)

LED—light-emitting diode, SBL—sludge blanket level

REFERENCES

Artinger, R., G. Buckau, S. Geyer, P. Fritz, M. Wolf, and J.I. Kim. 2000. "Characterization of Groundwater Humic Substances: Influence of Sedimentary Organic Carbon." *Applied Geochemistry*. 15(1):97-116.

Chin, Y., G. Aiken, and E. O'Loughlin. 1994. "Molecular Weight, Polydispersity, and Spectroscopic Properties of Aquatic Humic Substances." *Environmental Science & Technology*. 28(11):1853-1858.

Dentel, S., and K. Kingery. 1988. *An Evaluation of Streaming Current Detectors*. Denver: AWWA and AWWA Research Foundation.

Edzwald, J.K. 1993. "Coagulation in Drinking Water Treatment: Particles, Organics, and Coagulants." *Water Science & Technology*. 27(11):21-35.

Edzwald, J.K., W.C. Becker, and K.L. Wattier. 1985. "Surrogate Parameters for Monitoring Organic Matter and THM Precursors." *Journal AWWA*. 77(4):122-132.

El Khorassani, H., F. Theraulaz, and O. Thomas. 1998. "Application of UV Spectrophotometry to the Study of Treated Wastewater Discharges in Rivers." *Clean Soil Air Water*. 26(5):296-299.

Hargesheimer, E.E., and C.M. Lewis. 1995. *A Practical Guide to Online Particle Counting*. Denver, Colo.: AWWA Research Foundation and AWWA.

Hargesheimer, E., O. Conio, and J. Popovicova, eds. 2002. *Online Monitoring for Drinking Water Utilities*. Denver, Colo.: AWWA Research Foundation and CRS PROAQUA.

ISO (International Organization for Standardization). 2016. Water Quality—Determination of Turbidity. ISO 7027-1. Geneva: ISO.

Kalbitz, K., S. Geyer, and W. Geyer. 2000. "A Comparative Characterization of Dissolved Organic Matter by Means of Original Aqueous Samples and Isolated Humic Substances." *Chemosphere*. 40(12):1305-1312.

Kavanaugh, M.C. 1978. "Modified Coagulation for Improved Removal of Trihalomethane Precursors." *Journal AWWA*. 70(11):613-620.

Pernitsky, D.J., and J.K. Edzwald. 2003. "Solubility of Polyaluminum Coagulants." *Journal of Water Supply: Research and Technology–AQUA*. 52(6):395-406.

Potter, B.B., and J. Wimsatt. 2009. *Method 415.3, Rev. 1.2: Determination of Total Organic Carbon and Specific UV Absorbance at 254 nm in Source Water and Drinking Water*. Washington: U.S. Environmental Protection Agency.

Rigobello, E.S., A. Di Bernardo Dantas, L. Di Bernardo, and E. Maria Vieira. 2011. "Influence of the Apparent Molecular Size of Aquatic Humic Substances on Colour Removal by Coagulation and Filtration." *Environmental Technology*, 32(15): 1767-1777.

Sadar, M.J. 1998. *Turbidity Science: Technical Information Series – Booklet No. 11*. Loveland, Colo.: Hach.

Standard Methods for the Examination of Water and Wastewater. 2005 (21st ed.). Washington: APHA, AWWA, and WEF.

Tambo, N., and T. Kamei. 1978. "Treatability Evaluation of General Organic Matter. Matrix Conception and Its Application for a Regional Water and Wastewater System." *Water Research*. 12(11):931-950.

USEPA. 1999. *Guidance Manual for Compliance with the Interim Enhanced Surface Water Treatment Rule: Turbidity Provisions*. EPA 815R-99-010. Washington: USEPA.

USEPA. 2017. *Analytical Methods Approved for Compliance Monitoring Under the Long Term 2 Enhanced Surface Water Treatment Rule*. EPA 821-F-17-001. Washington: USEPA.

White, M.C., J.D. Thompson, G.W. Harrington, and P.C. Singer. 1997. "Evaluating Criteria for Enhanced Coagulation Compliance." *Journal AWWA*. 89(5):64-77.

This page intentionally blank.

Chapter 10

Case Studies

Managed by Rebecca Venot

This chapter consists of nine case studies. The case studies cover a range of topics included in this manual. These studies are intended to present illustrative examples that can be adapted to other water treatment facilities by users of this manual and to demonstrate how the concepts in some of the previous chapters can be applied. Titles and authors of the case studies in the order in which they are referenced and presented in Chapter 10 are as follows:

1. "Conversion From Alum to Ferric Sulfate at the Addison-Evans Water Treatment Plant, Chesterfield County, Va.," George Budd and George Duval
2. "Machine Learning to Optimize Water Treatment Plant Operations," Kim Ervin and Enoch Nicholson
3. "Optimizing Backwashing and Filter Ripening for Traditional and Biological Filters," William Carter, Jason Curl, and Kevin Linder
4. "Optimizing Coagulation for Recycled Water Filtration," Kristine Yung, David J. Pernitsky, Chathura Abeyrathna, Cassandra Butler, and Craig Mizutani
5. "Jar Test Calibration," George Budd and Paul Hargette
6. "Multiparameter Coagulation Optimization at Newton, N.C." Amir Alansari, James Amburgey, and Brian Wilson
7. "NOM Measurements for Coagulation Control," Tom Elford and David J. Pernitsky
8. "Net Charge Equals Positive Change," David Teasdale
9. "Streaming Current Monitor Pilot Study: The Detection of a Ferric Chloride Feed Failure," Michael Sadar

Case Study 1

Conversion From Alum to Ferric Sulfate at the Addison-Evans Water Treatment Plant, Chesterfield County, Va.

George Budd and George Duval

In the late 1990s, coagulation evaluations were conducted at the Addison-Evans WTP in Chesterfield County, Va., to improve organics removal and decrease the levels of disinfection byproducts (DBPs) in the finished water. This plant contains a conventional rapid-mix, flocculation, sedimentation, and filtration sequence with a design flow of 12 million gal per day (mgd) (45 ML/d). Source water is obtained from a reservoir that is characterized by low turbidity and alkalinity and relatively high total organic carbon (TOC). Evaluations that were performed included multidimensional coagulant characterizations that involved bench-scale testing of alum, ferric chloride ($FeCl_3$), ferric sulfate ($Fe_2(SO_4)_3$), and polyaluminum chloride to identify the regions of best performance for both particle and organic removal. As a result of the coagulation evaluations, a full-scale conversion from alum to ferric sulfate commenced in early 1999. Results of the full-scale conversion were reported by Budd et al. (2004).

Figure 10-1 illustrates the change in TOC removal that resulted from the full-scale conversion to ferric sulfate as the primary coagulant. Source water quality conditions during the time period evaluated (January to September 1999) were characterized by

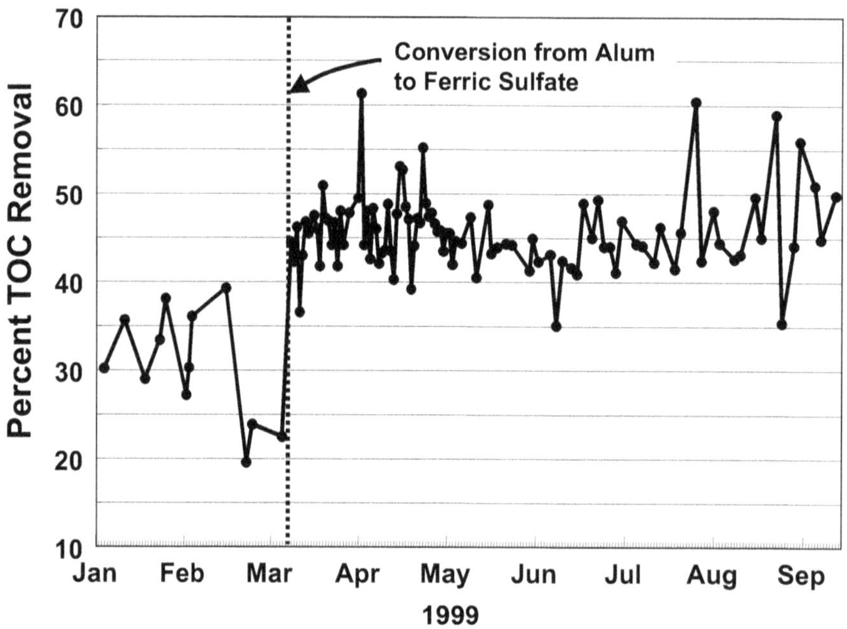

Source: Budd et al. 2004
TOC—total organic carbon

Figure 10-1 Effect of conversion from alum to ferric sulfate on TOC removal at Chesterfield County, Va.

alkalinity of 7–10 mg/L as calcium carbonate ($CaCO_3$), turbidity of 9–12 ntu, pH of 6.7–6.9, and source water TOC concentrations of 4.5–7.0 mg/L. At the time of conversion to ferric sulfate, a detailed ferric sulfate coagulant profile for settled turbidity was developed from bench-scale results. The detailed profile is contained in Figure 10-2 and indicates an optimal pH for coagulation with ferric sulfate in the range of 5.4–6.0, with a ferric sulfate dose of 30–40 mg/L. Full-scale operation with ferric sulfate in the pH range of 5.2–5.5 was subsequently found to provide effective floc formation for clarification and filtration during the initial period following the conversion. Although the change in coagulant chemical was driven by a goal of improved TOC removal, it was essential to use coagulation conditions that provided effective floc formation as well as increased TOC removal.

For the Addison-Evans WTP, the pH for effective operation with ferric sulfate has remained in the same general range because of the conversion to ferric sulfate. Ferric sulfate dosage has been increased when required by changes in water quality that can occur following storms and during periods when source water is obtained from the lower level of the reservoir that serves as the water source.

The conversion to ferric sulfate and the subsequent improved TOC removal have led to the reduced formation of trihalomethanes and sum of five haloacetic acids (Figures 10-3 and 10-4). Hurricanes and periods of bottom withdrawal from the source water reservoir (identified in these figures) represent special events of heightened DBP formation; reductions in DBP formation have been observed with ferric sulfate under these conditions as well.

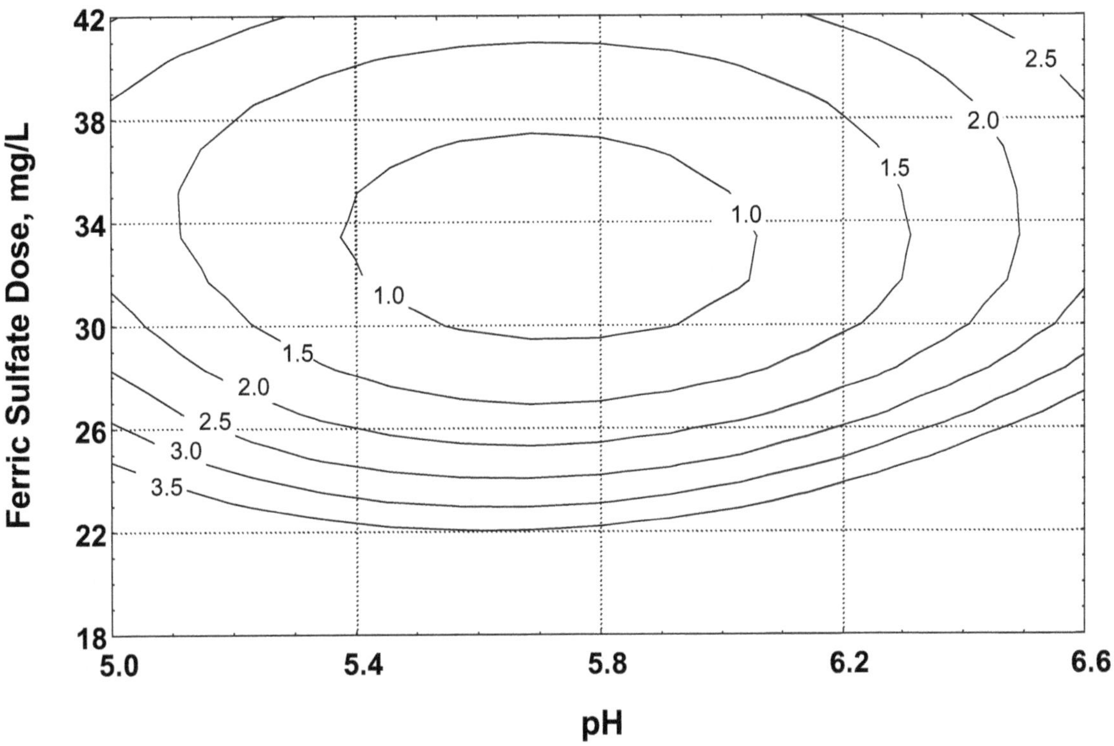

Source: Budd et al. 2004

Figure 10-2 Settled turbidity contours for ferric sulfate at Chesterfield County, Va.

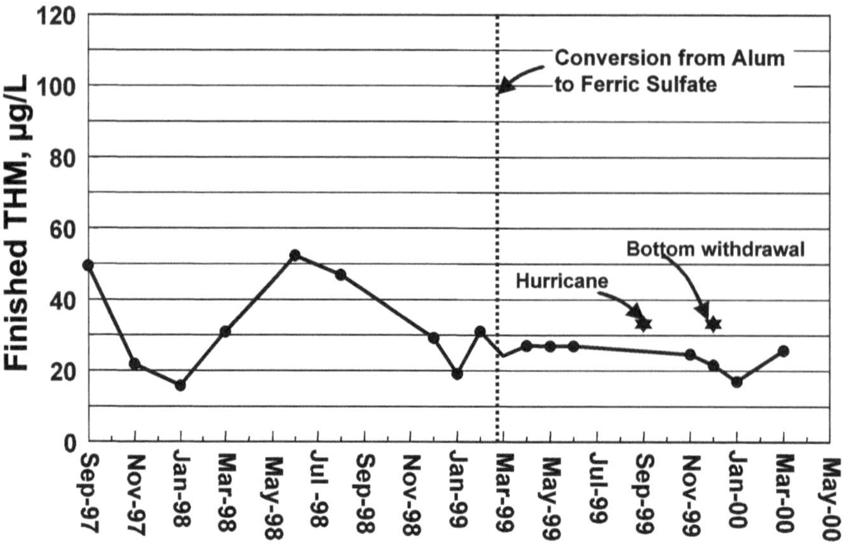

Source: Budd et al. 2004

THM—trihalomethane

Figure 10-3 Effect of conversion from alum to ferric sulfate on THM formation at Chesterfield County, Va.

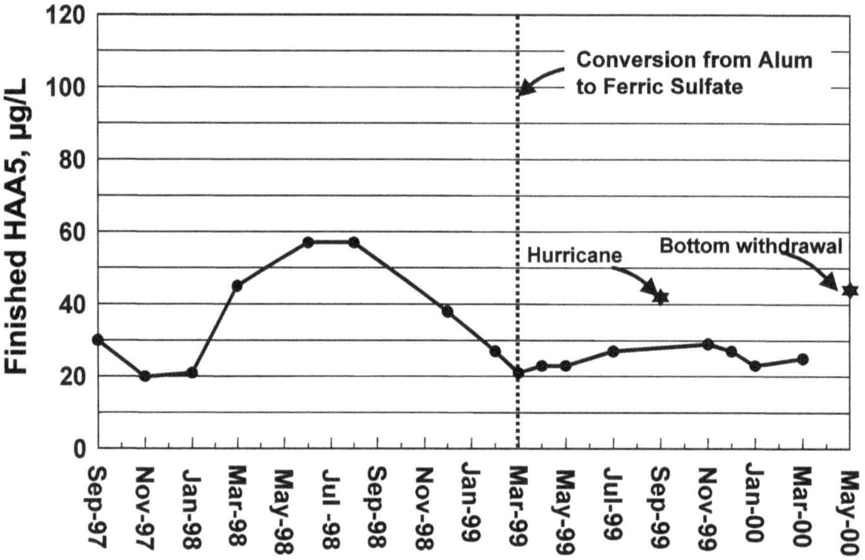

Source: Budd et al. 2004

HAA5—sum of five haloacetic acids

Figure 10-4 Effect of conversion from alum to ferric sulfate on HAA5 formation at Chesterfield County, Va.

REFERENCE

Budd, G.C., A.H. Hess, H. Shorney-Darby, J. J. Neemann, C.M. Spencer, J.D. Bellamy, and P.H. Hargette. 2004. "Coagulation Applications for New Treatment Goals." *Journal AWWA*. 96(2):102-113.

Case Study 2

Machine Learning to Optimize Water Treatment Plant Operations

Kim Ervin and Enoch Nicholson

Machine learning (ML) and artificial intelligence (AI) provide an opportunity to optimize WTP operations that result in efficiencies (such as chemical savings and solids reduction) and improved performance reliability. WTPs are equipped with sensors that continuously collect data. The traditional process of analyzing this collected raw data is computation heavy and relies on mechanistic modeling. With increased computational power, machines can analyze vast amounts of data and use algorithms to model the complex relations between the various parameters and their effect on the treatment process. ML and AI can be used to create more effective water treatment processes and provide early detection of problems.

Optimization performed by operators is largely based on experience, which accumulates over time as they become more experienced at controlling the treatment processes. ML and AI can accelerate this process by making the operational data, and insights derived from it, accessible to less experienced operators.

Coagulation is one of the most critical processes in surface water filtration plants, and modeling the coagulant dose is challenging because of its complexity, nonlinearity, and numerous contributory variables. When the source water quality changes, many operators are challenged to adjust chemical dosages, and improper dosages can result in plant upsets. Most frequently, this is the result of insufficient coagulant required to neutralize particles and organics in the source water. To address this risk, a natural tendency for operators is to overdose coagulant to provide more assurance of performance.

One case study of the application of ML and AI was a pilot project to optimize chemical coagulation for a 30-mgd WTP with sand ballasted flocculation in California in 2020. The solution developed for this application was a purely data-driven solution. This means that the actual operating data of the WTP were downloaded from supervisory control and data acquisition (SCADA) and then used to develop a mathematical relationship between the available parameters. The team developed a predictive neural model to optimize coagulant dose to maximize turbidity removal and minimize treatment cost for the given set of source water conditions. The model was developed on the basis of two years of available historical plant data taken at 15-min increments.

The data inputs to the model included the following and are shown in the green bubbles in Figure 10-5:

- Source water quality: temperature, turbidity, pH, ultraviolet transmittance (UVT)
- Plant operating conditions: flow, number of treatment trains on-line, chemical doses (coagulant, polymer)
- Settled water quality: turbidity

These data were used to build an ML algorithm that predicts the settled water quality on the basis of source water quality and plant operating conditions. Then, AI was used to optimize the dose on the basis of the source water and plant conditions. In every application, the optimization conditions may vary. In this example, the optimization was a function of

- meeting the maximum settled water turbidity limits—the targets were based on US Environmental Protection Agency guidance for settled water turbidity less

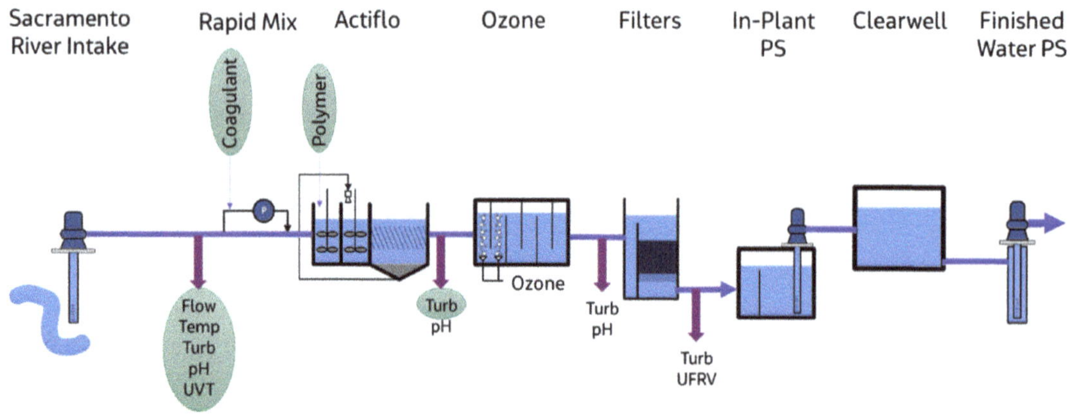

Source: Jacobs

PS—pumping station, UFRV—unit filter run volume, UVT—ultraviolet transmittance

Figure 10-5 WTP process flow diagram with samples used for model

than 1 ntu when source water is less than or equal to 10 ntu and less than 2 ntu when source water is greater than 10 ntu;

- minimizing the chemical cost—minimizing cost was the highest priority once the target turbidity goals were met. A similar model could also be developed around other parameters, such as minimizing finished water TOC; and
- minimizing the frequency of changes—because the operators make all chemical dose changes, the algorithm would not recommend frequent changes.

A comparison of the results of the optimization algorithm with actual plant data indicated that up to 10 percent savings in coagulant costs could be achieved by correcting marginal overdosing.

To achieve these savings, the information needs to be communicated to the operator. As part of this pilot project, a dashboard was created to display key operating parameters, recommendations, and potential cost savings, as shown in Figure 10-7. The dashboard operated as a web interface outside the plant SCADA system, beyond the firewall, to maintain plant security.

LESSONS LEARNED

The pilot study successfully demonstrated the potential for ML and AI to monitor and optimize coagulation. The study also identified several considerations for future applications.

- Model type—It is important to clearly define the objective of the study and select the algorithm best suited to the problem.

The original algorithm was based on time series and looking at the conditions immediately before the current condition to predict the required dose. As shown in Figure 10-6, this model was accurate at predicting settled water turbidity under normal conditions; it did not perform as well when water quality conditions changed suddenly.

In a subsequent pilot for another treatment plant, the team evaluated the effects of forest fires on water treatment. In this application, it was important to identify the unusual events rather than the normal events. The model was developed to identify the outliers and group them by the conditions that created them, such as hot and dry or storm runoff.

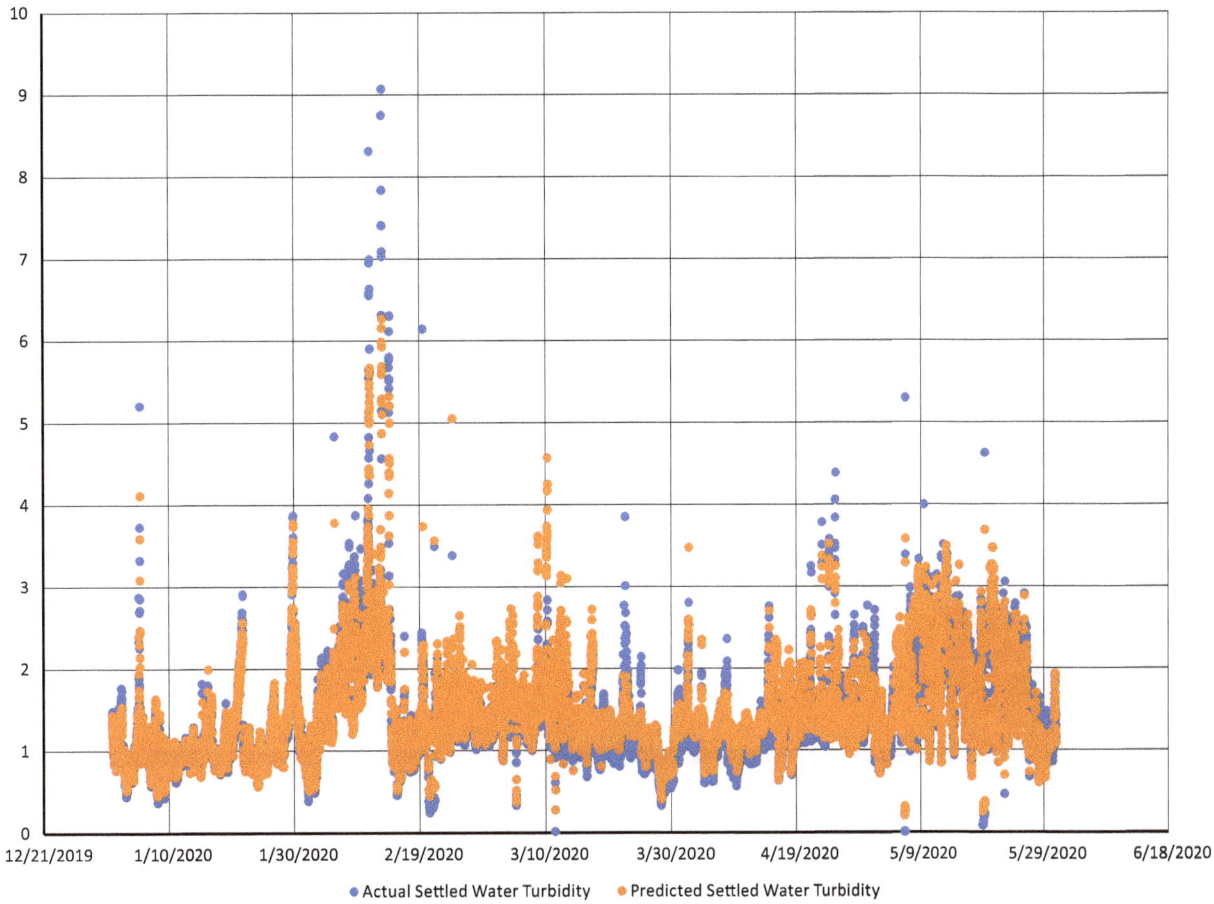

Source: Jacobs

Figure 10-6 Model predicted and actual settled water turbidity

The ultimate coagulation model will be able to correctly recommend the optimal coagulant dose under all conditions.

- Data quality and quantity—The data will determine the model options and accuracy.

The ML and AI models can be driven by any and all available data. SCADA data are generally preferred because they provide continuous data on frequent time steps. However, SCADA data can contain errors resulting from bad sensors, lack of calibration, maintenance, or off-line equipment. These data sets can require significant cleaning (identifying and treating values that are not representative of the conditions). Other data sources, such as manual data logs, can be beneficial because they are precleaned by the person entering the data, so there are fewer erroneous readings. However, these data sets can require a considerable effort to format and upload the data for use in the model.

Another challenge with WTP data for coagulation is the limited range of data available. Coagulation doses are not adjusted frequently, and treatment plants only operate at conditions that meet regulatory and treatment goals. Therefore, there is often insufficient information about the optimal coagulation dose outside of normal operating parameters. To address these limitations, the team has explored the creation of other data sets using jar tests or mechanistic models to create data sets that define the optimal dose.

- Cultural adoption—The ML and AI models rely on implementation by operators.

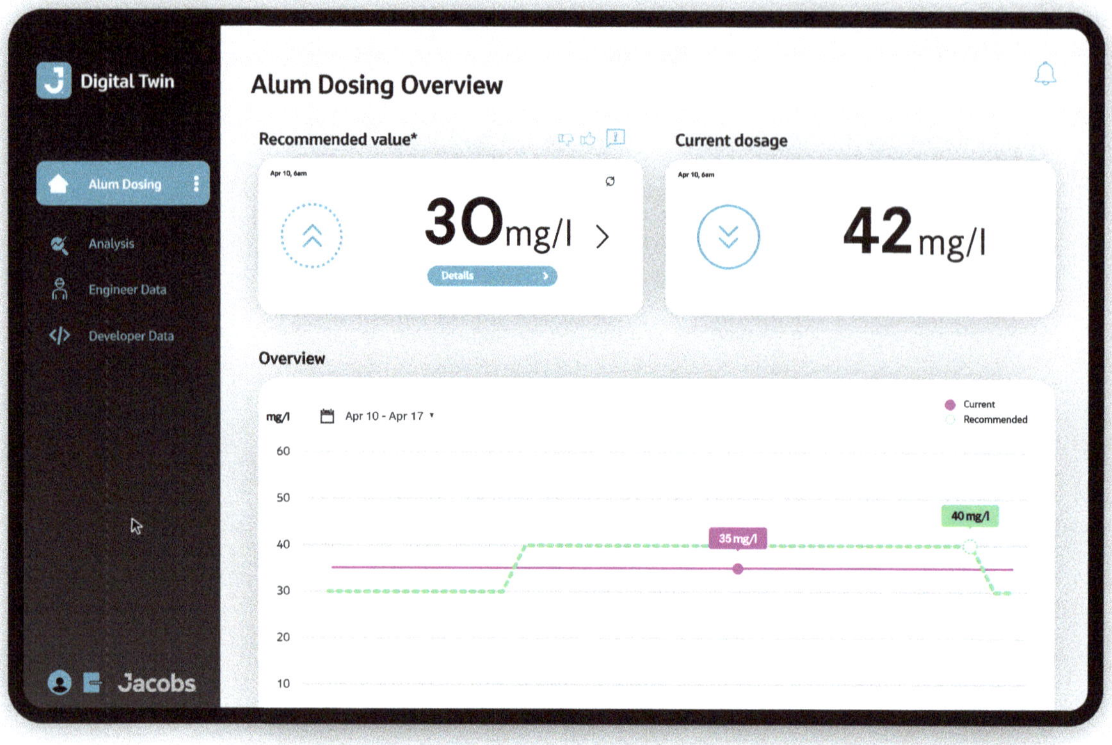

Source: Jacobs

Figure 10-7 Example of an operator optimization dashboard screen

Ultimately, the industry may reach a level of confidence in the ML and AI models to insert them directly into the control of the treatment plant. In the meantime, a dashboard tool like the one shown in Figure 10-7 can be useful to communicate the treatment recommendations to the operators. Other projects have used a text notification system to alert operators when a change is recommended. Any of these systems will only be effective if the operators have confidence in the algorithm and its recommendations. This trust takes time to establish and continued engagement by the development team to ensure the model provides beneficial information to the operators.

CONCLUSIONS AND RECOMMENDATIONS

The use of ML and AI algorithms will continue to evolve in the drinking water treatment industry. Coagulation presents one of the best opportunities for ML and AI application because it is a complex system, and the time to see the effects of a change is generally several hours. ML and AI can better predict when dose changes will be needed and recommend the optimal chemical doses to provide a more reliable and steady operation and opportunity for cost savings. These tools are still in the early development stage, and improvements are anticipated in the next few years.

Case Study 3

Optimizing Backwashing and Filter Ripening for Traditional and Biological Filters

William Carter, Jason Curl, and Kevin Linder

BACKGROUND

The Binney Water Purification Facility (BWPF) in Aurora, Colo., was placed into service in 2010. BWPF provides a unique opportunity for assessing biological filtration (biofiltration) optimization strategies because the plant includes two source waters with separate, parallel treatment trains. One of the treatment trains, the Aurora Reservoir train, is a conventional treatment train with biofilters that treats a pristine source water source from the mountains. The other treatment train, the South Platte train, is an advanced treatment train for indirect potable reuse and includes precipitative softening, an advanced oxidation process with UV light and hydrogen peroxide, biofiltration, and granular activated carbon adsorption. Before distribution, both trains are combined immediately upstream of free chlorine disinfection followed by chloramination. A process flow diagram is shown in Figure 10-8.

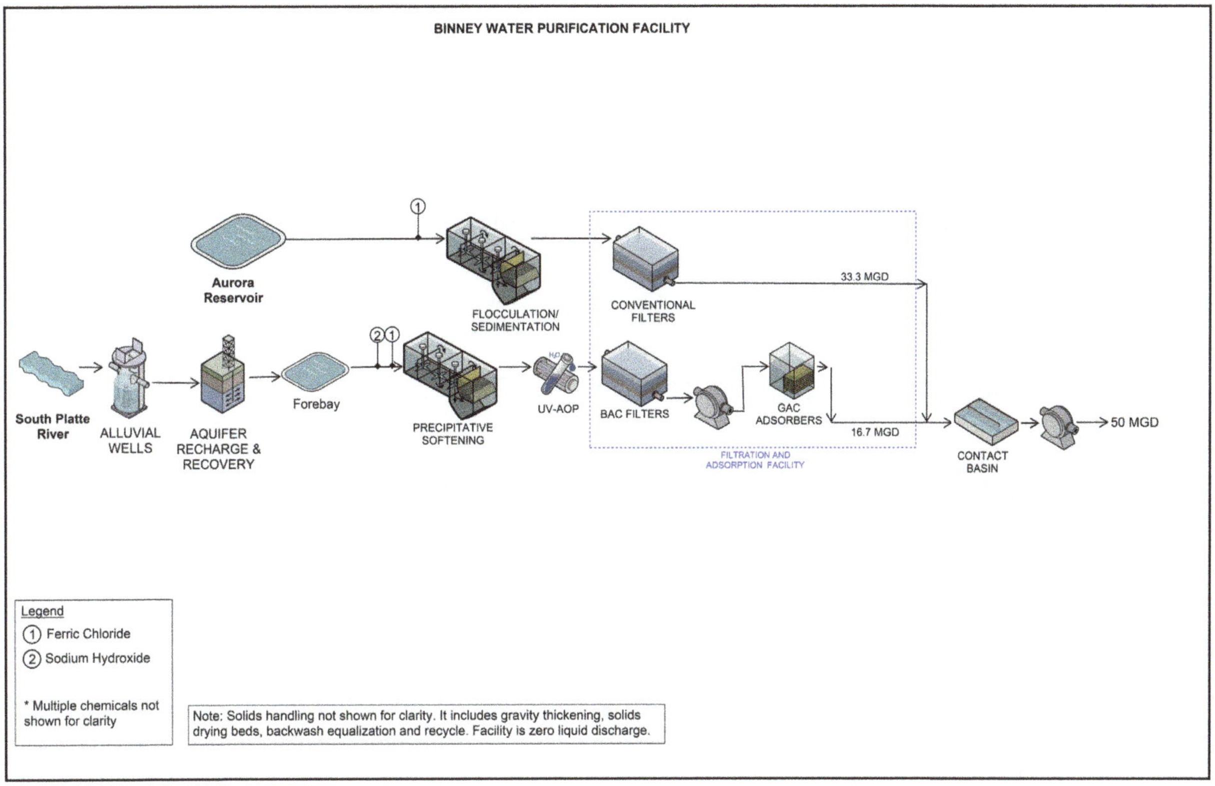

Source: Hazen & Sawyer

BAC—biological activated filters, GAC—granular activated carbon, UV-AOP—ultraviolet light with advanced oxidation process

Figure 10-8 Binney water purification flow process flow diagram

INTRODUCTION

In 2013, operations staff at the BWPF completed a self-assessment as part of their involvement in the AWWA Partnership for Safe Water (Linder 2015). One of the performance-limiting factors identified through this process was overall filter process capacity. Overall filter process capacity was addressed by commissioning four new biofilters that were deferred during the plant's original construction.

To improve filtration process capacity and reduce the filter-to-waste volume, all biofilters were equipped with programming to allow an extended terminal subfluidization wash (ETSW) step at the end of the normal backwash procedure. ETSW was intended to reduce the volume of water used during filter-to-waste by accelerating the filter ripening process. On-line media expansion and backwash turbidity monitors were also added to facilitate backwash optimization.

OBJECTIVES

The goals of the backwash optimization were as follows:

- Reduce the magnitude and duration of turbidity spike (ripening) when starting a filter run
- Not negatively affect unit filter run volume (UFRV)
- Increase filter operating efficiency

$$\text{Efficiency} = \frac{\text{Filter Production Volume}}{\text{Filter Production Volume} + \text{Backwash Volume} + \text{FTW Volume}} \quad \text{(Eq 10-1)}$$

- Mitigate occasional manganese (Mn) releases when filters are offline for extended periods

See Figure 10-9 for preoptimization design, operation, and performance summary for BWPF biofilters. See Figure 10-10 for a typical preoptimization filter run and ripening profile.

TYPICAL FILTER PROFILES BEFORE OPTIMIZATION

STEP 1: OPTIMIZING BIOFILTRATION TO PREVENT MANGANESE RELEASE

The first step in optimizing the filter backwashing process was to reduce the opportunity for manganese (Mn) release during startup following shutdowns. Aurora Water has learned that it is important to maintain adequate dissolved oxygen (DO) concentrations in the biofilters, even when the biofilter is not in service. If DO in the filter is depleted through biological activity, anoxic conditions cause manganese removed previously and retained in the media to be resolubilized. To mitigate these anoxic conditions, Aurora Water added a "DO backwash" to replenish DO after a filter has been offline for 150 h or when the filter DO is less than or equal to 2 mg/L. A "Prep" backwash was also added just before a filter coming on line. Both additions are intended to replenish the DO level before being placed in service; see Figure 10-11.

- DO backwash is used to replenish DO after a filter has been offline for 150 h or the filter DO is less than or equal to 2 mg/L.
- Prep backwash is used before a filter coming on-line (within 12 h) to replenish the DO level before being placed in service.

With the addition of the "DO" and "Prep" backwashes, the release of manganese was mitigated, but filter efficiency decreased slightly compared with the preoptimization condition because of the additional backwash water required.

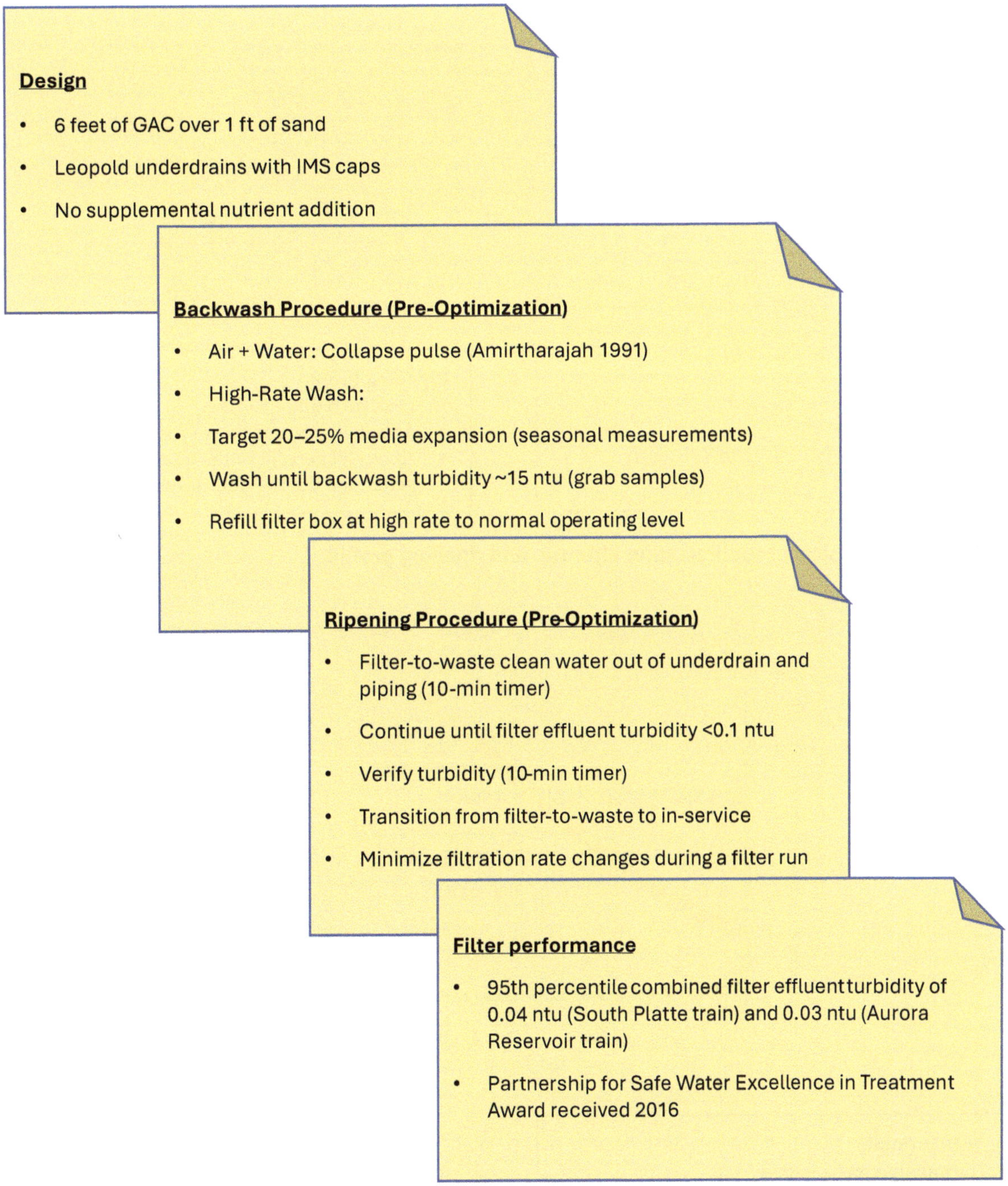

GAC—granular activated carbon, IMS—industrial material supply

Figure 10-9 Preoptimization design, operation, and performance summary for BWPF biofilters

STEP 2: ETSW

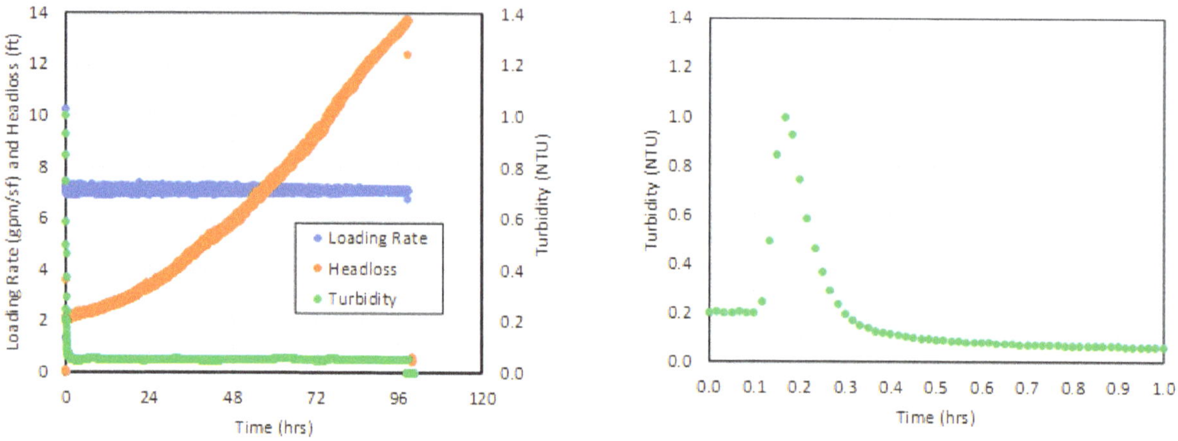

Filter Run Profile
- Terminated on head loss
- UFRV: 42,000 gal/ft^2
- Backwash and filter-to-waste: 635 gal/ft^2
- Filter Efficiency: 98.5 percent

Ripening Profile
- Peak turbidity of 1 ntu
- About 30 min to achieve <0.1 ntu
- About 70 percent of the total waste volume generated during filter-to-waste

Figure 10-10 A typical preoptimization filter run and ripening profile

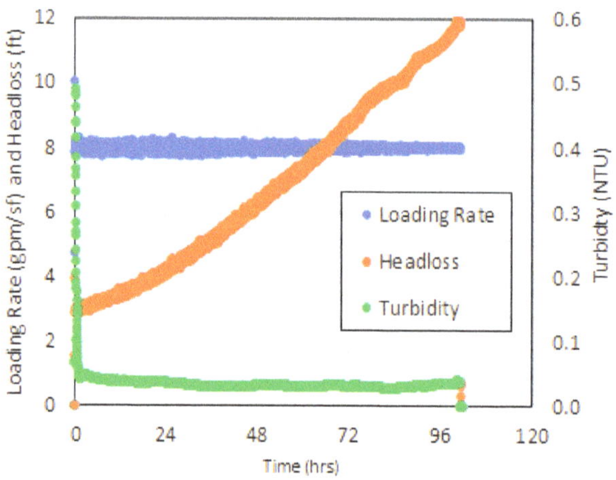

Filter Run Profile
- Terminated on head loss
- UFRV: 48,000 gal/ft2
- Backwash and filter-to-waste: 930 gal/ft2
- Filter efficiency: 98.1 percent

Figure 10-11 A typical filter run with use of DO and Prep backwashes for filters offline for extended periods

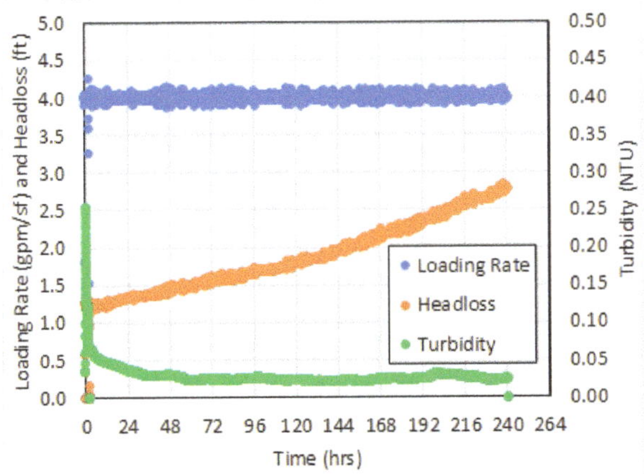

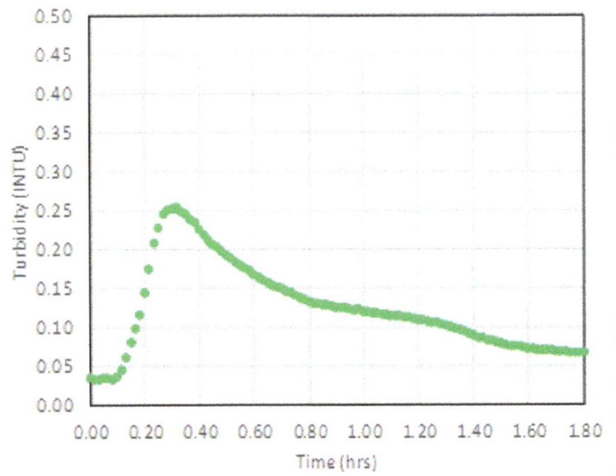

Filter Run Profile

- Terminated on time (250 h)
- UFRV: 57,000 gal/ft^2
- Backwash and filter-to-waste: 1,000 gal/ft^2
- Filter efficiency: 98.3 percent

Ripening Profile

- Peak 0.25 ntu
- More than 1 h to achieve <0.1 ntu (the increased length of time is because of the lower filter loading rate on that day)
- About 30 percent of the total waste volume created during filter-to-waste

Figure 10-12 A typical filter run and ripening profile with ETSW

The next step in the optimization process was to introduce an ETSW (Amburgey 2005) step into the backwash procedure. ETSW is an additional step after high-rate wash in which the backwash flow is reduced to a media subfluidized rate (low to no media expansion) for at least one filter-box volume (the volume within a single filter box, below the top of the backwash troughs) to evacuate all backwash remnant particles without dislodging more. This step displaces dirty water without abrading more dirt off the media grains. Odd-numbered filters used the ETSW step to allow for comparison with even-numbered filters with the ETSW step; see Figure 10-12.

Adding ETSW reduced the peak turbidity observed during ripening, and filters with ETSW showed improved performance compared with those without the ETSW step. Therefore, ETSW was implemented on all filters for subsequent optimization steps.

STEP 3: ADDING ETSW TO DO WASHES

The next iteration was to modify the DO wash using ETSW rates instead of fluidizing the bed. Because the DO and Prep backwashes are used on filters that are already clean, there is no need to fluidize media again and potentially fill the filter box with dirty water. The backwash sequence was again modified for half of the filters, and performance is summarized subsequently. This same subset of filters also initiated the use of instrumentation to inform the backwash sequence and monitor bed expansion to adjust high-rate backwash flows (targeting 20 percent bed expansion) and terminate high-rate wash on the target turbidity (45 ntu) that was selected on the basis of experience with these filters.

See Figure 10-13 for a typical filter run and ripening profile with ETSW, lower DO backwash flows, and modified high-rate fluidization and backwash waste turbidity targets.

Filter performance was moving in the right direction. The modifications to the backwash sequence improved filter efficiency but did not meet the operator's performance goals.

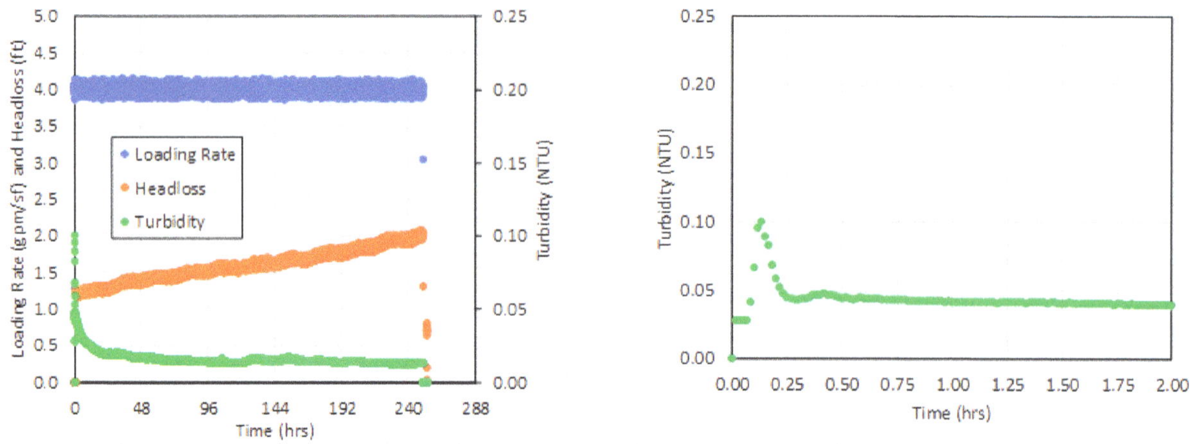

<u>Filter Run Profile</u>

- Terminated on time (250 h)
- UFRV: 60,000 gal/ft^2
- Backwash and filter-to-waste: 540 gal/ft^2
- Filter efficiency: 99.1 percent

<u>Ripening Profile</u>

- Peak 0.1 ntu
- 15 min to achieve <0.1 ntu
- About 15 percent of the total waste volume created during filter-to-waste

Figure 10-13 A typical filter run and ripening profile with ETSW, lower DO backwash flows, and modified high-rate fluidization and backwash waste turbidity targets

STEP 4: ADDING ETSW TO THE PREP WASH

Last, perform ETSW as part of the Prep wash. This iteration tested half of the filters using the ETSW as part of the backwash procedure after a run as well as part of the filter preparation step after being off-line for extended periods.

Ripening with ETSW during normal and Prep backwashes resulted in further improvement, as the subsequent graph and table illustrate.

See Figure 10-14 for a typical filter run and ripening profile with ETSW, lower DO and Prep backwash flows, and modified high-rate fluidization and backwash waste turbidity targets.

SUMMARY AND CONCLUSIONS

This implementation of the ETSW process in this case study was atypical in that it did not focus entirely on filtered water quality immediately after restarting a backwashed filter. ETSW was implemented into three separate types of backwashes, and one emphasis was on the overall water production efficiency of the filter (instead of just water quality). Backwash optimization at BWPF accomplished three intended goals:

- mitigated occasional manganese releases from biological filters;
- Increased filter efficiency from around 98.5 to greater than 99 percent, reducing total waste volume by around 30 percent; and
- maintained high UFRVs: routinely started a filter without a turbidity spike at greater than 0.1 ntu.

The ripening profiles shown subsequently in Figure 10-15 clearly illustrate the reduction in the turbidity spike following startup and the filter-to-waste (light blue) duration for each step of the optimization.

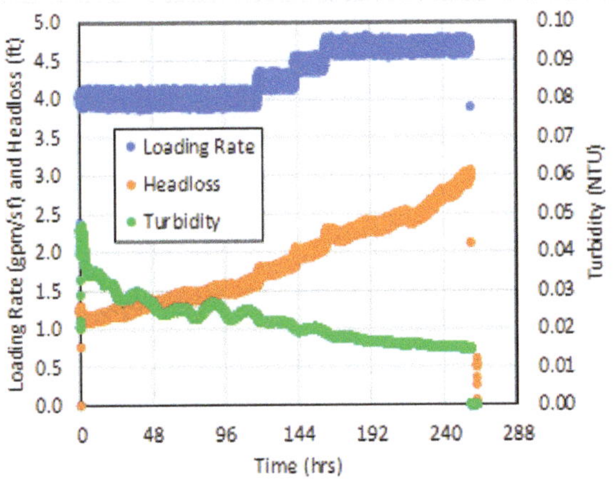

Filter Run Profile

- Terminated on time (250 h)
- UFRV: 67,000 gal/ft²
- Backwash and filter-to-waste: 572 gal/ft²
- Filter efficiency: 99.1 percent

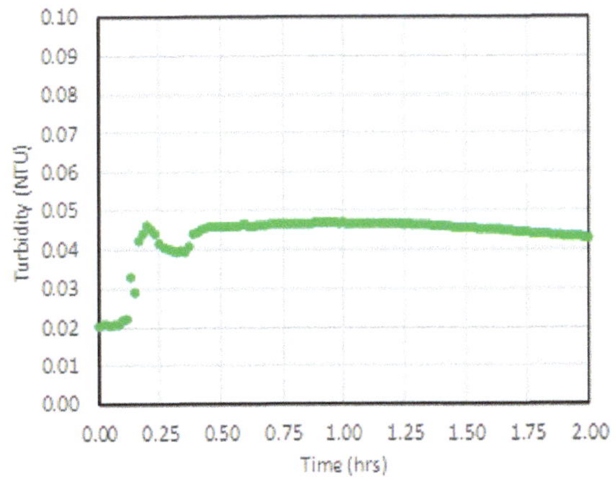

Ripening Profile

- Peak <0.1 ntu never greater than 0.1 ntu
- About 15 percent of the total waste volume created during filter-to-waste

Figure 10-14 A typical filter run and ripening profile with ETSW, lower DO and Prep backwash flows, and modified high-rate fluidization and backwash waste turbidity targets

At the completion of the optimization study, and after adjusting flow rates and durations following the approach outlined previously, the following backwashes were implemented as part of operations at the facility. The application rates are for single-cell filters, each 780 ft². Loading rates are provided in parentheses.

Backwash—occurs after a normal filter run and is normally initiated on the basis of filter run time or total filtered volume. It could also occur because of high turbidity or high head loss.

Setpoints

- Air-scour flow (scfm: 1,700 (2.2 scfm/ft²)
- Air-scour time (min): 3.0
- Low flow (with air at above rate) (gallons per minute [gpm]): 3,500 (4.5 gpm/ft²)
- Low flow time (min): 3.0
- High flow (gpm): 18,000 (23.1 gpm/ft²)
- High flow time (min): 5.0
- ETSW flow (gpm): 6,600 (8.5 gpm/ft²)
- ETSW time (min): 6.0

DO wash—used to replenish DO after a filter has been offline for 150 h or the filter DO is less than or equal to 2 mg/L.

Setpoints

- Air-scour flow (scfm): 1,700 (2.2 scfm/ft²)
- Air-scour time (min): 1.0
- Low flow (with air at the above rate) (gpm): 2,700 (3.5 gpm/ft²)
- Low flow time (min): 1.0

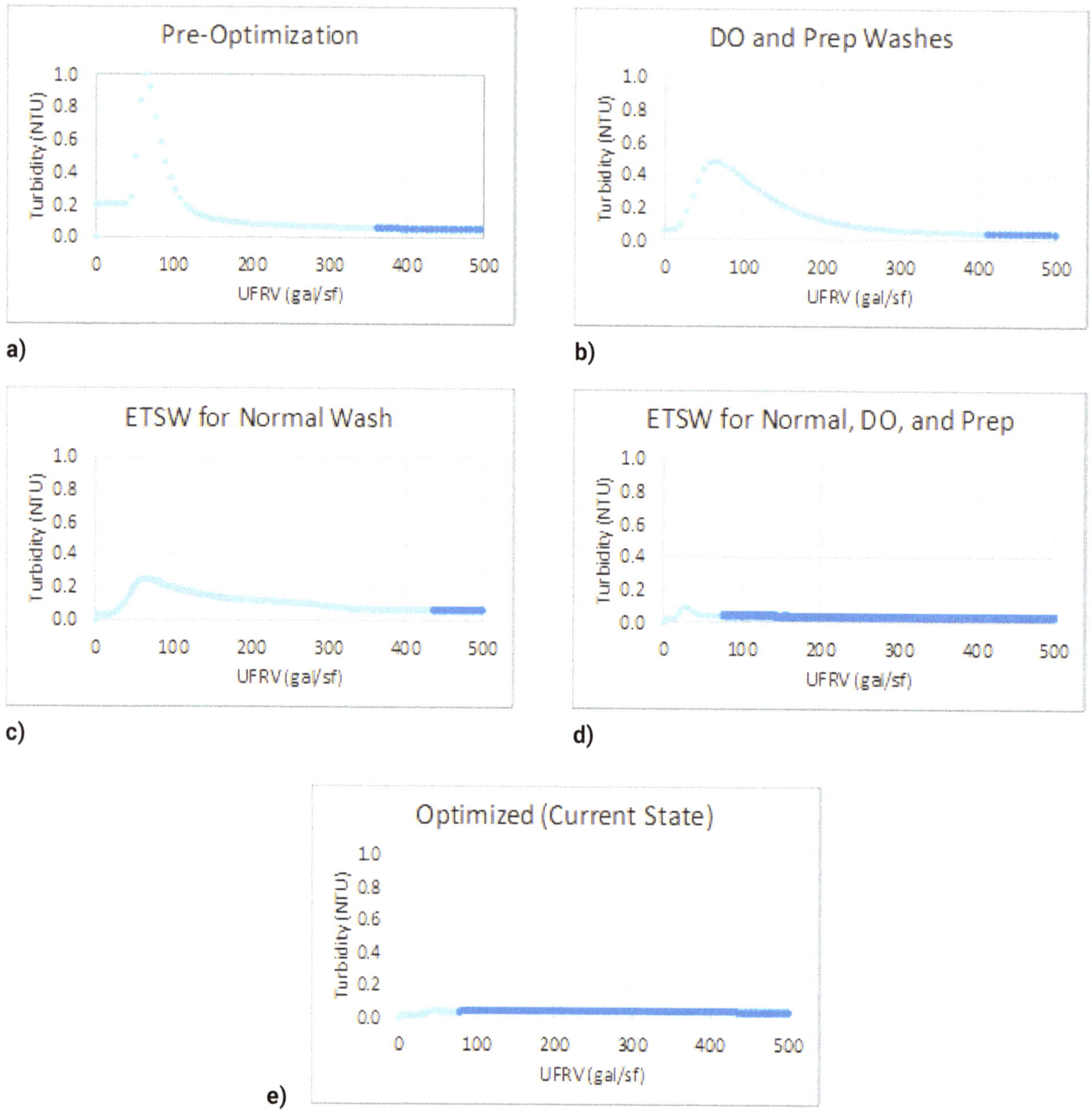

Figure 10-15 Ripening profiles: preoptimization (a); DO and Prep washes (b); ETSW for normal wash (c); ETSW for normal, DO, and Prep (d); and optimized (current state) (e)

- High flow (gpm): 17,000 (AR) (21.8 gpm/ft^2)18,000 (SP) (23.1 gpm/ft^2)
- High flow time (min): 1.0
- ETSW flow (gpm): N/A
- ETSW time (min): N/A

Prep wash—used before a filter coming on-line (within 12 h) to replenish the DO level before being placed in service.

Setpoints

- Air-scour flow (scfm): 1,700 (2.2 scfm/ft^2)
- Air-scour time (min): 1.0

- Low flow (with air at the above rate) (gpm): 2,700 (3.5 gpm/ft^2)
- Low flow time (min): 1.0
- High flow (gpm): 18,000 (23.1 gpm/ft^2)
- High flow time (min): 1.0
- ETSW flow (gpm): 6,600 (8.5 gpm/ft^2)
- ETSW time (min): 6.0

If operations staff identify areas for improvement and develop strategies to better understand the finest details of each unit process and how to address performance limiting factors, results like BWPF experienced are possible. The tenacity and dedication of BWPF operations and maintenance professionals resulted in improved biological filter performance with a well-documented, new and improved, optimized backwash protocol. The BWPF team has demonstrated that a dedication to continuous improvement results in measurable benefits in both improved water quality and reduced operational costs.

REFERENCES

Amburgey, J.E. 2005. "Optimization of the Extended Terminal Subfluidization Wash (ETSW) Filter Backwashing Procedure." *Water Research.* 39:314-330.

Amirtharajah, A., N. McNelly, G. Page, and J. McLeod. 1991. *Optimum Backwash of Dual Media Filters and GAC Filter-Adsorbers With Air Scour.* Denver: AWWA Research Foundation, AWWA.

Linder, K., and B. Martin. 2015. *Self-Assessment for Water Treatment Plant Optimization.* Denver: AWWA.

Case Study 4

Optimizing Coagulation for Recycled Water Filtration

Kristine Yung, David J. Pernitsky, Chathura Abeyrathna, Cassandra Butler, Craig Mizutani

The Central Contra Costa Sanitary District (Central San) Wastewater Treatment Plant (WWTP), in Martinez, Calif., achieves tertiary treatment through granular media filtration and disinfection to provide recycled water for landscape and industrial uses. This case study presents the results of bench and pilot-scale coagulation and filter testing to improve filter performance.

INTRODUCTION

The Central San WWTP tertiary filters are conventional sand/anthracite gravity filters. Alum is added to the UV-disinfected secondary effluent, and this coagulated water is applied to the filters without flocculation or sedimentation (i.e., an inline filtration approach). The primary coagulant, alum, is dosed in high quantities (100–200 mg/L) at the beginning of each filter run to load the filter bed with floc particles to improve particle removal, and then coagulant is only added intermittently to maintain turbidity less than 2 ntu. The filters consist of 12 in. of sand 0.4–0.6 mm in diameter and 24 in. of anthracite 0.7–1.3 mm in diameter and are typically operated at 1 gpm/ft^2. The Central San WWTP has a low solids retention time (SRT), which may affect the filtration efficiency of the secondary effluent because low-SRT plants are known to produce smaller particle sizes compared with higher-SRT activated sludge plants. Central San has a filter effluent turbidity target of 2 ntu, per California Uniform Statewide Recycling Criteria (California Code of Regulations, Title 22, Division 4, Chapter 3).

As discussed in Chapter 1, unlike membrane or cloth filters, which act exclusively by physical size exclusion, particles are removed in granular media filters by multiple forces, including electrostatic attachment of the particles to the surface of the media grains, detachment of the particles by hydraulic shear forces, and physical sieving mechanisms when the filter bed becomes heavily loaded with particles. Granular media filters require coagulation pretreatment to neutralize the charge on incoming particles so that they can approach and attach to the media grains. This is typically achieved by the addition of alum or ferric metal-salt coagulants or cationic "coagulant aid" polymers. However, in certain situations, even though incoming particles are near-neutrally charged and do attach to the filter media grains, they are not strongly attached and can be detached as a result of hydraulic shear forces. This can lead to short filter runs as a result of turbidity breakthrough. This phenomenon is most often seen for small particle sizes and at high filtration rates. In these cases, high-molecular-weight "filter aid" polymers can also be added to enlarge and/or strengthen the floc particles to improve particle removal and retention in the filter bed.

Because of the complex mechanisms present in wastewater filtration, determining the optimal coagulation conditions for inline filtration is especially difficult. Traditional jar tests rely on visual observations of floc size and settled water turbidity to aid in selecting the "best" coagulant dose. However, none of these indirect parameters provides insight into the filterability of the coagulated water—what will the filter effluent turbidity be, and how long will the run last before terminal head loss or turbidity breakthrough? Zeta potential measurements, on the other hand, provide a direct measurement of the charge of particles entering the filter, making zeta potential a possible tool for coagulation and filtration optimization.

RAPID JAR TESTING METHODOLOGY

For this study, a rapid jar test procedure was used to efficiently screen candidate coagulants and doses. Bench-scale rapid jar testing was initiated to evaluate the charge neutralization capacity of various inorganic coagulants and cationic polymers. Zeta potential measurements of the coagulated water were used as the key performance indicator for coagulation effectiveness. The rapid jar test procedure described here does not follow all the typical best practices for jar tests described in Chapter 7. For this application, there was no full-scale flocculation or clarification step before the filters, and the rapid jar test results were going to be validated with pilot-scale trials, so an abbreviated procedure that concentrated on generating floc charge data was sufficient.

Rapid jar tests were conducted as follows. Disinfected secondary effluent samples were added to a 2-L jar. For each test, coagulant chemicals (coagulant and/or polymer and/or hydrochloric acid [HCl]) were added simultaneously to the jar and subjected to 1 min of rapid mixing. After rapid mix, a sample was withdrawn from the jar with a 10-mL syringe and analyzed immediately for zeta potential using a zeta potential analysis instrument (Zetasizer Nano ZSP, Malvern Panalytical, Malvern, UK). No flocculation or settling step was included. In addition, zeta potential measurements of the full-scale WWTP secondary effluent were taken to compare bench-scale results with full-scale operation.

The exact zeta potential at which good filter performance occurs is somewhat site-specific; however, as mentioned in Chapter 1, research has shown that the zeta potential range of –5 to +5 mV is associated with good clarification and filter performance, and this range was used to evaluate coagulants.

PILOT-SCALE TESTING METHODOLOGY

To evaluate the filterability of the coagulated water, secondary effluent was passed through one of two identical 10-ft.-deep and 4-in.-diameter filter columns that were loaded with the same 12-in. of sand 0.4–0.6 mm in diameter and 24 in. of anthracite 0.7–1.3 mm in diameter as the full-scale filters. Filter loading rates of 2, 2.3, and 5 gpm/ft^2 were tested. Although higher than current full-scale filtration practices, a 5-gpm/ft^2 filter loading rate was used for most of the pilot test runs because the higher loading rate would increase hydraulic shear and exacerbate particle detachment effects, making for a worst-case test condition, and filter runs could be conducted during regular WWTP operating hours without the need for 24-h operation.

The target filtration column performance goals included the following:

- Filter effluent turbidity less than 2 ntu
- UFRV of 2,880 gal/ft^2. This assumes 24 h of operation at a filter loading rate of 2 gpm/ft^2 (this is significantly lower than would typically be achieved in drinking water treatment filtration).

Alum, ferric chloride, and aluminum chlorohydrate (ACH), as well as two classes of organic polymers, were tested. HCl was used for pH adjustment for some tests. The polymers were classified by the chemical suppliers on the basis of their intended purpose: coagulant aid polymers and filter aid polymers. The coagulant aid polymers (denoted as CA-1 through CA-3) were low- to medium-molecular-weight, high-cationic-charge polymers. Their primary purpose was to provide high concentrations of positive charge for charge neutralization and the production of near-neutral floc particles. The filter aid polymers (denoted as FA-1 and FA-2) were higher-molecular-weight, cationic polymers. The purpose of these polymers was to provide interparticle bridging between floc particles to create larger and stronger floc particles that would be retained better in the filter. In addition, two blended products were tested (denoted as Blend-1 and Blend-2).

These products contained a proprietary mixture of polyaluminum coagulant and cationic polymer. All products were dosed in accordance with the suppliers' recommendations.

RESULTS

During the test period, the secondary effluent applied to the filters was typically between 4 and 6 ntu, with a pH of 7–8, a UV absorbance at 254 nm between 0.09 and 0.19 cm^{-1}, and a temperature of 15–20°C.

Coagulant doses are reported as mg/L of dry alum (as $Al_2[SO_4]_3 \cdot 14H_2O$) for alum and mg/L of product for the proprietary blended coagulants, ACH, and the organic polymers.

The initial jar tests were aimed at optimizing coagulant doses to achieve near-neutral particle charge (–5 to +5 mV). Figure 10-16 shows the jar test results using alum, with and without pH adjustment, and ACH, along with a zeta potential measurement of a grab sample taken from the full-scale WWTP at the same alum dose. Figure 10-17 shows the jar test results for several cationic "coagulant aid" polymers.

The main conclusions drawn from the jar tests were the following:

- Very high doses of alum (approximately 180 mg/L) were required for near-neutral charge conditions. This alum dose dropped the pH to the mid-6s. There was good agreement between jar test data and plant data for the alum.

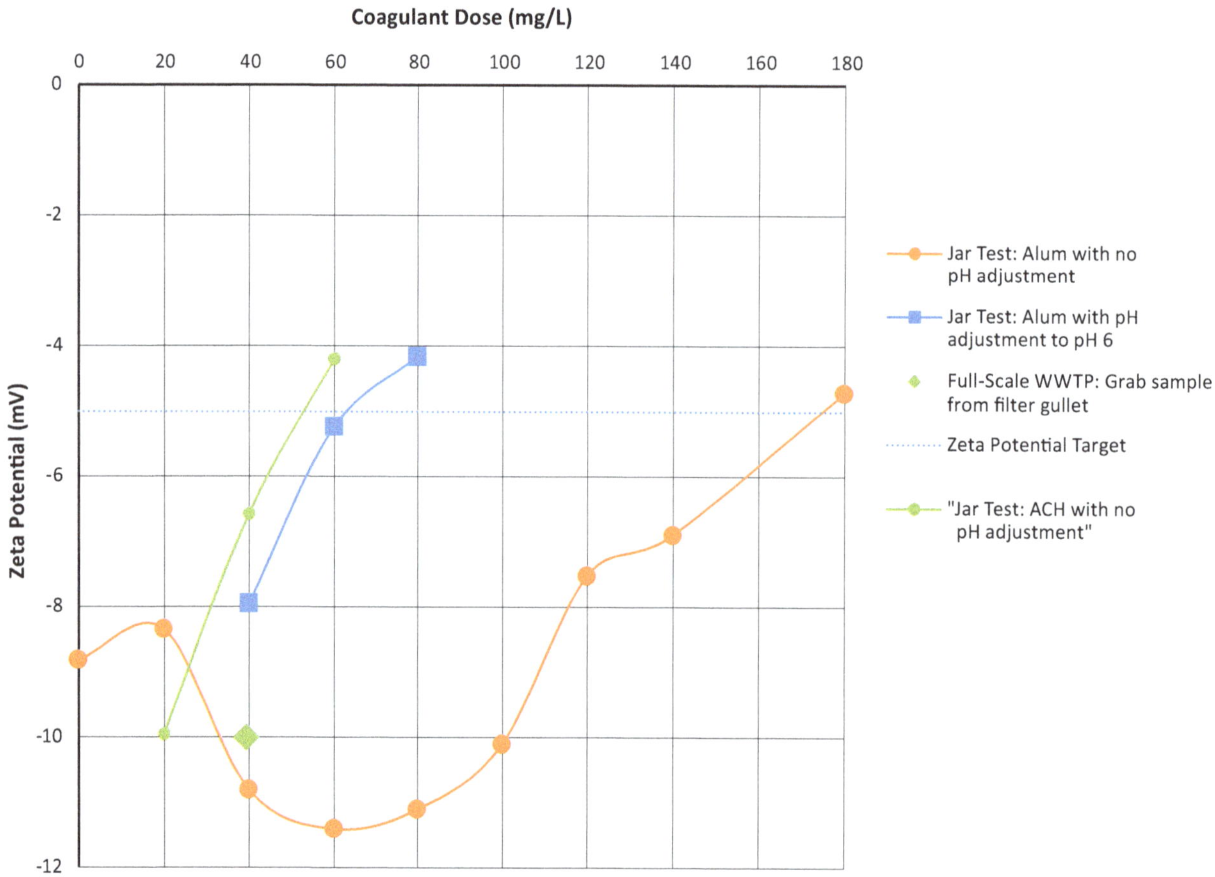

Source: Stantec

ACH—aluminum chlorohydrate, WWTP—wastewater treatment plant

Figure 10-16 Zeta potential versus coagulant dose for alum and ACH

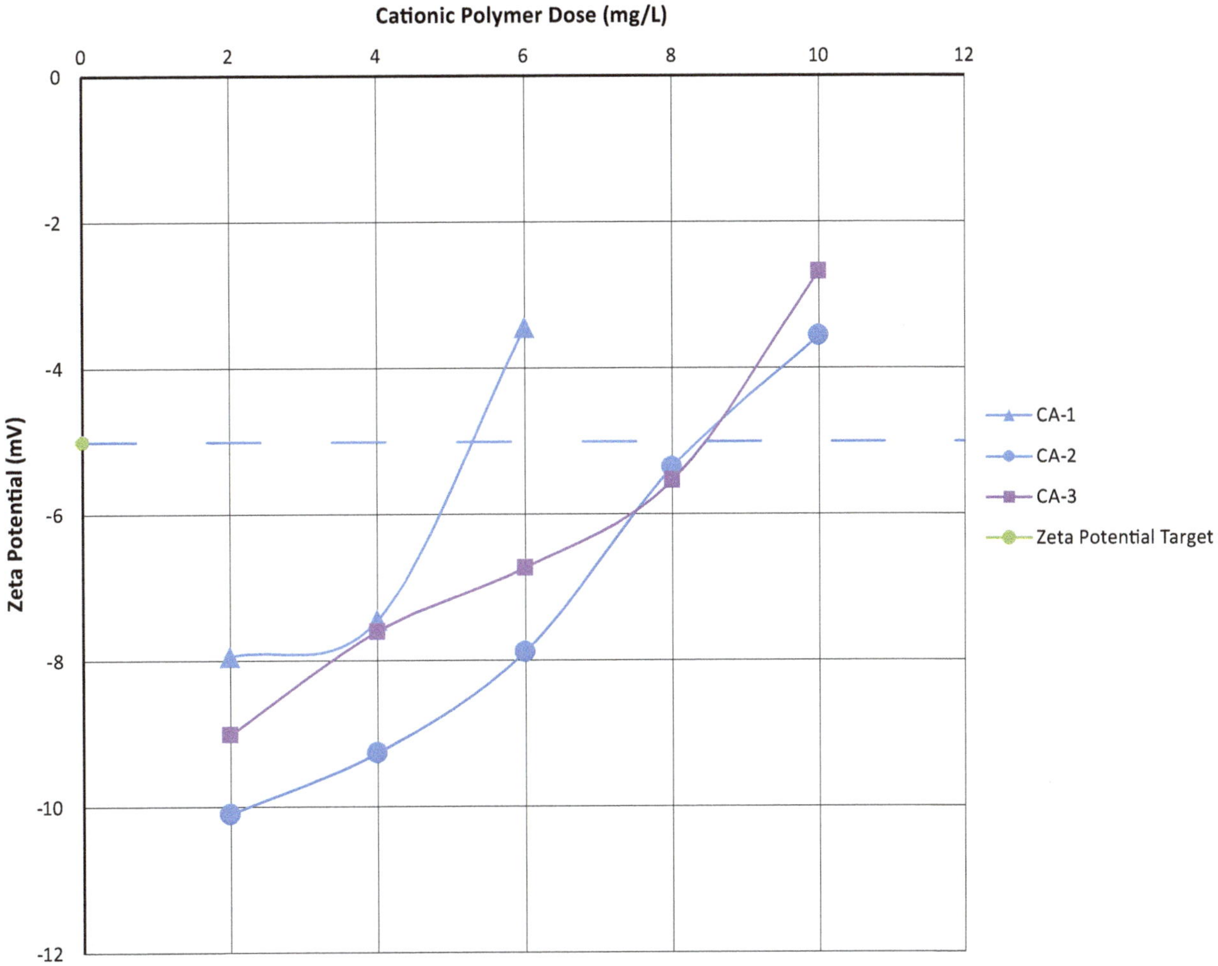

Source: Stantec

Figure 10-17 Zeta potential versus coagulant dose for cationic "coagulant aid" polymers

- Alum with pH adjustment was better at achieving near-neutral charge conditions than alum without pH adjustment; lower doses were required (approximately 70 mg/L) at pH 6.0.
- ACH was better at achieving near-neutral charge conditions than alum and was a viable option. Results showed that moderate ACH doses (approximately 60 mg/L) were required. This ACH dose dropped the pH to 7.0–7.5.
- Cationic "coagulant aid" polymer was a viable alternative for achieving near-neutral charge conditions at doses of approximately 8 mg/L. Polymer also generated significantly fewer solids than the alum or ACH.
- Ferric chloride had poor performance at achieving near-neutral charge conditions with this water (data not shown).

On the basis of these results, a series of full-scale tests was conducted with the medium-molecular-weight "coagulant aid" cationic polymer (CA-1). Over a range of doses between 10 and 20 mg/L, zeta potential ranged from –10.8 to –3.8 mV. Even at the higher doses in which the zeta potential was greater than the target of –5 mV, filter performance was poor, filter ripening was slow, and effluent turbidity was high. Because of variation in wastewater characteristics, more polymer was required in the full-scale tests than in the jar tests to achieve a similar zeta potential. These results suggested that filtration of this effluent would require more complex coagulation approaches; production of near-neutral floc particles was not sufficient to ensure good filter performance with these coagulant chemicals.

To investigate further, without affecting the operation of the full-scale WWTP, pilot columns were installed in the filter building. On the basis of the bench testing results, the following coagulants schemes were tested at pilot scale:

- alum alone (baseline performance),
- ACH alone,
- high-molecular-weight "filter aid" cationic polymer alone,
- ACH plus "coagulant aid" cationic polymer,
- ACH plus high-molecular-weight "filter aid" cationic polymer, and
- proprietary mixture of polyaluminum coagulant and cationic "blended" polymer.

On the basis of the poor full-scale performance and high doses tested, coagulant aid polymers on their own were not tested further.

Note that pH reduction with alum was not considered to be economical and was not tested further. Pilot-scale results are summarized for each coagulant combination in Table 10-1.

To illustrate the performance of various coagulant combinations, more detailed run data from several filter runs are presented subsequently. Figure 10-18 shows the results of a pilot filter run that started with a high-molecular-weight filter aid polymer (FA-2) followed by the addition of ACH halfway through the filter run. The dose callouts on the graph indicate the time a new coagulant dose was started. The filter aid polymer was not successful in achieving filter effluent turbidity reduction by itself, but soon after the addition of ACH, the filter ripened and turbidity in the filter effluent declined.

Figure 10-19 shows a filter run with an ACH dose of 25 mg/L. Notice the excessive filter ripening time to 2 ntu and high filter effluent turbidity after ripening. The filter turbidity never dropped to less than 1.7 ntu, and the UFRV was low. For comparison, Figure 10-20 shows a filter run with a dose of 18 mg/L of ACH and 0.003 mg/L of FA-2 high-molecular-weight polymer. In this case, the filter ripened rapidly to 2 ntu, and filter effluent turbidity dropped to less than 1 ntu for most of the run. The comparison between Figure 10-19 and Figure 10-20 shows the effect of the high-molecular-weight polymer in improving particle removal and increasing particle retention in the filter bed.

Table 10-1 Pilot filtration testing results

Coagulant	Coagulant Dose *mg/L*	UFRV *gal/ft²*	Average Zeta Potential of Filter Influent during Run *mV*	Average Filter Effluent Turbidity *ntu*
Alum	100	860	–10.6	0.52
ACH	25–35	854–1,952	–9 to –7	0.6–1.9
ACH plus FA-1	25 plus 10–15	639–1,797	–9.5 to –4.5	1.1–1.8
ACH plus FA-2	18–25 plus 0.001–0.01	1,495–1,816	–10.7 to –8.6	0.83–1.27
Blend-1	20–45	2,012–3,728	–3.0 to –0.22	0.53–1.4
Blend-2	20–35	1,526–1,776	Not recorded	0.91–1.4

Source: Stantec

ACH—aluminum chlorohydrate, FA—filter aid, UFRV—unit filter run volume

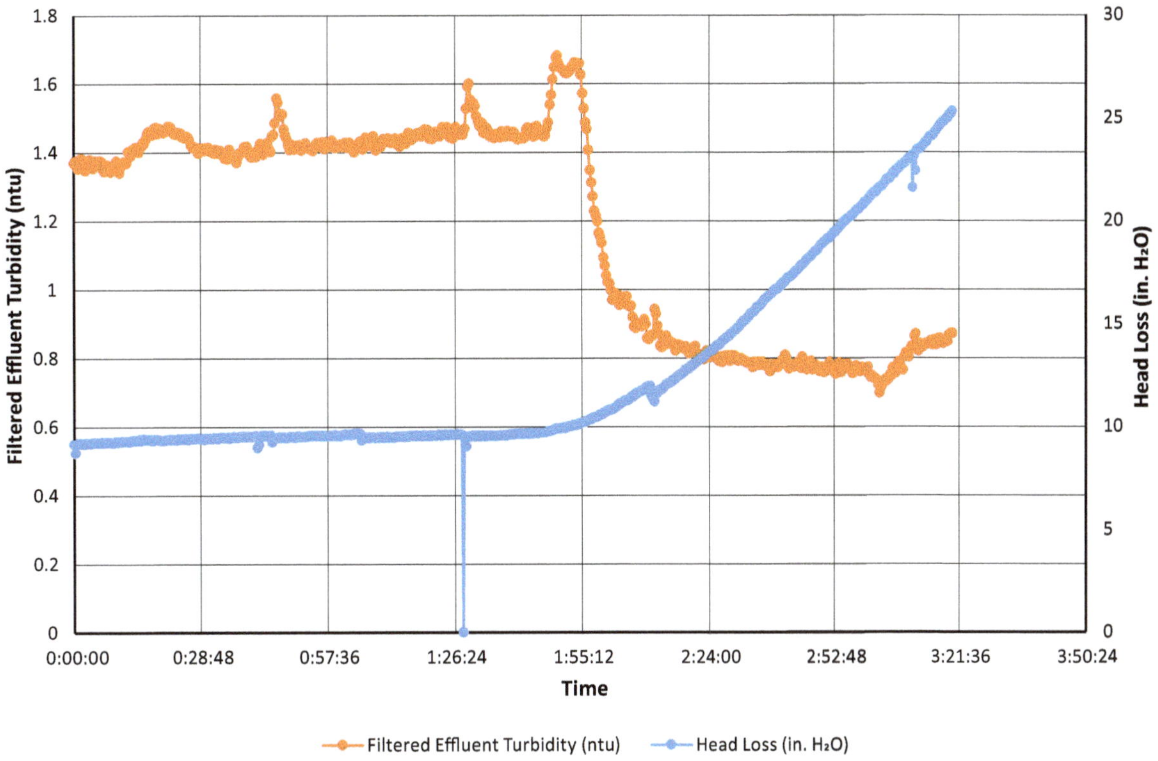

Source: Stantec

ACH—aluminum chlorohydrate, FA—filter aid

Figure 10-18 Pilot filter turbidity and head loss with high-molecular-weight polymer and later addition of ACH

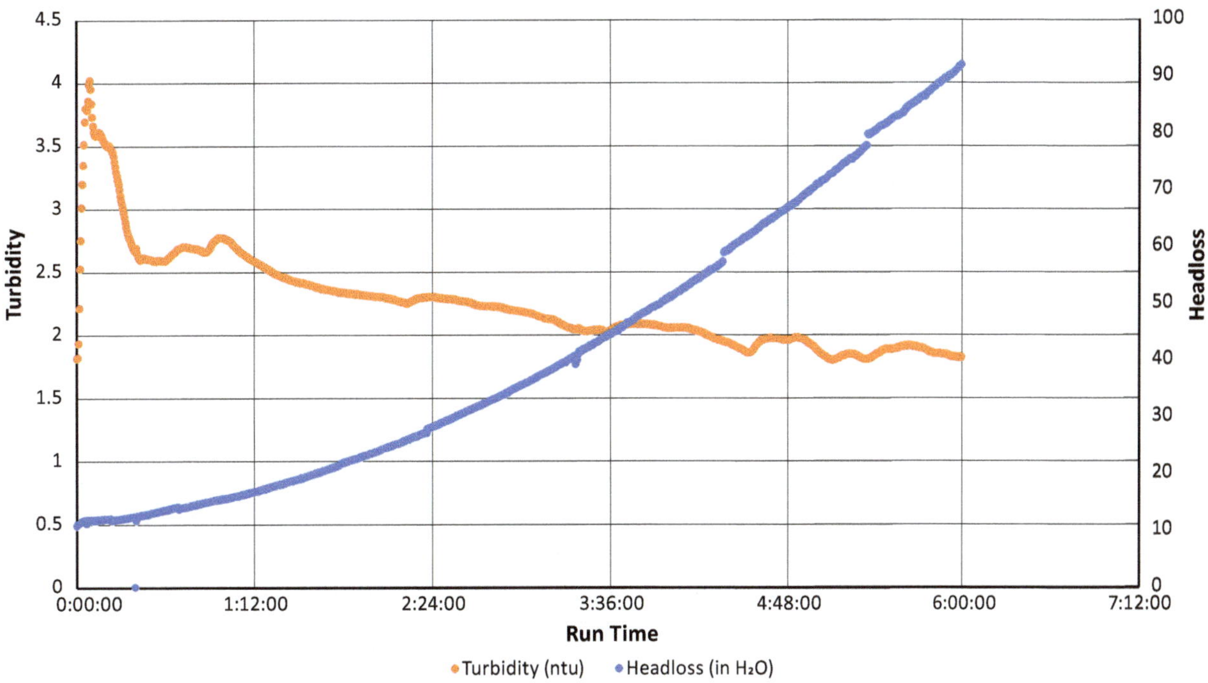

Source: Stantec

Figure 10-19 Pilot filter turbidity and head loss with 25 mg/L of ACH

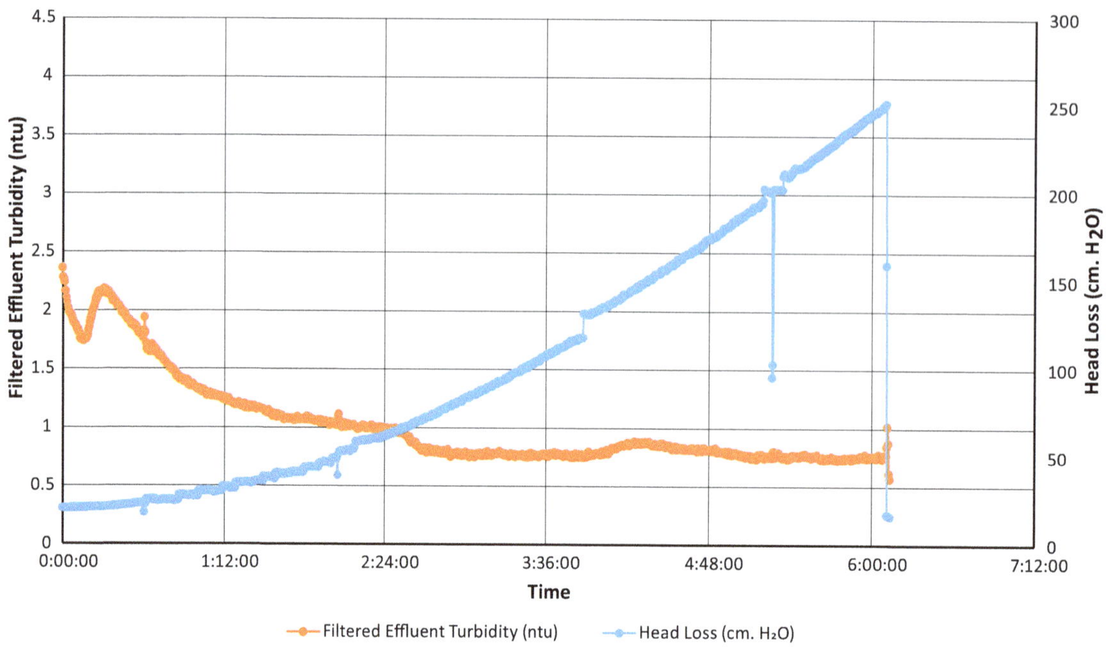

Source: Stantec
FA—filter aid

Figure 10-20 Pilot filter turbidity and head loss with 18 mg/L of ACH and 0.003 mg/L of FA-2

Figure 10-21 shows a screening exercise in which the dose of the blended coagulant Blend-1 was continually increased during a single filter run. Good, filtered water turbidity was seen with this coagulant; an increase in dose reduced the filter effluent turbidity, and an excellent correlation was observed with zeta potential.

The main conclusions drawn from the pilot-testing were as follows:

- Alum alone met the turbidity goal, but the UFRVs were very low. At the coagulant doses needed to generate near-neutral floc particles, solids loading to the filters was very high, resulting in high head loss and short run times.
- ACH alone met the turbidity goal and achieved higher UFRVs than alum, but the UFRVs were still low. At the coagulant doses needed to achieve near-neutral floc particles, solids loading to the filters was high, resulting in high head loss and short run times.
- The high-molecular-weight "filter aid" polymer (FA-2) alone was unable to meet the turbidity goal.
- The combination of ACH and "coagulant aid" cationic polymer (FA-1) showed no improvement compared with ACH alone.
- The combination of ACH and a high-molecular-weight "filter aid" polymer (FA-2) performed reasonably well. The ACH–FA-2 combination consistently achieved higher UFRVs than ACH alone, but the project UFRV goal was not met.
- The best performance was seen with Blend-1, a proprietary combined polyaluminum coagulant and organic polymer. This coagulant achieved the lowest consistent filtered water turbidity and achieved the UFRV project goal for more than 60 percent of the filter runs at 5 gpm/ft^2. When the pilot filters were run at a loading rate of 2 gpm/ft^2, the UFRV value was very near or exceeded the project UFRV goal for every trial with Blend-1. For this coagulant, the zeta potential values were correlated with filter performance.

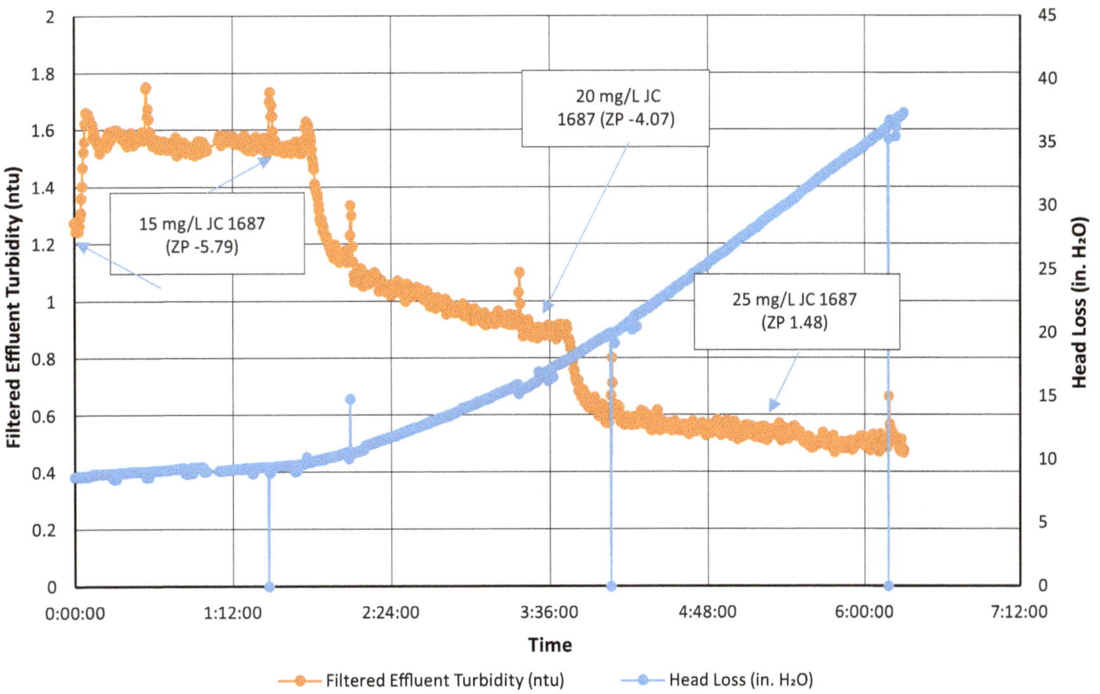

Source: Stantec

Figure 10-21 Pilot filter turbidity, head loss, and zeta potential with various doses of Blend-1

SUMMARY

The results of the bench and pilot-testing program indicated that optimization of coagulation for this wastewater filtration application required consideration of multiple coagulation mechanisms.

For the best-performing product, the proprietary polyaluminum–organic polymer product (Blend-1), the classic relationship between zeta potential and filter performance was seen. As coagulant dose increased, filter effluent turbidity decreased as particles became more neutral. Good UFRV results were seen when zeta potential measurements were within the range of –5 to +5 mV recommended in Chapter 1. Zeta potential measurements would be very useful in optimizing the dose of this product.

However, with other coagulants, the production of near-neutral particles was not necessarily sufficient to guarantee good filter performance. Primary coagulants like alum and ACH required very high doses to produce neutral particles, and although these particles were removed in the filter, the high solids loading resulted in short filter runs. When coagulant doses were decreased, head loss rates dropped, but early turbidity breakthrough then resulted. In some cases, the addition of a high-molecular-weight "filter aid" polymer with ACH reduced filter effluent turbidity and lengthened the UFRVs by suppressing turbidity breakthrough, even though zeta potential values were relatively low (–10.7 to –4.5 mV). Bench-scale zeta potential measurements would be insufficient for optimizing the dose for these products; pilot- or full-scale trials would be required.

Case Study 5

Jar Test Calibration

George Budd and Paul Hargette

Bench-scale testing (jar testing) is frequently conducted at treatment facilities to simulate coagulation conditions, often with a goal of determining optimal conditions for turbidity reduction. One problem that can occur with jar testing is that some jar test protocols are not representative of conditions in full-scale plants and are not adequate for assessing the effects that coagulation adjustments might have on full-scale turbidity reduction. When conducting jar testing, a key first step is a calibration of jar test conditions to the full-scale plant to provide a match of settled turbidity at settling times that reflect actual hydraulic loading within the plant. Procedures for using jar testing to provide this type of simulation have been described in this manual and in the literature (Hudson 1981).

An important aspect of the calibration process is the application of rapid-mix and flocculation times similar to those used at full-scale. Mixing speeds of the test apparatus are then adjusted until a reasonable match in results is achieved. Flocculation speeds are tapered as appropriate to approximate the sequence in the full-scale facility. Variation in floc formation with scale has been observed by other researchers (Clark et al. 1994). These types of differences that occur with scale of flocculation may be reflected in jar test results because mix conditions that provide for calibration at the jar test level of evaluation often produce a lower G value than is applied at full-scale, so mixing speeds may need to be adjusted accordingly.

Appropriate jar test settling times are developed on the basis of simulating hydraulic loading rates in the full-scale clarification basins. A hydraulic loading rate for conventional sedimentation basins is often in the range of 0.5 gpm/ft^2 (1.2 m/h), which equates to a settling velocity of 20 mm/min (0.8 in./min). If a standard 2-L jar test beaker is used in testing, the distance from the water surface to the sample port is 100 mm (4 in.). Given this distance, solids settling at a velocity of 20 mm/min (0.8 in./min) should be able to sink beyond the sample port within 5 min. Test protocols that allow settling times much in excess of this time will not reflect the settling conditions at most full-scale facilities, and marginal floc conditions that appear to be acceptable at significantly longer settling times will not be effective under actual full-scale operating conditions.

Table 10-2 shows a calibration testing sequence as presented by Budd et al. (2004). The calibration in these evaluations is based on achieving settled water turbidity in the jar test similar to that achieved in the full-scale facility. In this case, it was important to incorporate a simulation of the effect of prechlorination before an effective calibration was attained, as shown in the data for jar test 7 in the far-right column.

Table 10-2 Example of a calibration testing sequence

		Jar Test Identification Number						
Calibration Steps	Plant	1	2	3	4	5	6	7
Rapid mix								
Speed, rpm	—	100	100	150	150	150	150	150
Duration, s	—	180	180	180	180	180	180	180
Flocculation								
Speed (stage 1), rpm	—	40	25	40	35	35	35	35
Speed (stage 2), rpm	—	20	12	20	15	20	20	20
Speed (stage 3), rpm	—	—	—	—	—	10	10	10
Duration/stage, min	—	20	20	20	20	13	13	13
Settling time, min	—	7	7	7	7	7	7	7
Alum dosage, mg/L	—	23	22	23	23	23	23	18
Chlorine dosage, mg/L	—	—	—	—	—	—	4	4
Raw turbidity, ntu	—	32	46	32	22	20	20	20
Settled pH	6.6	6.7	6.6	6.6	6.7	6.7	6.9	6.7
Settled turbidity, ntu	2.0–2.5	7.1	13	6.4	3.8	3.6	1.4	2.5

Source: Budd et al. 2004

rpm—revolutions per minute

REFERENCES

Budd, G.C., A.F. Hess, H. Shorney-Darby, J.J. Neemann, C.M. Spencer, J.D. Bellamy, and P.H. Hargette. 2004. "Coagulation Applications for New Treatment Goals." *Journal AWWA*. 96(2):102.

Clark, M.M., R.M. Srivastava, J.S. Land, L.J. McCollum, D. Bailey, J.D. Christie, and G. Stolarik. 1994. *Selection of Mixing Processes for Coagulation*. Denver: AWWA Research Foundation and AWWA.

Hudson, H.E., Jr. 1981. *Water Clarification Processes: Practical Design and Evaluation*. New York: Van Nostrand Reinhold.

Case Study 6

Multiparameter Coagulation Optimization At Newton, N.C.

Amir Alansari, James Amburgey, and Brian Wilson

This case study presents results from a coagulation optimization investigation at a conventional treatment facility in Newton, N.C., that used a bench-scale filtration process as part of the jar test procedure instead of just settled water turbidity. The filtration step adds additional precision to coagulant dose selection that jar test settling alone cannot provide. The study's goal was to assess the treatability of the water and identify zones in which the plant could operate to reduce their chemical usage and costs while maintaining their stringent treatment goals. Table 10-3 shows the average source water quality parameters measured during this study. A jar testing procedure was used that used bench-scale granular media filters as the primary performance metric for optimization (Figure 10-22). A detailed step-by-step guide and videos of the jar test procedure are freely available online (Alansari 2020). Performance metrics included settled and filtered water turbidity, UV254, dissolved organic carbon (DOC), zeta potential, and total chemical costs ($/ mgal treated).

Jar tests results were visualized using contour plots of the performance metrics (e.g., filtered turbidity percent removal). Minimum target removals were achieved by operating within the selected optimal boundaries of a given metric (e.g., greater than 97 percent filtered turbidity removal). Multiple removal criteria were achieved for which the boundaries of different performance metrics overlapped. An optimal condition was defined as the least expensive condition that met a metric's percent removal criterion. The limits for settled water and filtered turbidity for the jar test results were set to less than 1.0 ntu (at least 84 percent removal) and 0.2 ntu (at least 97 percent removal), respectively. It is important to keep in mind that the full-scale turbidity goal of less than 0.1 ntu differed from the jar test goal of less than 0.2 ntu because the jar test filters only contained 3 in. of media and operated at greater than 5 gpm/ft^2. The jar test was intended to identify optimal coagulation conditions (dose and pH), whereas the treatment plant was intended to achieve optimal operating values under those conditions. The criteria for UV254 (at least 79 percent removal) and DOC (at least 50 percent removal) were selected on the basis of the third quartiles of the measured removals. All coagulant doses in this study are reported as mg/L as alum ($Al_2(SO_4)_3{\cdot}14H_2O$).

Table 10-3 Source water quality characteristics

Parameter	Study Average (Range)
Turbidity (ntu)	5.8 (3.0–7.1)
UV254 (1/cm)	0.043 (0.039–0.051)
DOC (mg/L as C)	0.8 (0.7–0.9)
SUVA (L/mg-m)	5.4 (4.5–6)
pH	7.1 (6.9–7.3)
Temperature ()	20.1 (19.3–21)
Alkalinity (mg/L as $CaCO_3$)	18*
Zeta potential (mV)	–12.4 (–17.1 to –4.1)

* From plant datasheet

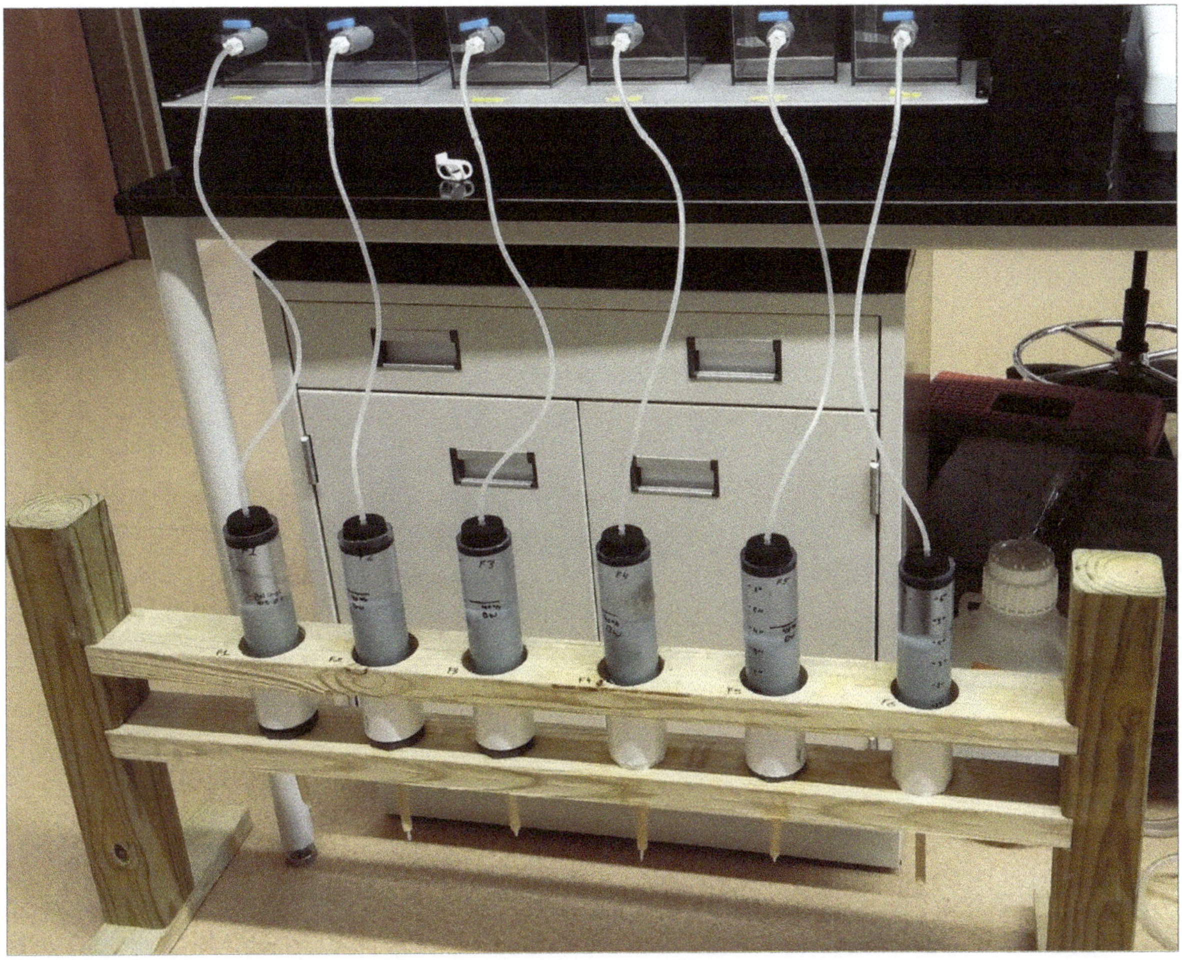

Figure 10-22 Bench-scale jar test filter stand with six identical granular media filters

ZETA POTENTIALS AND TURBIDITY REMOVALS

Zeta potential samples of the coagulated water were collected after the jar test rapid-mix stage and analyzed immediately (Figure 10-23). The data points represent actual jar test coagulation conditions (i.e., dose and pH). The alum titration curve (white dots in Figure 10-23) represents the postcoagulation pH for a series of alum doses without pH adjustment—a function of the applied coagulant dose and alkalinity for this water. The overlaid boundaries on the zeta potential contour plot represent zones of greater than 90 percent filtered turbidity removal (dashed line), optimal filtered turbidity removal (diagonal grid pattern), and optimal settled removal (grey fill). Also shown in Figure 10-23 is the pH and coagulant dose range within which the treatment plant operated during one year before the study. The grayscale intensity of the historical operating range corresponds to the frequency of the data points (i.e., darkest regions correspond to the highest frequencies).

The zeta potential of the particles was found to be predominantly negative at greater than pH 6.5 and predominantly positive at less than pH 6.5. This pattern appears to coincide with the theoretical distribution of aluminum hydrolysis species (Pernitsky and Edzwald 2006). In the optimal filtered removal region (i.e., 97 percent removal), the zeta potentials range from –13.5 to 13.1 mV with an average of –6.2 mV. Similarly, the zeta

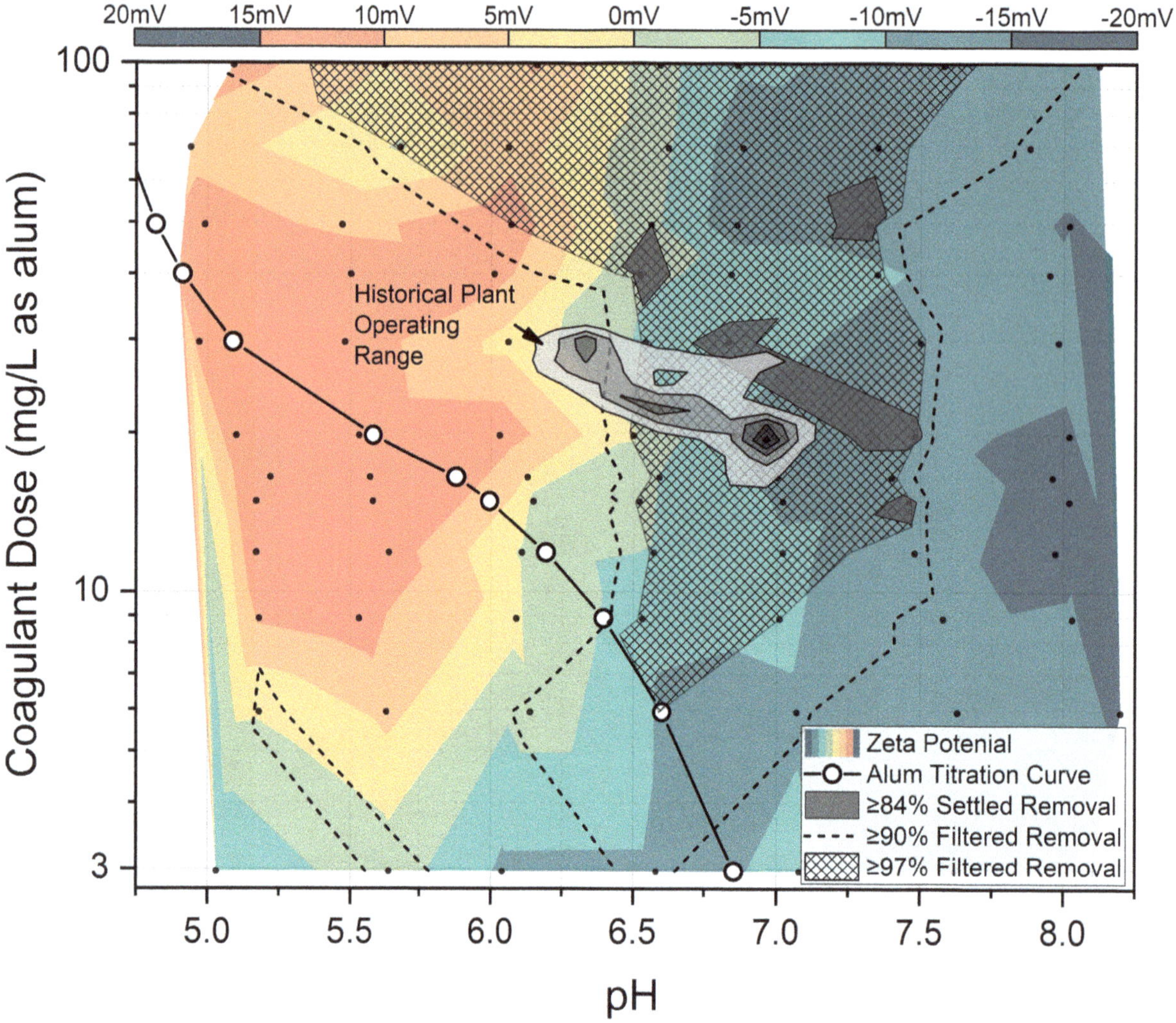

Figure 10-23 Zeta potential measurements and turbidity removals in jar tests (Newton, N.C.)

potentials in the greater than 90 percent boundary ranged from –15.1 to 13.1 mV with an average of –5.9 mV. The optimum filtered turbidity removal zone was a single broad and continuous zone that coincided with a significant proportion of the treatment plant's operating conditions, but the optimal settled water turbidity results occurred in small, scattered subsets of the optimal filtered zones. Most optimal settled water turbidities also occurred at higher dosages (typically more than 20 mg/L of alum). The possibility of missing optimal coagulation conditions and overfeeding coagulant would be considerably higher using settled water turbidity results from the jar tests, as is common practice at the time of this writing.

MULTIPARAMETER OPTIMIZATION

Figure 10-24 shows a plot of the filtered turbidity removals (in color) overlaid with the overlapping optimal removal zones for settled water turbidity, UV254, and DOC (all in grey) extracted from their respective contour plots (data not shown). Only "optimal" coagulant conditions less than 30 mg/L as alum-14 were considered in the selection process because this represented the historical worst-case (in terms of chemical costs) scenario

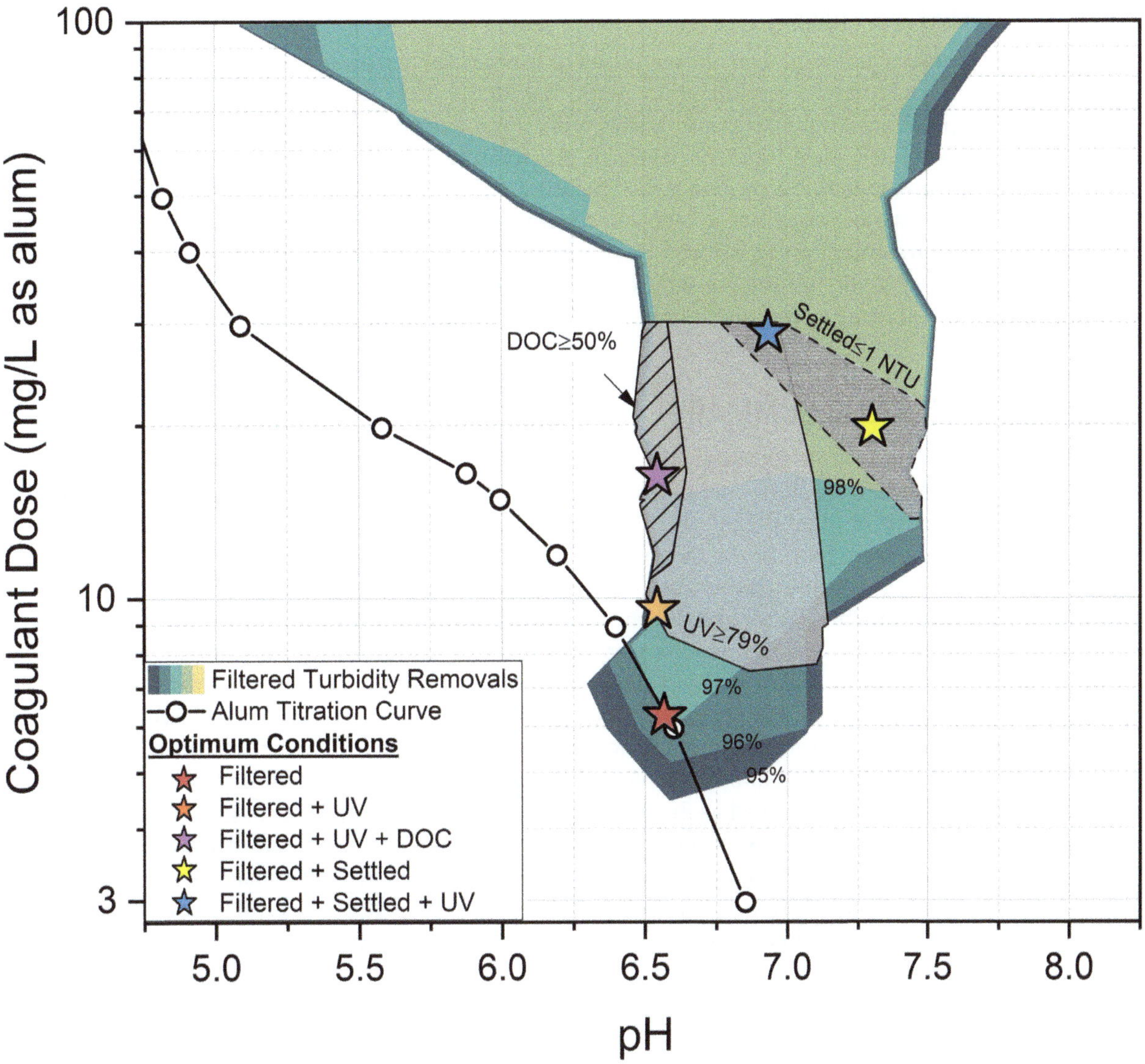

DOC—dissolved organic carbon, UV—ultraviolet

Figure 10-24 Multiple-criteria optimization alternatives with filtered jar testing

for this treatment plant. The star symbols in Figure 10-23 show the optimal coagulation conditions for up to five different optimization scenarios, beginning with filtered only (red star). The settled turbidity removal boundary did not overlap with the DOC removal boundary; therefore, no coagulation condition existed that optimized both. In other words, the treatment plant likely would have to sacrifice optimal DOC removals to achieve optimal settled turbidity removals (on the basis of these jar test results). At this treatment plant, optimizing DOC removals was not vital because the source water DOC levels were very low, and the plant did not have any issues meeting DBP regulatory limits. Table 10-4 compares the optimal coagulation conditions for different sets of optimization parameters. The chemical costs at the combined optimal filtered, UV254, and DOC (purple star) were $20 per mil gal of water treated. However, when settled turbidity removals were added as an optimization parameter, the chemical costs more than doubled with no added benefit to filtered turbidity removals.

CHEMICAL COST SAVINGS

These experimental results were used in an open discussion and mutual learning opportunity on coagulation optimization with the staff. The treatment plant's preoptimization operational range was compared with their postoptimization range (one year after the study) in Figure 10-25. On average, the treatment plant reduced their coagulant dose by approximately 50 percent after the study. This was achieved without any effect on the full-scale performance in terms of settled and filtered water turbidity removals or UFRVs. The treatment plant reduced their annual chemical spending by approximately 25 percent (~$18,000) by optimizing their coagulation conditions on the basis of the filtered turbidity jar test results presented here.

FILTERED JAR TESTING SUMMARY

The bench-scale filtration step adds additional precision to coagulant dose selection that jar test settling alone cannot provide. The benefits of optimizing coagulation conditions on the basis of filtered water turbidity instead of settled water turbidity as determined by jar testing were found to be threefold: filtered turbidity removals were generally independent of settled water turbidity removals, which could lead to both false optimums and missed optimums by sedimentation; the possibility of overfeeding the coagulant on the basis of jar test sedimentation results was minimized; and the jar test results with filtration were immediately usable without needing to "calibrate" the jar test procedure to the facility design as is typically recommended for sedimentation-based jar tests. However, the jar test filters cannot completely replicate the performance of a full-scale filter with respect to filter ripening, early breakthrough, and run length (UFRV), so like all jar tests, the bench-scale results should be used as the basis for confirmatory full-scale testing.

Although it was not necessary to further optimize the removal of dissolved organics for DBP control at this facility, multiparameter optimization was shown to be possible according to the previous filtered jar test results. Overall, chemicals savings of approximately 25 percent annually in this study made the jar test optimization case study a long-term benefit for this facility.

Table 10-4 Performance versus cost comparisons for multiple optimums

Optimization Criteria	pH	Coagulant Dose*	Filtered Removal %	Settled Removal %	UV Removal %	DOC Removal %	Cost/mgal $
Filtered	6.6	6	97	66	77	45	9
Filtered + UV	6.5	10	98	76	79	49	15
Filtered + DOC	6.6	12	98	74	81	50	20
Filtered + UV + DOC	6.6	12	98	74	81	50	20
Filtered + settled	7.3	20	98	84	61	35	43
Filtered + settled + UV	6.9	29	98	84	80	42	56

DOC—dissolved organic carbon, UV—ultraviolet

* mg/L as alum

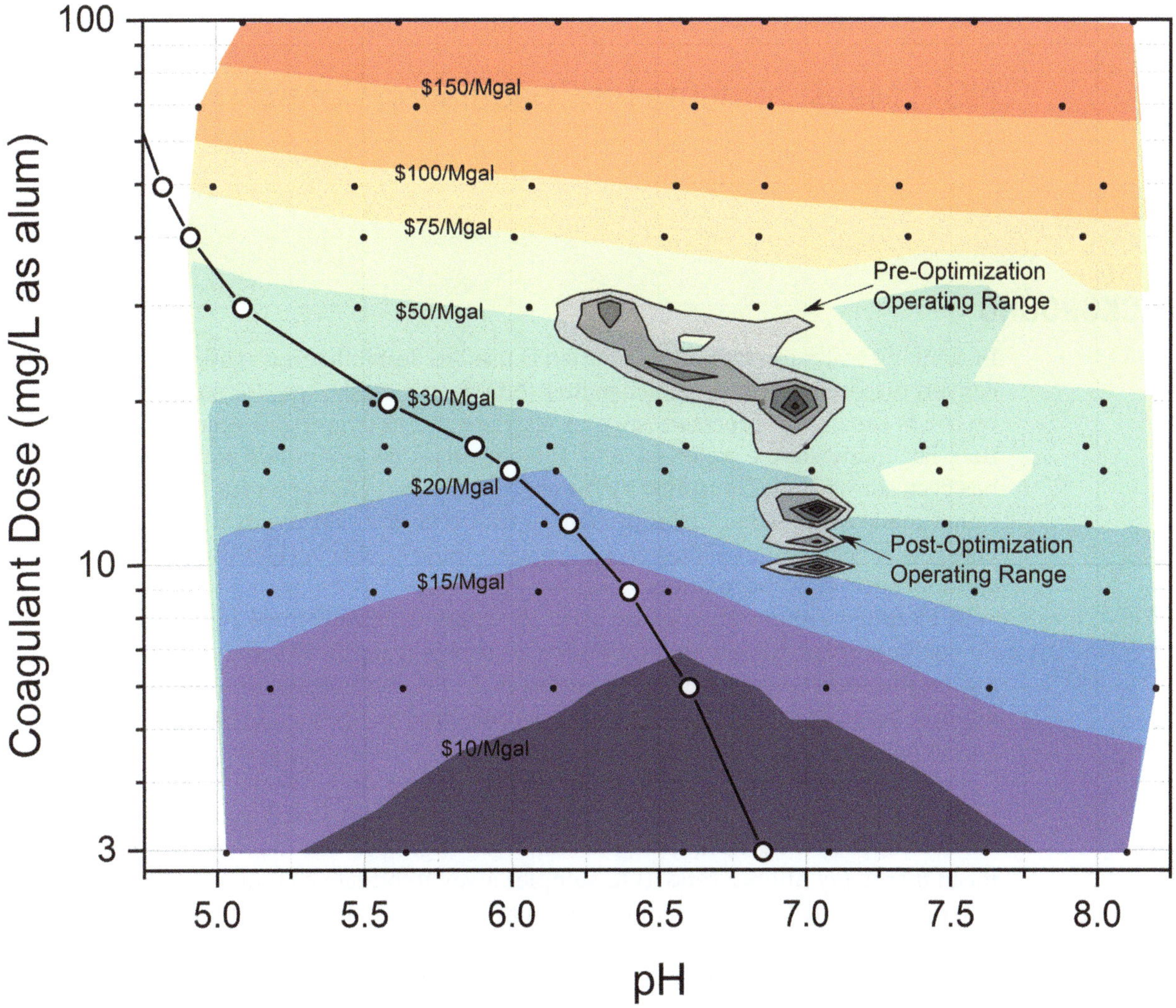

Figure 10-25 Comparing full-scale chemical costs before and after optimization study

REFERENCES

Alansari, A. 2020. "The Next-Generation Jar Test Procedure." www.amiralansari.com/jartest (accessed August 7, 2025).

Pernitsky, D.J., and J.K. Edzwald. 2006. "Selection of Alum and Polyaluminum Coagulants: Principles and Applications." *Journal of Water Supply: Research Technology–AQUA.* 55(2):121-141.

Case Study 7

NOM Measurements for Coagulation Control

Tom Elford and David J. Pernitsky

BACKGROUND

In drinking water treatment, coagulation is used to destabilize suspended particles and to convert dissolved natural organic matter (NOM) to a solid phase for subsequent removal by solids-separation processes such as clarification, flotation, and granular media filtration. For many source water sources, the coagulant dose required for proper treatment has been shown to be controlled by the concentration of NOM present in the source water rather than source water turbidity. This is true even if NOM removal is not the primary objective of coagulation. This makes measurement of NOM concentration a critical parameter for controlling coagulation processes.

Despite its importance, many plant operators, especially those with smaller systems, do not have access to current information on source water NOM concentrations. Depending on the source water source, increases in NOM concentration can occur without an increase in turbidity. When this occurs, operators may not be aware of the resulting coagulant underdose until filtered and/or clarified water turbidity begins to increase.

TOC measurements are typically used in the water treatment industry to quantify NOM concentrations. Although high-quality, reliable on-line and bench-top instruments suitable for WTP use are available, their relatively high costs and maintenance requirements mean that many utilities send TOC samples away to off-site commercial labs. The delays inherent with off-site analysis limit the use of this information for making operational decisions.

UV254 has also been shown to be an excellent measure of NOM concentration in source waters, and the analytical procedure is simple, fast, and inexpensive. The exact relationship between UV absorbance and NOM concentration is unique for each source water source, however. For a given source water source, increases in UV absorbance indicate increasing NOM concentrations and increasing coagulant demands.

This case study describes the use of UV absorbance and TOC measurements to aid coagulant dose determinations at the city of Calgary Glenmore WTP.

GLENMORE WTP

The Glenmore WTP is one of two plants serving the city of Calgary, which is located approximately 100 km (60 mi) east of the Rocky Mountains, in Alberta, Canada. The plant is a 350-ML/d (92-mgd) conventional treatment plant treating water from the Glenmore Reservoir, which is an inline impoundment on the Elbow River.

Turbidity levels are typically low throughout the year, although two runoff events in the spring provide challenging treatment conditions: a lowland runoff and a mountain runoff. The lowland runoff consists of a small turbidity spike but a relatively large increase in TOC because snowmelt carries NOM from forested and agricultural lands. This TOC increase results in a large increase in the coagulant dose at the WTP. The mountain runoff occurs later, once the snowpack in the mountains melts. This runoff event results in high flows and high turbidity because of the erosion of silts in the mountains.

Several bench-top and on-line instruments are used at the Glenmore WTP to monitor source water quality, as shown in Table 10-5. The plant coagulant dose is selected on the basis of jar tests conducted by plant staff. The dose is verified by monitoring settled and filtered water turbidity. The filters are operated to maintain a filtered water turbidity of less than 0.1 ntu.

Table 10-5 Analytical instrumentation used at the Glenmore WTP for monitoring source water quality

Parameter	Instrument Type	Instrument Model
Source water UV absorbance	Bench-top	Hach DR 4000
Source water TOC	On-line	Hach TOC 1950 plus
Source water turbidity	On-line	Hach 1720D (low range) and Hach Surface Scatter 6 (high range)

Source: Tom Elford

TOC—total organic carbon, UV—ultraviolet

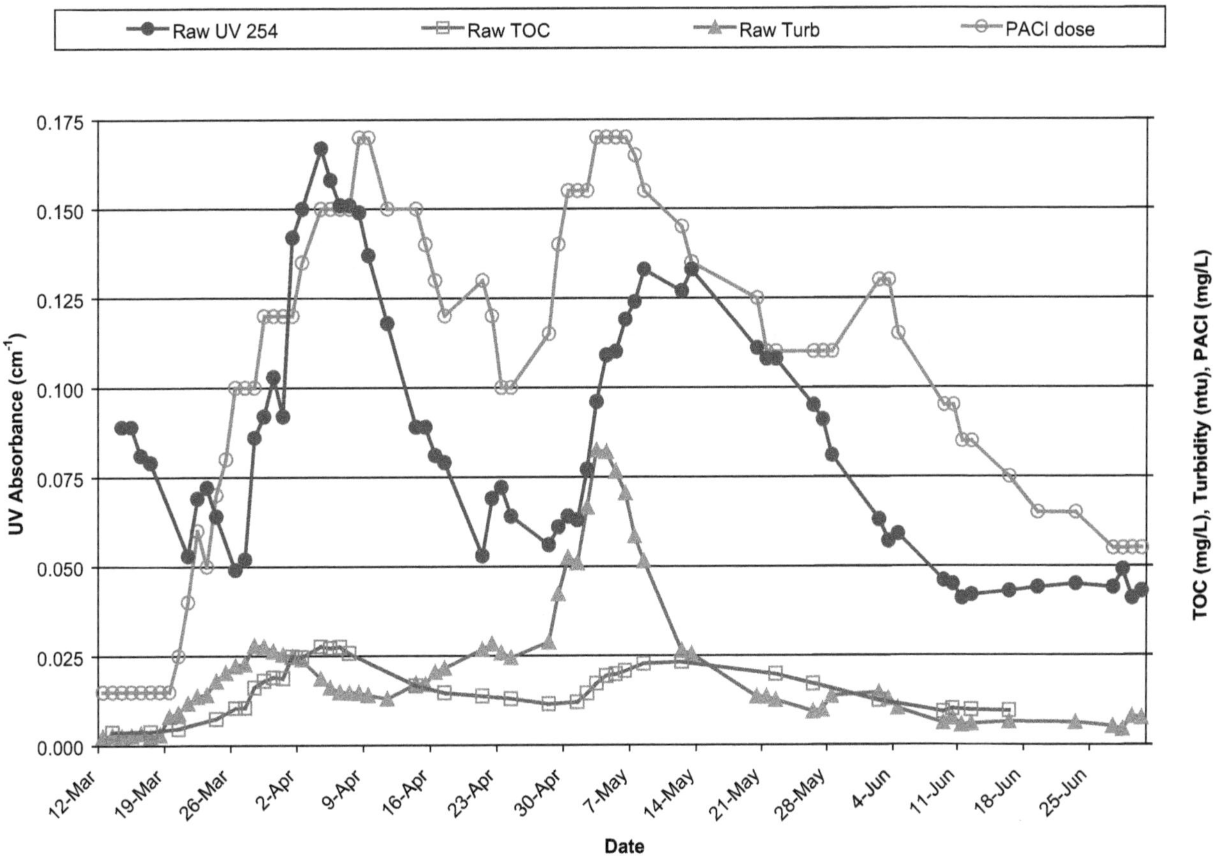

Source: Tom Elford

PACl—polyaluminum chloride, TOC—total organic carbon, Turb—turbidity, UV—ultraviolet

Figure 10-26 Relationship between source water quality parameters and coagulant dose for the Glenmore WTP

Comparison of Source Water NOM Concentration to Coagulant Dose

To investigate the relationship between source water quality and coagulant dose, historical plant records were analyzed, and daily source water turbidity, TOC, and UV absorbance data were plotted against the actual coagulant dose used in the plant for a period of several months in 2003.

Data from a typical spring runoff period are shown in Figure 10-26. As described previously, a clear increase in source water TOC and UV absorbance can be seen in April, corresponding to the lowland runoff, followed by an increase in TOC, UV absorbance, and turbidity during the May mountain runoff. Although operators were using changes in source water turbidity as the primary trigger for increasing coagulant dose at the time these data were collected, the figure shows the close relationship between source water UV absorbance and TOC and the actual coagulant dose used in the plant. The increased coagulant doses required for proper treatment plant performance (filtered water turbidity was maintained at less than 0.1 ntu during this entire period) can be seen to be related to the elevated NOM concentrations in the source water. The April data also illustrate that an increase in coagulant demand can occur in the absence of a large turbidity increase (the peak turbidity during the April runoff period was only 5 ntu).

CONCLUSIONS

From an operational perspective, UV absorbance and TOC measurements have proved to be useful tools for predicting changes in source water quality at the Glenmore WTP. Increases in source water UV absorbance and TOC provide operators with an early warning of NOM increases, which enable operations staff to anticipate changes in coagulant dose rather than responding to treated water turbidity excursions after the fact. Furthermore, Glenmore operations staff has observed that when there is an increase in source water NOM, a slightly higher coagulant dose is needed in the plant to maintain filtered water turbidity than the dose predicted from the settled turbidity results from jar tests. Again, source water TOC and UV absorbance measurements allow the plant operators to anticipate this situation and to better determine the optimal coagulant dose for the plant.

Case Study 8

Net Charge Equals Positive Change

David Teasdale

In February 2005, Canada's largest low-pressure enhanced coagulation membrane WTP began producing water. Located in Kamloops, B.C., the plant supplies up to 42 mgd (160 ML/d) of drinking water to more than 80,000 residents. Before the plant was completed, pilot studies were done to examine the use of streaming current monitors (SCMs) for optimizing water quality. The results validated the technology's implementation in a distinct and innovative arrangement. Because an SCM was incorporated in the treatment process in May 2006, a 30 percent increase in membrane performance was noted over the following calendar year. As of December 2008, baseline coagulant dosing has seen a quantifiable reduction of 25 percent overall.

SOURCE WATER

The South Thompson River, the treatment plant's source water source, is typical of many rivers in the British Columbia interior. It has relatively low alkalinity, low TOC–DOC, fairly consistent pH, with potential for short-term turbidity spikes primarily during events of spring runoff (up to 600 ntu). Like most surface water sources in Canada, the river undergoes substantial temperature fluctuations during a calendar year. Inorganic clay makes up the largest proportion of the river's particulate matter.

Traditionally, coagulation has been used to achieve particle destabilization, and most often feed rates are based on source water turbidity levels. As a rule, the higher the turbidity, the higher the coagulant dosage required for effective particle removal and destabilization. At the Kamloops plant, a primary consideration for full-scale operation was to examine whether an on-line dosing strategy could be developed that would compensate for fluctuating levels of turbidity and NOM in the source water—that is, whether it would be possible to develop some sort of on-line coagulant feed system that automatically compensated for variables that can be seen (turbidity) and variables that cannot be seen (NOM).

NOTE: *Reprinted from Opflow, July 2007.*

Industry Changes

Municipalities treating source waters high in NOM have often needed to use coagulant dosages that exceeded the amounts normally required for particle charge reduction. NOM usually indicates the presence of decaying vegetation in the raw source water. With NOM present, turbidity may have little correlation to effective coagulant dosing, and the percentage of NOM removal often becomes the performance benchmark toward determining adequate coagulant feed rates.

WTPs across North America often use chemical coagulants without a way to quantify proper dosing levels. Other frequently overlooked factors include mixing kinetics and chemical reactions involved in the coagulation process. An on-line SCM was incorporated at the Kamloops Centre for Water Quality to help quantify proper coagulant dosing limits.

How It Works

Over the years, instrument advances similar to zeta potential or electrophoretic mobility measurements have emerged in the water treatment industry. The most common is the SCM, which uses an electric sensor to determine when charge neutralization has been reached in a suspension. The theory is similar to the principle behind a zeta meter in that it is a charge-measuring device that has a measurement which produces a relative value. The device measures the net ionic and surface charges of colloids in suspension between two electrodes. A piston moves the water back and forth in the chamber, and positive and negative charges are moved downstream to the electrodes where they are measured and generate a streaming current value (SCV). The streaming current amplitude and polarity are a function of the sampling location and the type of coagulant used.

Previous piloting work at the plant had identified the "minimum" coagulant feed rate for process optimization in average source water conditions to be 3 mg/L of ACH, the coagulant of choice. That dosage resulted in the best overall membrane performance and optimized organics reduction. Two peristaltic pumps were installed to achieve this optimal dosing level. A flow-based coagulant pump feeds 2 mg/L of ACH into the plant's incoming source water, and a second peristaltic pump runs exclusively from a signal generated by the SCM and acts in a "top up and trim" fashion. A target streaming current set point is chosen that equates to adding a 1-mg/L dosage of additional coagulant. This system is depicted in the diagram in Figure 10-27. The "top up and trim" concept is shown in Figure 10-28. Using the aforementioned example, the sum of coagulant, added under ideal water conditions—a key consideration—is 3 mg/L, or the optimal dosing value determined from pilot studies.

This setup allows operators to respond to source water fluctuations in two ways. Adjustments to coagulant feed rates can be made when the SCM-controlled coagulant pump starts to approach maximum output based on a preset SCV. This would result from either a visible change in source water quality, such as a turbidity spike, or an invisible influence, such as an influx of dissolved organics after a heavy rainstorm. The SCM continuously analyzes changes in source water, whereas the base coagulant dosing pump maintains a feed rate slightly less than that required for optimal plant performance. Therefore, influxes of organics or invisible influences can be detected and compensated

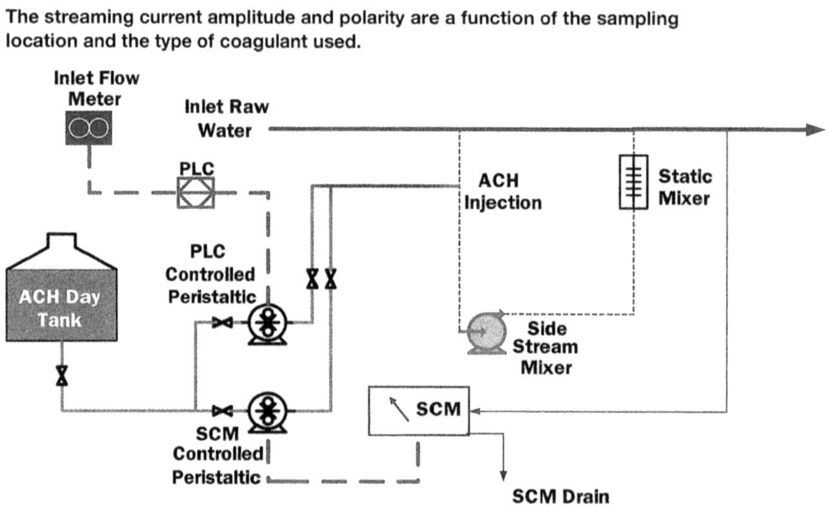

ACH—aluminum chlorohydrate, PLC—programmable logic controller

Figure 10-27 Kamloops SCM configuration

A target streaming current value set point is chosen that equates to a "top up and trim" dose of 1 mg/L of additional coagulant being added.

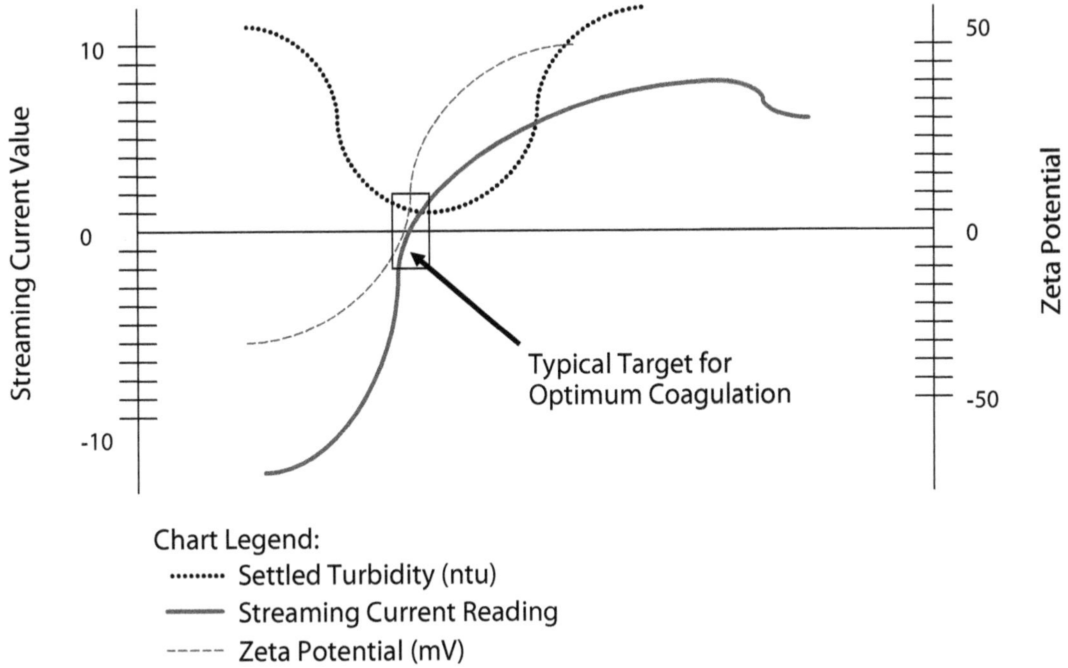

Figure 10-28 Optimal plant performance

for through an automatic increase in coagulant feed. Without an SCM in place, invisible influences are more likely to go undetected.

The other obvious benefit is that if either pump should stop operating, there would be a second unit running at all times. If the base-level pump would stop, the SCM would detect it and automatically compensate its feed rate accordingly. In the event the monitor unit failed, there would always be a minimum base coagulant dosage being fed into the source water stream—although less than ideal. Alarm set points on the SCM unit would indicate a problem with the coagulant feed system within minutes. Often in many larger plants, coagulant feed systems can malfunction for several hours before detection.

Unit Optimization

Choosing the "optimal" operating condition for an SCM can be complicated. The unit produces an on-screen value—usually negative, depending on where the unit is measuring—that corresponds with the level of particulate charge present, the SCV. The larger or more positive the value, the more coagulant or positively charged ions are present in the sample. One of the major operational challenges is determining the proper value or setpoint. The objective is to find the optimal SCV that corresponds with optimal organics removal, particle destabilization, and plant performance. SCM values are not typical of most on-line instruments because they must be calibrated to a baseline measurement that will vary from one source water source to another and from one treatment facility to another. Seasonal water changes may also produce variances in the "target" set point. Jar tests can help operators determine optimal base-level coagulant dosages when significant changes to source water conditions are encountered.

Other factors that need to be monitored include unit flow rates and piston wear. Both are imperative to the SCM unit operation and the determination of a baseline SCV. When source waters are prone to variable conditions, it may be prudent to consider a unit that is equipped with an automatic flushing mechanism.

Although SCMs are effective and simple in principle, the factors that influence their operation can be complex. This is especially true when operating conditions change or dramatic changes to source water quality are noted. It is important for operators to understand how such confounding variables may influence the unit's operation, and these limitations should be considered when determining whether an SCM is appropriate for a particular plant.

REFERENCE

Teasdale, D. 2007. "Net Charge Equals Positive Change" *Opflow.* Denver, CO. AWWA

Case Study 9

Streaming Current Monitor Pilot Study: The Detection of a Ferric Chloride Feed Failure

Michael Sadar

INTRODUCTION

In September 2007, an SCM was placed at a local surface WTP for field evaluation. The chief operator at this facility was considering the evaluation of SCM technologies for the monitoring and confirmation of coagulant feed.

The drinking water facility practices conventional treatment for surface water with a production capacity of between 15 mgd (57 ML/d) in winter and 35 mgd (132 ML/d) in summer. Treatment includes chemical injection to adjust pH and the addition of ferric chloride as the primary coagulant. After sedimentation, the water is filtered through dual-media anthracite filtration, followed by disinfection.

The source water originates in the central Rocky Mountains and is primarily from snowmelt. This surface water flows approximately 60 mi (about 100 km) before being captured in an impoundment of approximately 1,000 surface acres (400 ha) that is located in an urban area approximately 3 mi (5 km) from the plant. The reservoir is considered to be a very stable source water source but is susceptible to flashing from the rare strong thunderstorm.

Three SCM instruments from various manufacturers were installed in the plant to continuously sample the flash mix effluent. SCM signals were logged directly into the plant SCADA system, but because this was only a trial, the SCM data were visually monitored only and were not wired to any alarms within the SCADA system. The SCM readings were displayed for each of the three instruments, with the values updated each minute. Every 24 h, the staff would clean each SCM measurement cell and re-zero the instrument. Through the re-zeroing of the instrument, the operators can, at a glance, determine if the SCM value is in its expected range. This value is expected to remain close to zero because of the typical stability of the source water. If the value deviated strongly negative, it is a potential signal that the coagulant feed was lower than the previous optimum, or if it was to deviate positively, that the coagulant feed was higher than optimum.

Initial Operation

The first three months involved the installation of the sensors, understanding the features of each sensor, and getting used to looking at SCM data as a process performance tool. This was also the time for the plant operators to become familiar with the technology and begin the process of understanding and using the information in their operation practices.

When integrating any new on-line monitor into plant operations, patience is required. The operators must first understand what the measurement is and how it can be useful in their day-to-day operations. They must understand the critical functions, operational protocols, and maintenance requirements of the instrument to ensure the generated data are reliable. Once the data are proved to be reliable, operators must learn how different changes in the source water and chemical dosages in the treatment processes can influence the response of each of these instruments. Having learned this, they can also learn how to use the SCM information to better monitor their treatment processes, optimize their processes, and troubleshoot unexpected events. Over the first three months of the SCM trial, the plant staff learned to understand streaming current technology, its respective strengths and limitations, the maintenance requirements of the instruments, and how to adjust coagulant doses in response to changes in streaming current output.

Detection of Chemical Feed Failure

Overall, the integration of the SCMs went well. However, it was not until the staff had an unexpected chemical feed failure that was detected by the SCMs that they fully embraced the use of SCM technology in plant operation.

On February 1, at approximately 1900 h, an abrupt increase in flash mix pH was seen, as shown in Figure 10-29. There was no change in the source water turbidity or the source water pH at this time. The operator initially suspected an overfeed of lime. It was not until the operator keyed on the SCM values and examined the SCM trend that he suspected a coagulant feed problem. The coagulant feed pumps were checked, and it was found that one of the coagulant feed pumps had partially lost its siphon, reducing the ferric chloride dose by approximately 50 percent. After the cause was identified and remedied, the operators fully bought into SCM technology as a valuable process-monitoring tool.

The value of streaming current technology can be seen by examining the data in Figures 10-29 and 10-30 in more detail. In addition to the immediate decrease in the SCM reading after pump failure, Figure 10-29 also shows the eventual effect this had on the settled water turbidity. Before the partial feed failure, the turbidity of the settled water was about 1.2 ntu. After the coagulant failure, the resulting settled water turbidity increased to 2 ntu. The peak in the settled water turbidity was seen approximately 3 h after the identification and correction of the coagulant feed failure, emphasizing the early detection provided by the SCM.

A similar trend is shown in Figure 10-30 for the filter effluent turbidity. The coagulant feed failure resulted in an eventual increase in the filter effluent turbidity from a base

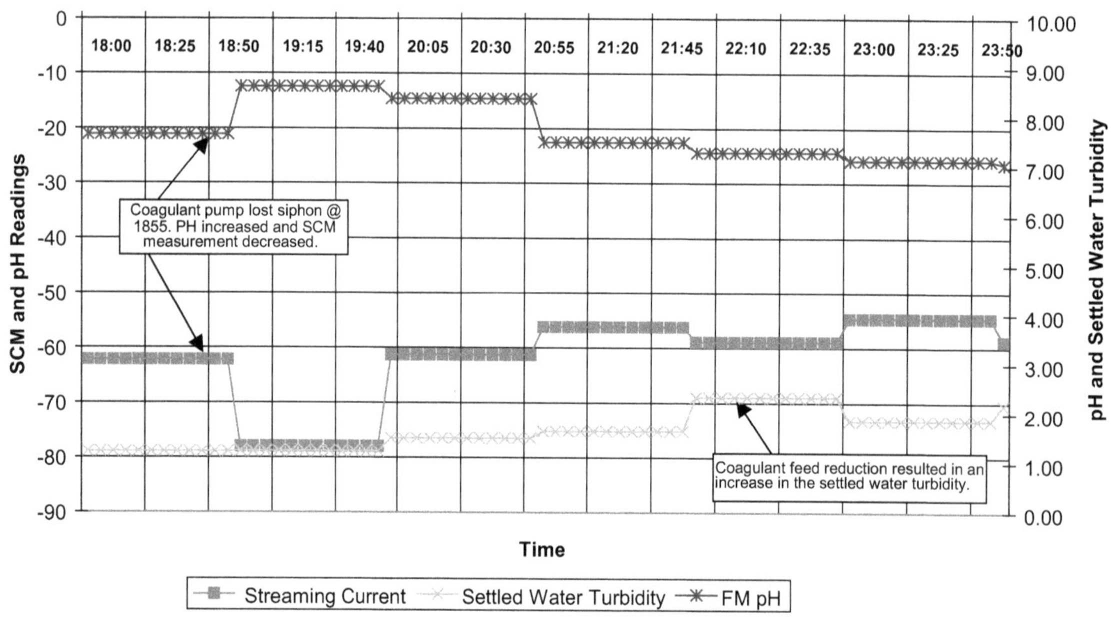

Source: Michael Sadar
FM—flash mix

Figure 10-29 SCM, pH, and settled water turbidity data during coagulant feed pump failure

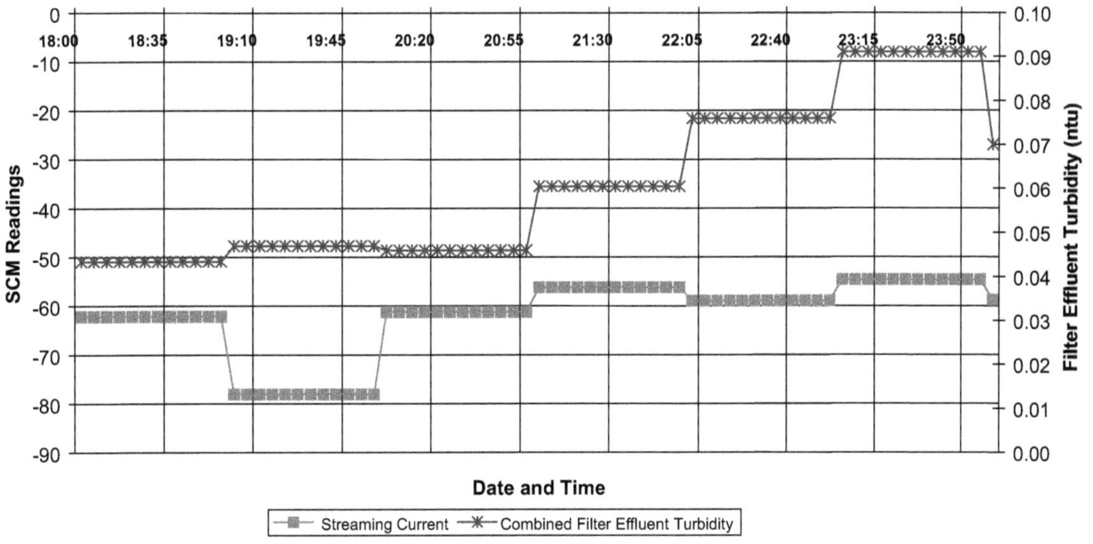

Source: Michael Sadar

Figure 10-30 SCM and filter effluent turbidity data during coagulant feed pump failure

value of 0.04 to a maximum value of 0.09 ntu. This peak value in filter effluent turbidity occurred 3.5 h after the coagulant underfeed event.

In summary, the underfeed incident caused the finished water turbidity to increase from a minimum of 0.04 to a maximum of 0.09 ntu. The settled water turbidity increased from about 1.3 to 2.3 ntu before returning to baseline. Although the finished water never approached regulatory limits for turbidity, the correlation between the decrease in SCM reading followed by the sequential increase in settled water and then the finished water turbidity is undeniable. The SCMs provided excellent early warning detection but were being used only as pilot study instruments at the time and were not integrated into the SCADA alarm system. If they had been integrated, it is probable that the subtle falloff in the ferric-metered dosage would have been caught and corrected even more quickly with a correspondingly smaller effect on both the settled and finished water turbidity.

Streaming Current Detector Pilot Study Conclusion

The coagulant feed failure event at the WTP was easily detected by SCM technology. The instrument was very stable and generated a stable baseline. The instrument can be easily re-zeroed, and this feature provided a consistent recognizable baseline value for the operator. The operators monitor over a hundred parameters concurrently for the operation of the treatment process, so they need a simple reference to determine if the SCM value is good or bad. Being able to set that reference to zero provides a simple means of problem identification, literally at a glance. This feature allows a single-point evaluation of the coagulant dosing system without the need to generate graphs or trend lines.

Following this underfeed event, the operators at this water plant have gained a significant amount of trust and confidence in the SCM instrument and now use it to monitor real-time stability of the chemical dosing system. The plant staff implemented and continues to practice a standard operating procedure for cleaning and re-zeroing of the SCM instrument on a daily basis. The plant staff also knows that if there is a pH increase combined with an SCM decrease, this indicates a coagulant failure. Without the SCM data, the operator would still have a level of uncertainty and would have to spend additional time troubleshooting the underfeed event. Time is of great importance, as this case has shown because events upstream can influence downstream processes throughout the entire plant.

This page intentionally blank.

Appendix A

Filter Index Test

The purpose of this test is to determine the filterability of the treated water resulting from the mixing process. The test involves measuring the time required to filter a known quantity of water through a micropore filter and comparing it with the time required to filter distilled water. The particles that are properly treated with the correct chemistry and mixing will not plug up the filter and therefore will allow the water to pass through quickly. Both the time and filtered turbidity are measured to determine the optimal treatment scheme. The apparatus is shown in Figure A-1 and involves the use of a Millipore filter, filter paper, and vacuum pump.

CONDUCTING THE FILTER INDEX TEST

On the jar test data sheet, place additional columns to record the time to filter 200 mL of sample from the jar test and the resulting filtered turbidity. The same size filter media and vacuum must be used for all tests to compare treatment schemes.

1. Measure 200 mL of deionized (DI) water and pour it into the Millipore filter. Turn on the vacuum pump, and record the time it takes to filter the entire volume through the filter paper and the filtered turbidity.
2. Immediately after stopping the mixing sequence in the jar test, collect 200-mL samples from each jar.
3. For each sample, replace the filter paper, record the time to filter 200 mL, and measure the resulting filtered turbidity.
4. Calculate the filter index (FI) by dividing the sample filter time by the DI filter time.

$$\text{Filter index} = \frac{\text{Time to filter sample}}{\text{Time to filter DI water}} \tag{Eq A-1}$$

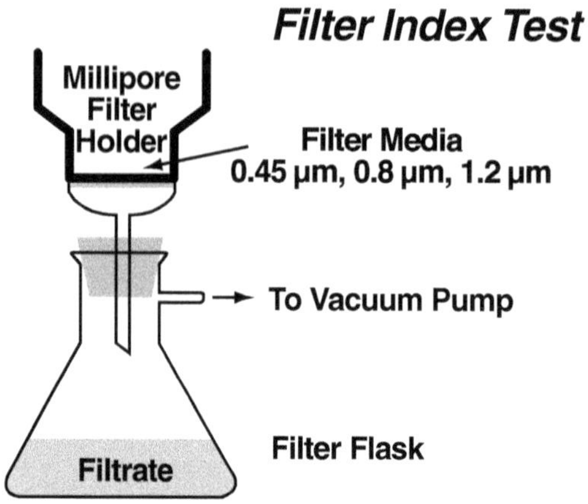

Source: ITT Water & Wastewater

Figure A-1 Filter index test apparatus

5. The comparison of the sample's FI number and filtered water turbidity will result in an optimal treatment scheme for the number of sample schemes evaluated.
6. For additional tests, the treatment scheme that yielded optimum results from all of the previous tests should be the treatment scheme in the first jar to serve as a benchmark for comparing the results of new schemes that are tested in the remaining jars.

The best treatment scheme will provide water that has the lowest FI number as well as the lowest filtered water turbidity. The water sample that has low filtered turbidity and high FI indicates that the particles in the water were not properly conditioned and the plant would experience filter plugging and short filter runs. The water sample that has high turbidity and low FI indicates that the particles are small enough to pass through the filter but are not coagulated and conditioned properly to be captured in the filter and would cause high turbidity in the finished water.

Appendix B

Velocity Gradient Curves for Magnetic and Marine Propeller Impellers

Figures B-1 and B-2 present laboratory velocity gradient curves for the magnetic impellers and marine propeller impellers in 11.5-cm square jars, respectively.

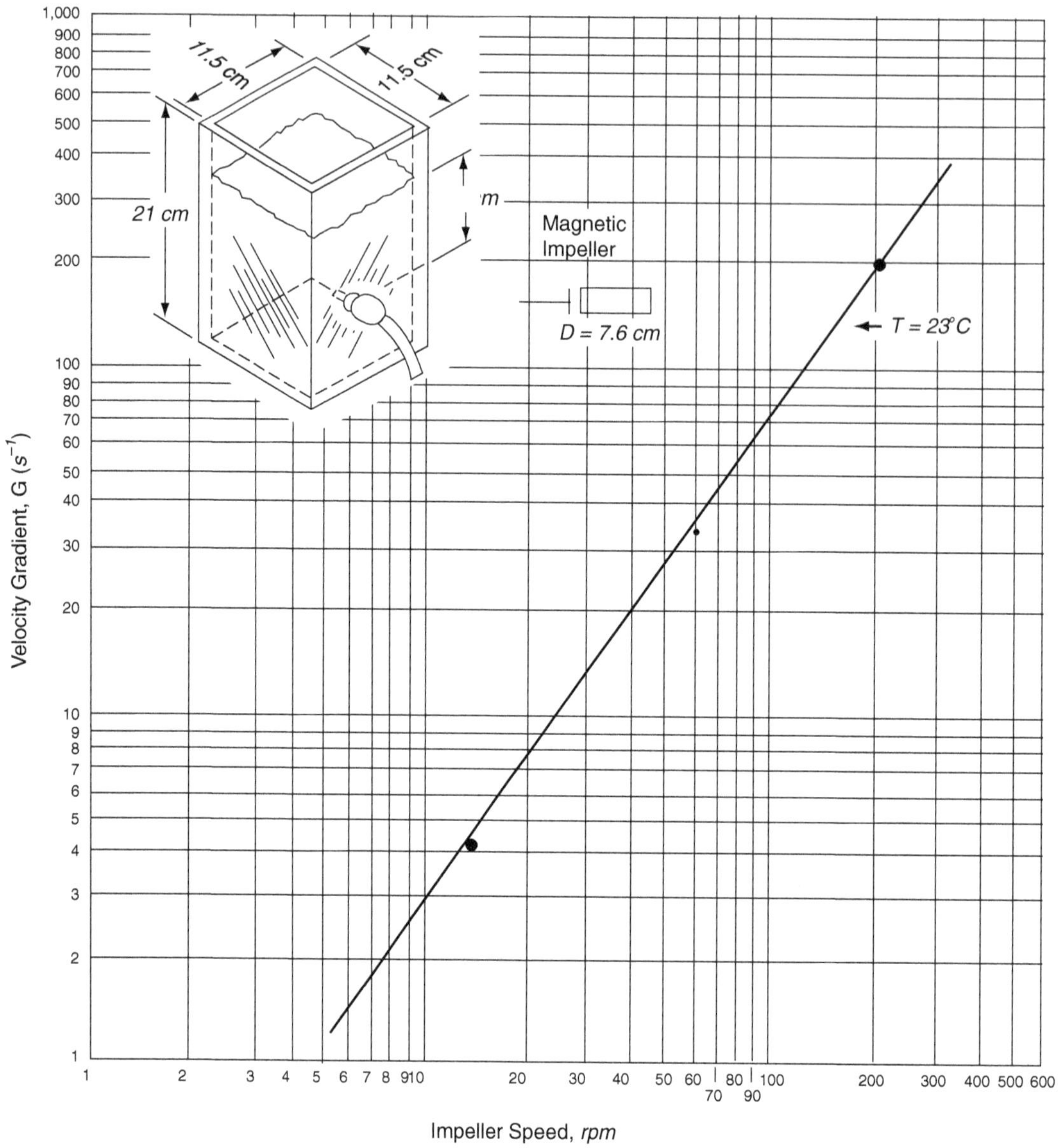

Source: Environmental Engineering & Technology, Inc.

Figure B-1 Laboratory *G* curve for magnetic impeller

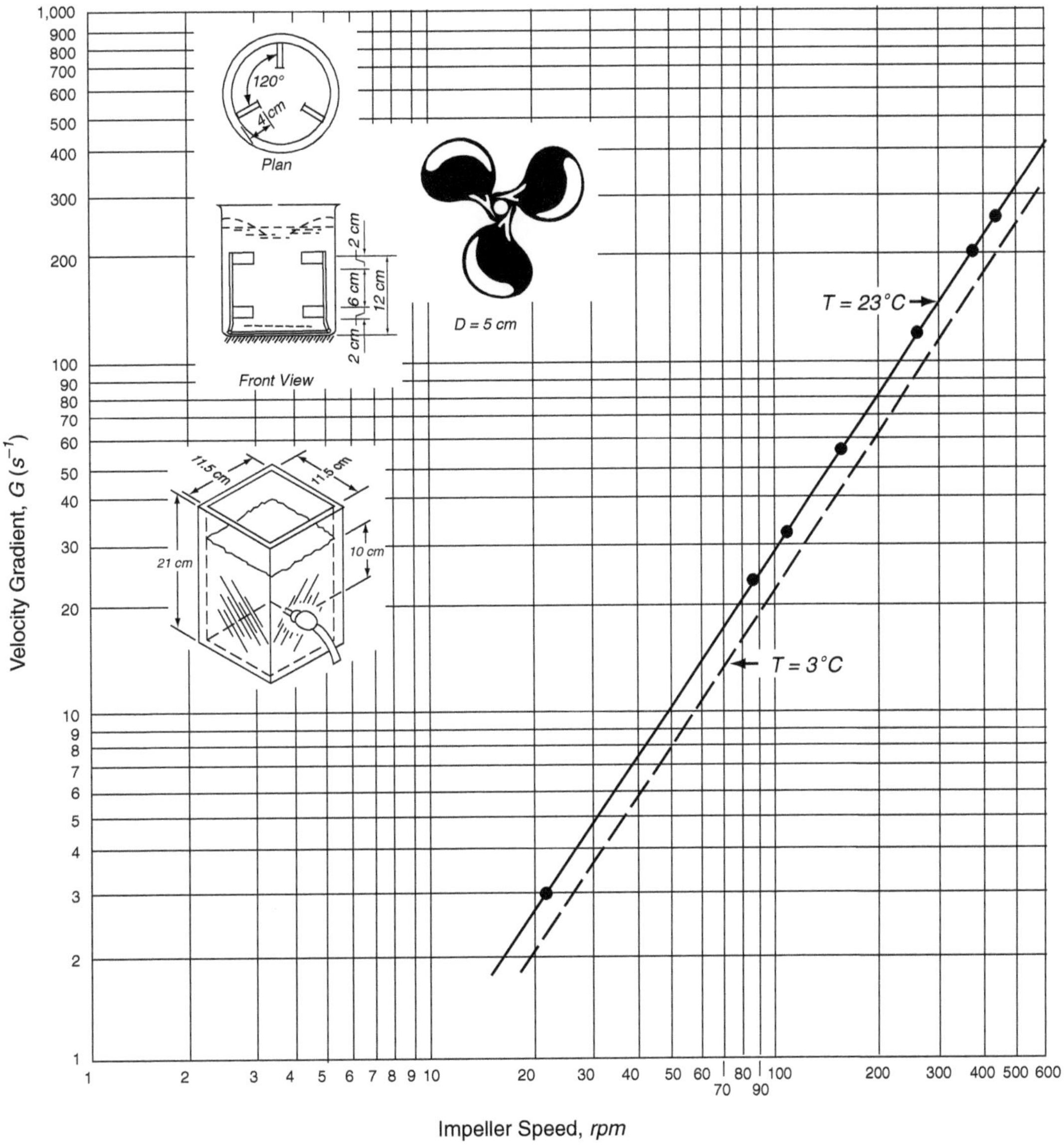

Source: Cornwell and Bishop 1983

Figure B-2 Laboratory *G* curve for marine propeller

REFERENCE

Cornwell, D.A., and M.M. Bishop. 1983. "Determining Velocity Gradients in Laboratory and Full-Scale Systems." *Journal AWWA*, 75:9:470–475.

This page intentionally blank.

Appendix C

Examples of Standard Operating Procedures (SOPs)

Bull Creek Surface Water Treatment Plant
Coagulation–Flocculation Jar Test

1. Scope

1.1. This procedure covers a general procedure for the evaluation of a treatment to reduce dissolved, suspended, colloidal, and nonsettleable matter from water by chemical coagulation–flocculation, followed by gravity settling and filtration. This procedure may be used to evaluate color and turbidity removal along with residual aluminum content remaining after completion.

2. Summary

2.1. A coagulation–flocculation test is carried out to determine the chemicals, dosages, and conditions to achieve optimal results. The primary variables to be investigated are as follows:

- type of chemicals added,
- chemical dosages added,
- pH,
- order of additions of chemicals and mixing conditions,
- temperature, and
- aluminum residuals in samples after coagulation.

3. Interferences

3.1. There are some possible interferences that may make the determination of optimal jar test conditions difficult. These include the following.

- Temperature changes during the test: Thermal or convection currents may occur that interfere with the settling of coagulated particles. (A water bath should be used to reduce this interference.)
- Gas release during test: Flotation of coagulated floc may occur as a result of gas bubble formation caused by mechanical agitation, temperature increase, or chemical reaction.

4. Reagents

4.1. Purity of reagents—Plant grade chemicals should be used in all tests.

4.2. Source water—2-L samples should be used in each beaker.

4.3. Jar test chemicals—Use graduated cylinders for water in the chemical makeup shown subsequently.

Coagulants:

(A) Alum—Add 6.2 mL of raw plant alum into 200 mL of distilled water; 1 mL of this solution will be 10 mg/L in a 2-L source water sample. Make a new batch every day used.

(B) Liquid sodium aluminate—Add 8.6 mL of raw liquid sodium aluminate into 200 mL of distilled water; 1 mL of this solution will be 10 mg/L in a 2-L source water sample. Make a new batch every day used.

(C) Dry sodium aluminate—Add 4.2 g of dry sodium aluminate into 200 mL of distilled water; 1 mL of this solution will be 10 mg/L in a 2-L source water sample. Make a new batch every day used.

Coagulants Aids:

(A) Liquid polymer—Add 0.102 g of raw liquid polymer into 500 mL of distilled water; 1 mL of this solution will be 0.1 mg/L in a 2-L source water sample. Make a new batch every two weeks.

(B) Dry polymer—Add 0.102 g of dry polymer into 500 mL of distilled water; 1 mL of this solution will be 0.1 mg/L in a 2-L source water sample. Make a new batch every two weeks.

Taste and Odor Control:

(A) Powder activated carbon—Add 1 g of powder activated carbon into 500 mL of distilled water; 1 mL of this solution will be 1 mg/L in a 2-L source water sample. Make a new batch every month.

(B) Potassium permanganate—Add 0.206 g of potassium permanganate to 500 mL of distilled water; 1 mL of this solution will be 0.2 mg/L in a 2-L source water sample. Make a new batch every month.

pH Adjustments

(A) (Caustic soda) sodium hydroxide—Add 1.3 mL of 50 percent plant caustic soda into 500 mL of distilled water; 1 mL of this solution will be 1 mg/L in a 2-L source water sample. Make a new batch every month.

(B) Hydrated lime—Add 1.01 g of hydrated lime into 600 mL of distilled water; 1 mL of this solution will be 1 mg/L in a 2-L source water sample. Make a new batch every month.

5. Procedures

5.1. Refer to section 4.3 to determine the correct amounts of all chemicals to be used in the jar test. Measure out all proper dosages into the plastic syringes. Note: 1 mL is equal to 1 cc on the syringe.

5.2. Calculate proper mix times for the three revolutions per minute (rpm) mixing speeds. For the "flash mix block" on the jar test form. Take the source water flow on the highest side of the plant and convert the flow to gallons per minute (gpm) then divide by 16,830. The answer will be the mixing time at 150 rpm. Next, take the same gpm reading on the highest side of the plant and divide it by 13,600. This answer will be the minutes of mixing at 50 rpm. Next, take the same highest raw flow in gpm and divide it by 3,141. This answer will be minutes of mixing at 25 rpm. Insert answers on sheet.

Example: Total Raw Flow = 13.5 million gal per day (mgd)

East Flow = 7.0 mgd

West Flow = 6.5 mgd

$$\frac{7{,}000{,}000}{1{,}440} = 4{,}861 \text{ gpm}$$

$$\text{Flash Mix} = \frac{4{,}861}{1{,}440} = 0.29 \text{ min at } 150 \text{ rpm}$$

$$\text{1st Stage} = \frac{4{,}861}{13{,}600} = 0.36 \text{ min at } 50 \text{ rpm}$$

$$\text{2nd Stage} = \frac{4{,}861}{3{,}141} = 1.5 \text{ min at } 25 \text{ rpm}$$

Table C-1 Jar test data sheet

	Flash Mix	Slow Mix 1st Stage	Slow Mix 2nd Stage	Source Water Amount Tested *Liters*	Source Water Characteristics									Date Analyst
Minutes	2	3	10	1	Color	Turbidity *ntu*	Alkalinity *mg/L* $CaCO_3$	Temperature °C	pH	TDC *mg/L*	TSS *mg/L*	TDS *mg/L*	Hardness *mg/L* $CaCO_3$	Objective: Find correct alum dosage
rpm	150	50	25	2	135	5.8	19	21.7	6	12	—	60	18	—

$CaCO_3$—calcium carbonate, rpm—revolutions per minute, TDC—total dissolved solids, TSS total suspended solids, TDS-total dissolved solids
Jar test flash mix (mixing time in minutes) + flow into plant flash mix (gpm)/16,830 gal.
1st stage mixing time in minutes = flow into plant flash mix (gpm)/15,600 gal.
2nd stage mixing time in minutes = flow into plant flash mix (gpm)/3,141 gal.

Table C-2 An example of a jar test data sheet

	Chemicals Used *mg/L* in order of addition			Floc Characteristics			Filtrate						
Jar No.	PAC	Alum	Polymer	Minutes to Form First Floc	Description of Floc	Settling Rate *cm/s*	Color	Turbidity *ntu*	Alkalinity *mg/L* $CaCO_3$	pH	Temperature °C	Al^{3+} RES *mg/L*	TOC *mg/L*
1	5	30	0.2	Did not form	None	N/A	—	—	—	—	—	—	—
2	5	35	0.2	30 s	Size = 5.1 mm, normal floc	0.2	30	1.5	—	—	—	—	—
3	5	40	0.2	30 s	Size = 5.1 mm, normal floc	0.2	15	0.23	—	—	—	—	—
4	5	45	0.2	30 s	Size = 5.1 mm, normal floc	0.2	12.5	0.2	—	—	—	—	—
5	5	50	0.2	30 s	Size = 5.1 mm, normal floc	0.2	13	0.27	—	—	—	—	—
6	5	55	0.2	30 s	Size = 5.1 mm, normal floc	0.2	13	0.2	—	—	—	—	—

N/A—not applicable, No.—number, PAC—powdered activated carbon, RES—residue, TOC—total organic carbon

ALAMEDA COUNTY WATER DISTRICT Standard Operating Procedure WTP2	NUMBER 34	EFFECTIVE DATE: October 1, 1997 REVISED: August 30, 2002
SUBJECT: **Jar Test Procedure for WTP2, using the Pipette Method**		PAGE 1 OF 21

1. Scope and Application

1.1 The jar test is used in the water treatment industry to simulate the treatment process. The test should be conducted whenever there is a significant change in water quality or as a tool for coagulant dosage optimization.

1.2 The jar test is performed in the laboratory to determine optimum dosage for chemical coagulants (usually ferric chloride or aluminum sulfate) and coagulant aids (cationic polymer) needed to achieve overall efficient plant operation.

1.3 The optimum coagulant dosage for a plant balances several factors:

- Settled water turbidity.
- Long filter runs.
- Organic removal.

Jar tests can be used to qualitatively evaluate the effects of different coagulant doses on these trends.

1.4 Jar tests are also used to compare and evaluate different coagulants and/or combinations of coagulants and coagulant aids.

1.5 This Pipette method saves time in preparing reagents. Also no dilution occurs when rinsing the reagent containers as in the conventional Beaker method.

1.6 Each test run may take 2-3 hours to complete.

2. Apparatus and Materials

2.1 Phipps & Bird Stirrer model 7790-400 (figure 1).
2.2 Variable speed control (0-300 rpm).
2.3 6 @ 2 liter square jars, B-KER^2.
2.4 HACH turbidimeter 2100AN.
2.5 Vacuum pump.
2.6 Stopwatch.
2.7 6 @ Plastic bottles, 500 mL.
2.8 Thermometer.
2.9 Filter funnel and filter holder base (figure 5 - 6).
2.10 Membrane filter, Millipore, cat # HAWG 047 S3, filter type HA, 0.45 µm.
2.11 Vacuum flask (figure 5-6).
2.12 Septa, (12 one for each coagulant and coagulant aid used per jar) (figure 2.)
2.13 Illuminator, to provide light needed for a clear visual inspection.
2.14 Jar-test form (figure 10).
2.15 Graduated cylinder 20 and 250 mL.
2.16 1000 mL volumetric flasks (one for each stock solution).
2.17 Magnetic stir bars.
2.18 Magnetic stirrer.
2.19 Calculator.
2.20 Micro-pipette, two sizes: 10 to 100 µL (use yellow tips) and 100 to 1000 µL (use white tips) (figure 3.)

Figure C-1 An example of standard operating procedure (SOP)

ALAMEDA COUNTY WATER DISTRICT Standard Operating Procedure WTP2	NUMBER 34	EFFECTIVE DATE: October 1, 1997 REVISED: August 30, 2002
SUBJECT: **Jar Test Procedure for WTP2, using the Pipette Method**		PAGE 2 OF 21

3. Reagents and Dilutions

3.1 **Ferric Chloride $FeCl_3$** coagulant

Stock solution

- When to obtain a new sample:
- A new stock solution should be collected at least monthly or whenever a new delivery is received. The date of the most recent delivery can be found in the Chemical Delivery Binder found in the operators' console. **NOTE:** Ferric Chloride shipments are received often, so it is very likely that a week old stock solution is not from the most recent shipment and should be re-sampled so the date should always be checked.
- Obtaining a new sample:
- Dispose of previous month's storage in the laboratory sink.
- Rinse out the old bottle or a new one from the cabinet under the jar testing machine. Label the jar with the current date.
- If not an operator, ask the operator to obtain a ferric sample for you.

NOTE: Always notify the operator when taking a sample.

- Go to the ferric storage tank in the chemical bay with the empty bottle and a bucket for flushing the ferric line (use appropriate protective gear; gloves, goggles, raincoat).
- Slowly turn the sampling tap on the side of the tank and flush the line for 15 - 30 seconds into a bucket.
- Collect about 100 mL of ferric chloride

Calculation of uL of ferric chloride to add to the septa for each 2 liter jar using an Eppendorf pipette

- The specific gravity of the Ferric solution as well as the percent activity of the solution is needed for the calculation. This information is found from the most recent Ferric delivery in the Chemical Delivery Binder in the operators' console.
- The calculation for the uL of ferric chloride to add for each mg/L of ferric chloride is as follows:

$$\frac{\text{Specific gravity}}{\text{\% activity}} = \left(\frac{1\ mg\ FeCl_3}{L\ water}\right)\left(\frac{L\ Ferric}{kg\ Ferric}\right)\left(\frac{kg}{10^6\ mg}\right)\left(\frac{kg\ Ferric}{kg\ FeCl_3}\right)\left(\frac{2\ L\ water}{beaker}\right)\left(\frac{10^6\ uL}{L}\right)$$

Which simplifies to :

$$\frac{2}{S.G * \%\ active} = \text{\# of } \mu\text{L Ferric solution to add per mg/L } FeCl_3 \text{ dose.}$$

ALAMEDA COUNTY WATER DISTRICT Standard Operating Procedure WTP2	NUMBER 34	EFFECTIVE DATE: October 1, 1997 REVISED: August 30, 2002
SUBJECT: **Jar Test Procedure for WTP2, using the Pipette Method**		PAGE 3 OF 21

EXAMPLE:

If the Ferric chloride most recent shipment is about 43% active and its specific gravity is 1.47:

$$\frac{2}{1.47*0.43}=\frac{2}{0.6321}=3.16\ \mu\text{L Ferric per mg/L FeCl}_3\text{ per jar}$$

In this case, the 3.16 μL is the amount to add to the 2 L beaker for each mg/ L of the ferric chloride desired in the jar test. So if you want to simulate a dose of 15 mg/L Ferric Chloride:

$$15\frac{mg}{L}*3.16\frac{uL}{mg/L}=47.4\ \mu\text{L Ferric Chloride should be placed on the septa.}$$

- **If you choose to vary the Ferric dose:**
- **If you want to optimize Ferric, you usually vary it by 2-5 parts per jar.**

EXAMPLE:

Using the same number for specific gravity and % activity as in the previous example:

	Jar 1	Jar 2	Jar 3	Jar 4	Jar 5	Jar 6
Desired ferric dose mg/L	15	20	25	30	35	40
μL of ferric to add to each septa	47.4	63.2	79.0	94.8	111	126

3.2 **Aluminum Sulfate (Alum)** $Al_2(SO_4)_3$ coagulant

Stock solution

- When to obtain a new sample:
- A new stock solution should be collected at least monthly or whenever a new delivery is received. The date of the most recent delivery can be found in the Chemical Delivery Binder found in the operators' console.
- Obtaining a new sample:
- Dispose of previous month's storage in the laboratory sink.
- Rinse out the old bottle or a new one from the cabinet under the jar testing machine. Label the jar with the current date.
- If not an operator, ask the operator to obtain an alum sample for you. **NOTE: Always notify the operator when taking a sample.**

ALAMEDA COUNTY WATER DISTRICT Standard Operating Procedure WTP2	NUMBER 34	EFFECTIVE DATE: October 1, 1997 REVISED: August 30, 2002
SUBJECT: **Jar Test Procedure for WTP2, using the Pipette Method**		PAGE 4 OF 21

- Go to one of the alum feed pumps in the chemical bay with the empty bottle and a bucket for flushing the alum line (use appropriate protective gear; gloves, goggles, raincoat).
- Slowly turn the sampling tap on the feed pump and flush the line for 15 - 30 seconds into a bucket.
- Collect about 100 mL of alum

Calculation of uL of Alum to add to the septa for each 2 liter jar using an Eppendorf pipette

- This calculation is the same as the Ferric Chloride calculation and requires the same information: specific gravity and percent activity, which are obtained from the most recent Alum delivery in the Chemical Delivery Binder in the operators' console.
- Substituting Ferric with Alum, the equation simplifies to:

$$\frac{2}{S.G * \% \, active}$$ **# of µL Alum to add per mg/L Alum dose.**

EXAMPLE:

If the most recent shipment of Alum were 48% active and its specific gravity was 1.33:

$$\frac{2}{1.33 * 0.48} = 3.13$$ L Alum per mg/L alum per jar

In this case, the 3.13 µL is the amount to add to the 2 L beaker for each mg/L of the alum desired. So if you want to simulate a dose of 40 mg/L alum:

$$40\frac{mg}{L} * 3.13\frac{uL}{mg/L} =$$ 125 µL Alum should be placed on the septa.

- **If you choose to vary the Alum Dose:**
- **If you want to optimize Alum, you can vary it by 2-5 parts per jar.**
- Each jar is calculated as shown for the Ferric.
  NOTE: Alum is rarely used in the plant. Ferric Chloride is the primary coagulant and Alum would be used in case of an emergency.

3.3 **Cationic Polymer (PE-C, polyDADMAC) poly(dimethyl diallyl ammonium chloride) n($C_8H_{16}ClN$)** coagulant aid

Stock solution

- When to obtain a new sample:

ALAMEDA COUNTY WATER DISTRICT Standard Operating Procedure WTP2	NUMBER 34	EFFECTIVE DATE: October 1, 1997 REVISED: August 30, 2002
SUBJECT: **Jar Test Procedure for WTP2, using the Pipette Method**		PAGE 5 OF 21

- A new stock solution should be collected at least monthly or whenever a new delivery is received. The date of the most recent delivery can be found in the Chemical Delivery Binder found in the operators' console. **NOTE:** The PE-C deliveries occur much less frequently than the Ferric

and

therefore, a stock solution close to a month old may still be from the most recent delivery but the dates should always be checked.
- Obtaining a new sample:
- Dispose of previous month's storage in the laboratory sink.
- Rinse out the old bottle or a new one from the cabinet under the jar testing machine. Label the jar with the current date.
  If not an operator, ask the operator to obtain a PE-C sample for you.
  NOTE: Always notify the operator when taking a sample.
- Go to the PE-C storage tank in the chemical bay with the empty bottle and a bucket for flushing the PE-C line (use appropriate protective gear; gloves, goggles, raincoat).
- Slowly turn the sampling tap on the side of the tank and flush the line for 15 - 30 seconds into a bucket.
- Collect about 100 mL of PE-C

Intermediate Solution

- The cationic polymer used as a coagulant aid is very viscous, and should be diluted before using in a jar test.
- To make a **1:50 dilution** of the polymer (2%)
- use a 20 mL graduated cylinder to measure 20 mL of PE-C. Add to a 1000 mL volumetric flask. Rinse out the graduated cylinder with DI water into the flask to ensure you have removed all of the PE-C.
- Fill to the 1000 mL mark on the flask with deionized water. Mix the solution well using a stir bar if necessary.

Calculation of uL of PE-C working solution to add to the septa for each 2 liter jar using an Eppendorf pipette

- This calculation is similar to that of the Ferric but differs in the details.
- Although polymers from different manufacturers and batches contain varying amounts of active ingredients, for the purpose of calculations, we assume 100% active ingredient for the polyDADMAC product used at both treatment plants (as the operators assume when calculating plant doses.) As a result, the only information necessary from the Chemical Delivery binder is the specific gravity of the most recent shipment of PE-C

- Ass umi ng $\left(\frac{1\,mg\ PE-C}{L\ water}\right)\left(\frac{L\ PE-C}{kg\ PE-C}\right)\left(\frac{kg}{10^6\ mg}\right)\left(\frac{1000\ L\ working\ solution}{20\ L\ PE-C}\right)\left(\frac{2\ L\ water}{beaker}\right)\left(\frac{10^6\ uL}{L}\right)$

a 100 % active stock solution and taking into account the dilution to the working solution and specific gravity, the equation for PE-C is as follows:

Specific gravity Working solution

ALAMEDA COUNTY WATER DISTRICT Standard Operating Procedure WTP2	NUMBER 34	EFFECTIVE DATE: October 1, 1997 REVISED: August 30, 2002
SUBJECT: **Jar Test Procedure for WTP2, using the Pipette Method**		PAGE 6 OF 21

$$\frac{100}{(s.g.)} =$$

Which simplifies to:

# of µL PE-C working solution to add per mg/L PE-C dose.

EXAMPLE:

If the most recent shipment of PE-C has a specific gravity of 1.08:

$$\frac{100}{1.08} = 92.59 \text{ L PE-C working solution per mg/L PE-C per jar}$$

In this case, the 92.59 µL is the amount to add to the 2 L beaker for each mg/L of the PE-C desired. So if you want to simulate a dose of 1.3 mg/L PE-C:

$$1.3\frac{mg}{L} * 92.59\frac{uL}{mg/L} =$$ 120 µL PE-C working solution should be placed on the septa.

- **If you choose to vary the PE-C dose:**
- **If you want to optimize PE-C, you usually vary it by about 0.2 parts between jars.**

EXAMPLE:

Using the same number for specific gravity as in the previous example:

	Jar 1	Jar 2	Jar 3	Jar 4	Jar 5	Jar 6
Desired PE-C dose mg/L	0.9	1.1	1.3	1.5	1.7	1.9
µL of PE-C working solution to add to each septa	83.3	102	120	139	157	176

4. How to Use The Eppendorf Micro-Pipette (figure 3)

4.1 **Size of Micropipette**

- Yellow top micropipette: This pipette is to measure over a range from 10 - 100 µL and is accurate to one decimal place. The clear tips in the yellow boxes should always be used with this pipette.

- Blue Top micropipette: This pipette is for measuring amounts ranging from 100 - 1000 μL and is accurate to the ones place. The larger, clear tips in the blue boxes should always be used with this pipette.

ALAMEDA COUNTY WATER DISTRICT Standard Operating Procedure WTP2	NUMBER **34**	EFFECTIVE DATE: October 1, 1997 REVISED: **August 30, 2002**
SUBJECT: **Jar Test Procedure for WTP2, using the Pipette Method**		PAGE 7 OF 21

4.2 **Volume Setting**

- Press the gray locking button found on the left side of the volume window down and hold.
- Set the volume by turning the control button.
  NOTE: Always set higher volumes first. When changing the volume from a lower to a higher setting, turn the knob past the desired volume and then back again.
- Release the locking button. The set volume is now secured against inadvertent adjustment.

4.3 **Filling**

- Attach pipette tip securely.
- Press the control button down to first the stop and hold.
- Hold pipette vertical and immerse the tip approximately 3 mm into the liquid.
  NOTE: Do not place a micro-pipette into liquid without a tip in place and do

  not immerse it further than the removable tip.
- Let the control button glide back slowly and smoothly.
- Slide the tip out of the liquid along the inside wall of the vessel, wiping off any droplets that remain on the outside of the tip.

4.4 **Dispensing**

- Hold tip upright and press the control button slowly down to the first stop to empty the tip onto the septa.
- Press the control button to the second stop to empty the tip completely.
- Let the control button glide back.

4.5 **Removing Tip**

- To remove a tip, hold the control button all the way to the third stop where it will release the tip. Throw the used tip in the trash.
- Replace the tip when you change between chemicals or if it becomes contaminated. Always throw away the tips after each jar test.

5. Vacuum Pump / Filter Apparatus Operation (figure 5-6)

5.1 **Set up the filter apparatus**

- Open a 0.45 um Millipore HA membrane filter and carefully place it grid side up on the filter holder base that fits inside the flask.
- Place the magnetic filter funnel on top of the filter.

5.2 **Filter Operation**

- Turn on the vacuum pump and add a few drops of deionized water to just soak the filter paper. Wait for the vacuum to stabilize around 25 Hg.
- Using a graduated cylinder, measure out 200 mL of the water you are testing.

- Pour the 200 mL of test water into the filtration unit. Start the stopwatch

as

soon as the first drop of the water has been poured into the filtration unit. Stop the watch when the first dry spot appears on the filter paper. Record the time, in seconds, as the filter time.
- **Blank:** This process should always first be done with deionized water as a reference. Ideally the DI water should filter in 20-30 seconds. If necessary, the vacuum pressure should be adjusted to achieve this DI water time. The **Filter Index** = sample time (seconds) / DI water time (seconds).
- During the jar test, this filter process is done several times on the settled water from the test. The blank must be done with the DI water and the applied water and finished water from the sample taps in the laboratory are also tested.

6. Special Requirements

6.1 This procedure must be modified every time it is used to reflect conditions in the plant. Before beginning, record as much information on the jar test form as possible. Information such as raw water quality, number of flocculation and sedimentation basins, plant flow, plant dosages, and specifications for the most recent chemical deliveries will change with each test and are necessary to ensure that the jar test reflects the current plant conditions. Figure 10 contains a sample jar test form as well as where you can obtain all of the necessary data.

6.2 The doses to be tested will vary every time you do a jar test. Generally, routine jar tests are done to determine the best coagulant dose to be used in the plant. Ask the operator in charge his/her opinion about the current chemical dose. Generally, only one chemical is varied at a time and
two complete jar tests are run. Ask the operator which chemical to vary first and have him/her choose 6 doses such that the first beaker is under-dosed and the last is over-dosed. The middle beaker should contain the current chemical dose that the plant is using.
EXAMPLE: If the current plant doses are 19 mg/L Ferric and 1.2 mg/L PE-C, the first set of jars might test ferric doses of 10, 13, 16, 19, 22, and 25 mg/L with a constant PE-C dose at 1.2 mg/L. Once the test is run and an optimum ferric dose is determined, run the second test to determine the optimum PE-C dose. For example, if the optimum ferric dose was found to be 16 mg/L, the next test may keep the ferric dose constant at 16 mg/L and vary the PE-C doses of 0.6, 0.8, 1.0, 1.2, 1.4, and 1.6 mg/L.
NOTE: The doses that are obtained from the SCADA screens are not always accurate. If the operator has done a catch recently, use the calculated catch doses to decide on the doses to simulate in the jar test.

7. Procedure

7.1 Ask an operator to turn on an **ozonated water pump. Flush** the sample line for at least 10 minutes before filling the beakers.

ALAMEDA COUNTY WATER DISTRICT Standard Operating Procedure WTP2	NUMBER **34**	EFFECTIVE DATE: October 1, 1997 REVISED: **August 30, 2002**
SUBJECT: **Jar Test Procedure for WTP2, using the Pipette Method**		PAGE 9 OF 21

7.2 Complete the preliminary information on the jar test form. Record raw water quality data, reason for testing, plant doses, etc. on the jar test form. The data that is collected in the laboratory (such as RW quality and DI and AW filter times) can also be done during the flocculation time to save some time.

7.3 Calculate the **flocculation time** using the following formula and record the time in the appropriate spot on the jar test form.

$$\textit{floc time, min} = \frac{\textit{number of floc basins}}{\textit{flow rate, mgd}} \times 122$$

7.4 Calculate the **settling time** using the following formula and record the time in the appropriate spot on the jar test form.

$$\textit{settling time, min} = \frac{\textit{number of sed basins}}{\textit{flow rate, mgd}} \times 17.9$$

7.5 Set two timers such that the first one goes off after flocculation is complete (1 minute + flocculation time) and the second beeps at the end of the test (1 minute + flocculation time + settling time.)

7.6 Calculate the **mixing speed (rpm)** to simulate the floc basins. First find the Velocity Gradient using the appropriate Velocity Gradient vs. Plant Flow rate graph (figure 7 or figure 8) (it depends on the number of open tubes per cell.) Then use the Laboratory G Curve graph (figure 9) to

determine the mixing speed on the x-axis from the velocity gradient on the y-axis. Record all times and speeds on the correct portion of the jar test form.

EXAMPLE for 7.3 through 7.5**:**

Plant flow = 21 MGD, 6 floc basins, 4 sed basins, and 4 tubes are in use.

$$\text{Floc Time} = \frac{6}{21} * 122 = 34.8 \approx 35 \text{ minutes}$$

$$\text{Settling Time} = \frac{4}{21} * 17.9 = 3.4 \approx 3.5 \text{ minutes}$$

Velocity gradient: Use figure 7. Find the velocity by finding where the rate intersects with the appropriate curve. In this case, 21 MGD crosses the 6 basin curve at 17.5 ft/sec.

Speed: Use figure 9. Where the velocity gradient (y-axis) is 17.5 ft/sec, this crosses the solid line at an impeller speed (x-axis) slightly above 20 rpm. We will use 20 rpm for simplicity.

ALAMEDA COUNTY WATER DISTRICT Standard Operating Procedure WTP2	NUMBER **34**	EFFECTIVE DATE: October 1, 1997 REVISED: **August 30, 2002**
SUBJECT: **Jar Test Procedure for WTP2, using the Pipette Method**		PAGE 10 OF 21

On jar test form:

Time and Speed Sequence			
	Start Time	**Duration**	**Speed**
Rapid Mix	0 sec	30 sec	300 rpm
Slow Mix	30 sec	30 sec	200 rpm
Flocculation	1 min	35 min	20 rpm
Settling	36 min	3.5 min	0 rpm

End 39.5 min

Set the timer for 36 minutes and 39.5 minutes.

7.7 **Decide on the doses for the chemical and calculate the amount of chemicals to be added to each jar** to simulate those doses (Section 3; most likely the chemicals added will be only ferric chloride (3.1) and the cationic polymer (3.3).)

7.8 Using the Eppendorf pipette as described in section 4, **dispense the calculated amount of ferric & PE-C onto the septa** (12 total septa) (figure 2.)

7.9 Fill the 6 square jar test containers to the 2 liter mark with **ozonated water** from the laboratory sink. Collect all 6 samples within a few minutes to avoid variability in the test results. Hooking a longer hose onto the sample tap makes this easier.

7.10 **Place the jars onto the Phipps and bird Stirrer, lower and tighten the stir paddles into the jars, and turn on the light on the apparatus** (figure 1.) Measure the temperature in one of the jars and record on the jar test form.

7.11 Turn on the mixer to **300 rpm**, make sure that the jars are all placed correctly such that the stir paddles are not hitting the jars and they are centered so that the septa will fall into the jars properly.

7.12 **Crank the septa bar such that the septa fall into the jar containers and start the timer as you crank the bar.**

7.13 At **30 seconds**. reduce the mixing speed to **200 rpm**.

7.14 At **1 minute** reduce the mixing speed to the **calculated flocculation speed** (section 7/6.)

7.15 **At the first timer (1 minute + flocculation time) turn off the mixer and allow the jars to settle.**

7.16 **Place the half liter square plastic bottles labeled 1 through 6 upright with the sample taps** (located on the side of the jars) inside the plastic bottles so that when the settling time is up, you can open all the taps at once and collect all samples.

7.17 **Important parameters to observe** during the jar test and record on the jar test form

- Relative speed for the first floc formation in each jar (first=1 and last=6).
- Relative floc size between the jars (VS=very small, S=small, M=medium, L=large, VL=very large).
- Relative settling rate between the jars (fastest=1 and slowest=6).
- Relative clarity of the water between floc particles between the jars (best=1 and worst=6).

Record all observations and significant differences between jars on the bottom of the jar test form (flocculation example, figure 4.)

NOTE: It is a good idea to have an operator, a senior operator, or a supervisor observe the jars with you to determine which jars look the best.

ALAMEDA COUNTY WATER DISTRICT Standard Operating Procedure WTP2	NUMBER **34**	EFFECTIVE DATE: October 1, 1997 REVISED: **August 30, 2002**
SUBJECT: **Jar Test Procedure for WTP2, using the Pipette Method**		PAGE 11 OF 21

7.18 Immediately **after settling**, after the second timer is complete (1 minute + flocculation time + settling time), **collect approximately 500 mL of each jar sample** into the half liter square plastic bottles labeled 1 through 6. It is critical that the samples be collected from the sample tap,
located in the side of the jar test containers, all within a few minutes to avoid variability in the test results. If you have set up the plastic bottles as previously explained (section 7.16) you can open all the sample taps at once and simultaneously collect the samples from all jars.

7.19 **Analyze samples for turbidity** and record them as settled turbidity on the jar test form.

7.20 Determine the filtration time for deionized water, the applied water and filter effluent from the laboratory sink using the vacuum pump and the filter apparatus (section 5.) (This step can also be done earlier, while the jars are mixing to speed up the process.)

7.21 **Determine the filtration time for each jar sample** and record on the jar test form as filtered water filter time and calculate the filter index (section 5.)

7.22 **Analyze the filter effluent (filtrate) for turbidity for each jar sample** and record on the jar test form.

7.23 Make sure you have recorded all other necessary information on the jar test form and done all of the necessary laboratory analysis.

8. **Analysis and Conclusions**

8.1 Upon completion of the test, go over the results and come to a conclusion by choosing the most optimal coagulant and doses.

8.2 Confer with the operator and discuss each aspect of the test results together.

- **Floc Characteristic:** The ideal floc particle will be the 2nd or 3rd rank in speed of forming. The flock will be large but no so large that it sinks to the bottom of the beaker during the flocculation stage of the test. It should hold together and not break apart as it is stirred. The rank for settling should be 2nd or 3rd. If the floc settles too quickly there is a chance that settling will occur in the floc basins before it reaches the sedimentation basins.
- **Settled Water:** The operators want to see very clear water between the floc particles as they are forming. The clearer the water, the cleaner the end product will be. The optimal beaker will be rank 1. The settled turbidity should coincide with the amount of chemical added. It will most likely increase as the dose decreases.
- **Filtered Water:** The optimal filtrate water will be the one that filters the fastest and produces the cleanest filtrate turbidity. It is common to have all the jars produce very similar filtrate turbidity, but the distinguishing factor will be the filter time.

ALAMEDA COUNTY WATER DISTRICT Standard Operating Procedure WTP2	NUMBER 34	EFFECTIVE DATE: October 1, 1997 REVISED: August 30, 2002
SUBJECT: **Jar Test Procedure for WTP2, using the Pipette Method**		PAGE 12 OF 21

8.3 In the conclusion and comments on the jar test form, write a few sentences to summarize the test results. Note the jar number that produced the most optimal results and a brief description of why you and the operator selected it. Also discuss why the other jars failed, i.e., too much or too little coagulant or heavy floc.

8.4 The decision to change the plant's doses after the test is up to the operator. He/She may wish to examine the jar test form further.

8.5 Make sure a copy of the form is added to the jar test binder stored in the Laboratory.

9. Quality Control

9.1 The stock solution of alum, ferric chloride and PE-C should not be kept longer than one month and should be re-sampled if there has been a new shipment since the sample was taken.
9.2 The prepared working solutions should be dated and kept no longer than 2 days.
9.3 Water samples should be collected immediately before the analysis.
9.4 Wash volumetric flasks immediately after test is finished.
9.5 Wash beakers immediately after test is finished.
9.6 Clean jars with 0.02 N sulfuric acid once a month.
9.7 Dispose of previous month's storage in the laboratory sink.
9.8 Never lay pipettes down with liquid in the tip.
9.9 Never pipette liquid without attaching a tip to the pipette or submerge the pipette past the tip.

10. References

10.1 Introduction to Water Quality Analyses, Volume 4, American Water Works Association, 6666 W. Quincy Ave., Denver, CO 80235.
10.2 Water Quality, Principles and Practices of Water Supply Operations Series, Second Edition, 1995, American Water Works Association, 6666 W. Quincy Ave, Denver, CO 80235.
10.3 Water Treatment Handbook, Volume 1, Sixth Edition, 1991, Degremont, 11, rue Lavoisier-F 75384 Paris Cedex 08.
10.4 Operational Control of Coagulation and Filtration Processes, AWWA Manual M 37, 1992, American Water Works Association, 6666 W. Quincy Ave, Denver, CO 80235.

ALAMEDA COUNTY WATER DISTRICT Standard Operating Procedure WTP2	NUMBER **34**	EFFECTIVE DATE: October 1, 1997 REVISED: **August 30, 2002**
SUBJECT: **Jar Test Procedure for WTP2, using the Pipette Method**		PAGE 13 OF 21

ALAMEDA COUNTY WATER DISTRICT Standard Operating Procedure WTP2	NUMBER **34**	EFFECTIVE DATE: October 1, 1997 REVISED: **August 30, 2002**
SUBJECT: **Jar Test Procedure for WTP2, using the Pipette Method**		PAGE 14 OF 21

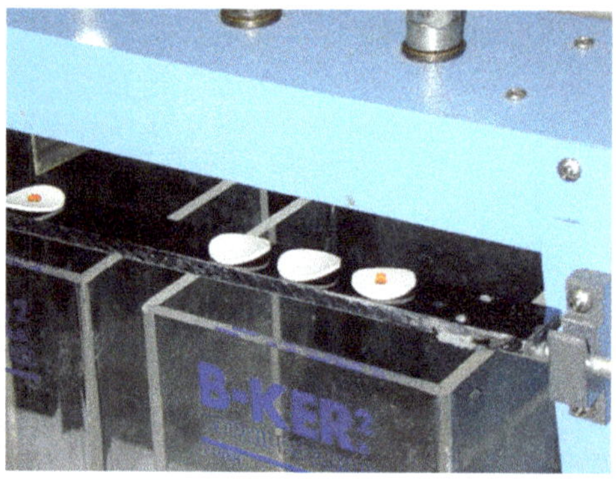

ALAMEDA COUNTY WATER DISTRICT Standard Operating Procedure WTP2	NUMBER **34**	EFFECTIVE DATE: October 1, 1997 REVISED: **August 30, 2002**
SUBJECT: **Jar Test Procedure for WTP2, using the Pipette Method**		PAGE 15 OF 21

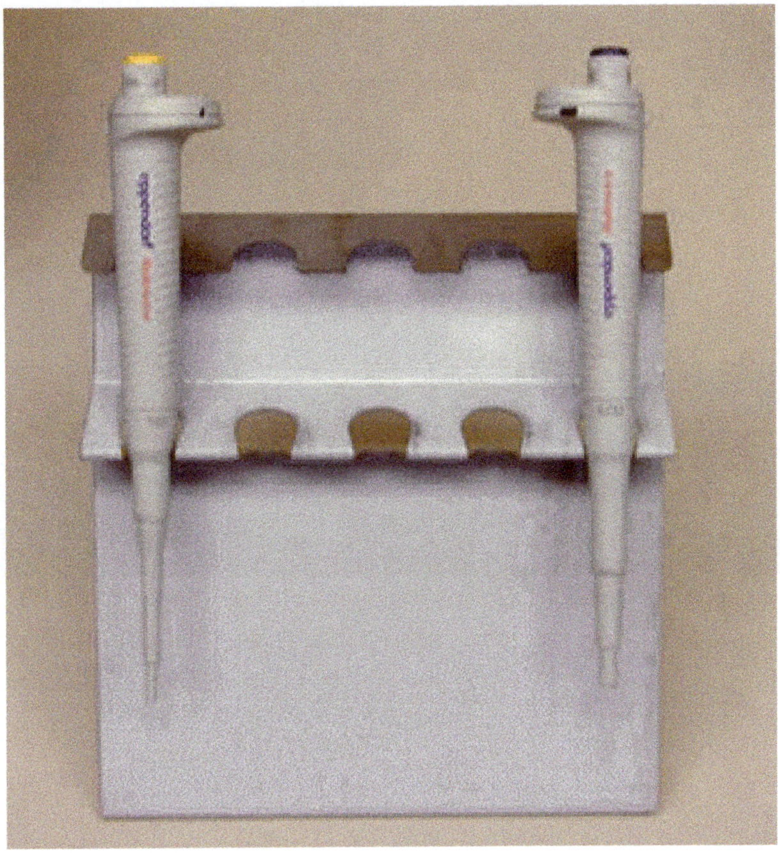

ALAMEDA COUNTY WATER DISTRICT Standard Operating Procedure WTP2	NUMBER 34	EFFECTIVE DATE: October 1, 1997 REVISED: **August 30, 2002**
SUBJECT: **Jar Test Procedure for WTP2, using the Pipette Method**		PAGE 16 OF 21

ALAMEDA COUNTY WATER DISTRICT Standard Operating Procedure WTP2	NUMBER 34	EFFECTIVE DATE: October 1, 1997 REVISED: **August 30, 2002**

SUBJECT: **Jar Test Procedure for WTP2, using the Pipette Method**	PAGE 17 OF 21

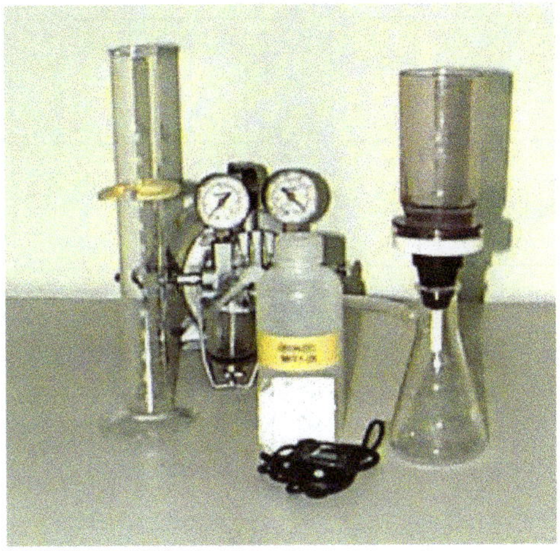

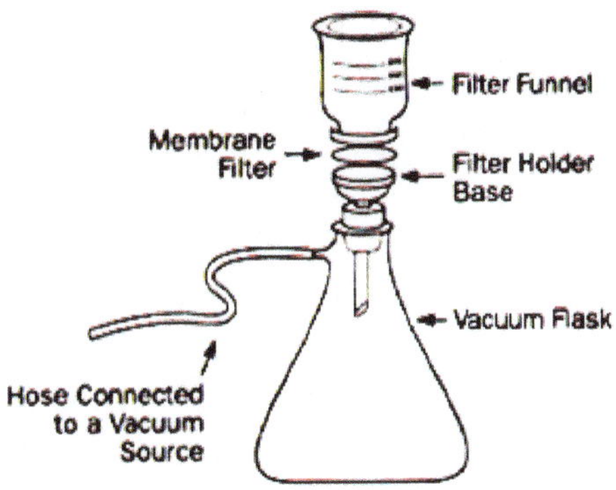

ALAMEDA COUNTY WATER DISTRICT Standard Operating Procedure WTP2	NUMBER **34**	EFFECTIVE DATE: October 1, 1997 REVISED: **August 30, 2002**

SUBJECT: **Jar Test Procedure for WTP2, using the Pipette Method**	PAGE 18 OF 21

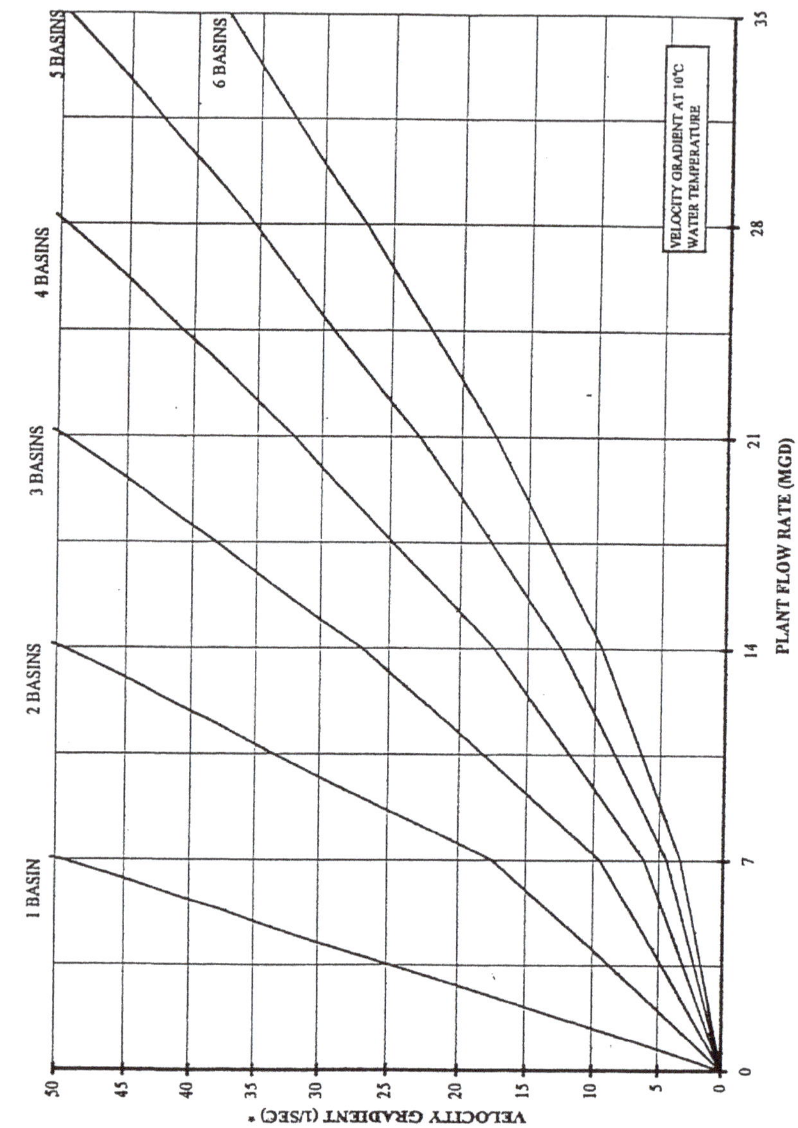

ALAMEDA COUNTY WATER DISTRICT Standard Operating Procedure WTP2	NUMBER **34**	EFFECTIVE DATE: October 1, 1997 REVISED: **August 30, 2002**

SUBJECT: **Jar Test Procedure for WTP2, using the Pipette Method**	PAGE 19 OF 21

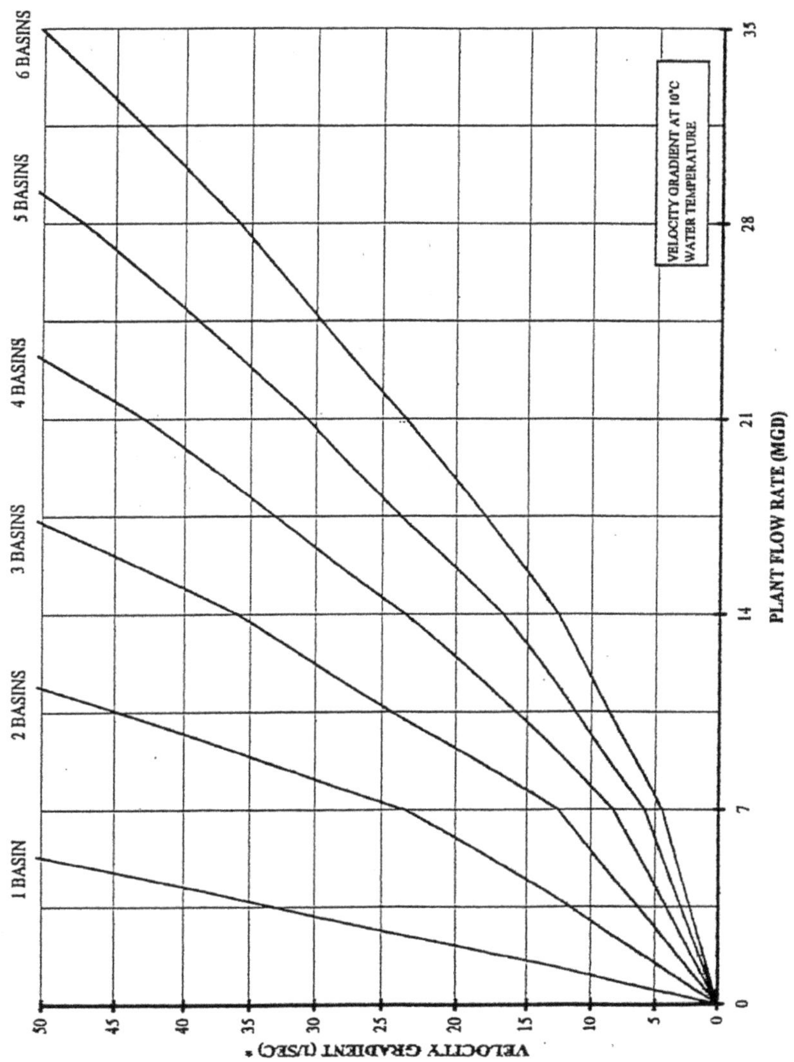

ALAMEDA COUNTY WATER DISTRICT Standard Operating Procedure WTP2	NUMBER **34**	EFFECTIVE DATE: October 1, 1997 REVISED: **August 30, 2002**
SUBJECT: **Jar Test Procedure for WTP2, using the Pipette Method**		PAGE 20 OF 21

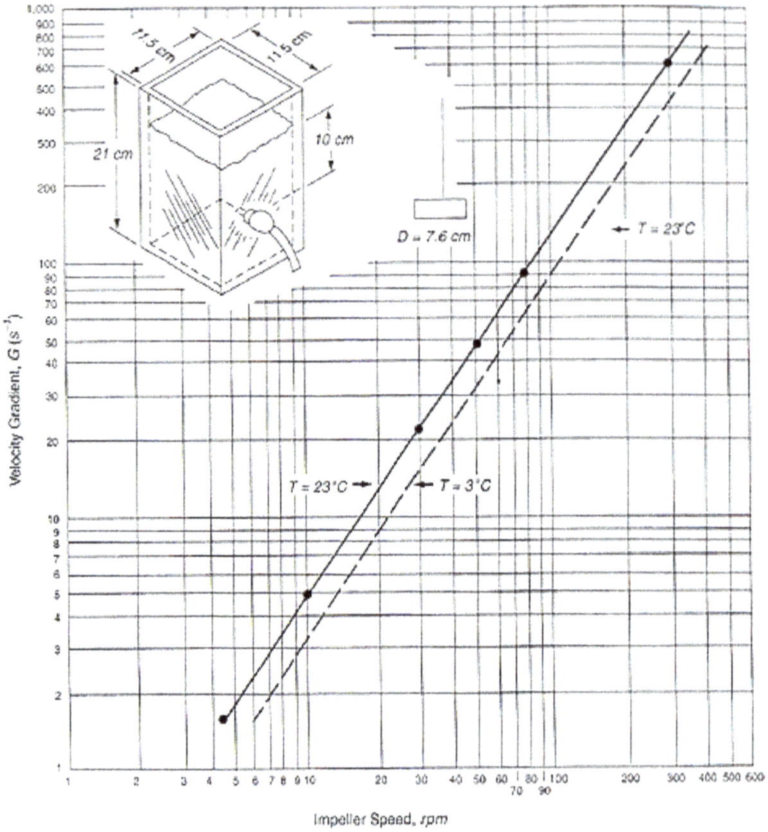

ALAMEDA COUNTY WATER DISTRICT Standard Operating Procedure WTP2	NUMBER 34	EFFECTIVE DATE: October 1, 1997 REVISED: August 30, 2002
SUBJECT: **Jar Test Procedure for WTP2, using the Pipette Method**		PAGE 21 OF 21

TREATMENT PLANTS JAR TEST FORM

Name:	Plant: TP1 ( ) TP2 ( )	Plant Rate: mgd	Time of Day:	Date:
Reason for Testing:		Method Used : (X)	Micro-pipette :	Beaker :

WTP Water Quality Data						Plant Process		TP 2 Only		Plant Dosages		Filter Index (FI) Test		
Raw Water	______% Delta	______% Del Valle				AW Turbidity	ntu	*# of Floc used*		Chlorine	mg/l	Filter Size/Type	0.45 um	Millipore-HA
Turbidity			Alkalinity			AW Filter Time	sec	*# of Seds used*		Ferric	mg/l	D. I. Water Time :		sec
pH			Spec Cond			AW Filter Index	(FI)	*# of Open Tubes*		PE-C	mg/l	Vacuum :		in Hg
Temp			Hardness			Filtered Water	ntu	*Ozone*	mg/l	Alum	mg/l	F I =	Sample Time / DI Time	

Chemicals Used (In Order of Addition)											Floc Characteristics			Settled Water		Filtered Water		
	Chlorine		Ferric		Alum		PE-C				Relative speed of forming	Relative Size	Relative Settling Rate	Relative Clarity Between Floc	Settled Turbidity	Filter Time	Filter Index	Filtrate Turbidity
Jar	mg/l	ul	mg/l	ul	mg/l	ul	mg/l	ul	mg/l	ul	(1-10)	(VS-VL)	(1-10)	(1-10)	(NTU)	(sec)	(FI #)	(NTU)

Ratings 1 = Best/Fastest 10 = Worst/Slowest S= Very Small S = Small M = Medium L = Large VL = Very Large

Time and Speed Sequence				Observations / Comments and Conclusions:
	Start Time	Duration	Speed	
Rapid Mix	sec	sec	rpm	
Slow Mix	sec	sec	rpm	
Flocculation	min	min	rpm	
Settling	min	min	rpm	

This page intentionally blank.

Index

D

J

K

L

M

N

O

P

R

S

T

U

V

W

Z

AWWA Manuals

- M1, *Principles of Water Rates, Fees, and Charges*, #30001
- M2, *Instrumentation and Control*, #30002
- M3, *Safety Management for Water Utilities*, #30003
- M4, *Water Fluoridation Principles and Practices*, #30004
- M5, *Water Utility Management*, #30005
- M6, *Water Meters—Selection, Installation, Testing, and Maintenance*, #30006
- M7, *Problem Organisms in Water: Identification and Treatment*, #30007
- M9, *Concrete Pressure Pipe*, #30009
- M11, *Steel Pipe—A Guide for Design and Installation*, #30011
- M12, *Simplified Procedures for Water Examination*, #30012
- M14, *Backflow Prevention and Cross-Connection Control: Recommended Practices*, #30014
- M17, *Fire Hydrants: Installation, Field Testing, and Maintenance*, #30017
- M19, *Emergency Planning for Water and Wastewater Utilities*, #30019
- M20, *Water Chlorination/Chloramination Practices and Principles*, #30020
- M21, *Groundwater*, #30021
- M22, *Sizing Water Service Lines and Meters*, #30022
- M23, *PVC Pipe—Design and Installation*, #30023
- M24, *Planning for the Distribution of Reclaimed Water*, #30024
- M25, *Flexible-Membrane Covers and Linings for Potable-Water Reservoirs*, #30025
- M27, *External Corrosion Control for Infrastructure Sustainability*, #30027
- M28, *Rehabilitation of Water Mains*, #30028
- M29, *Water Utility Capital Financing*, #30029
- M30, *Precoat Filtration*, #30030
- M31, *Distribution System Requirements for Fire Protection*, #30031
- M32, *Computer Modeling of Water Distribution Systems*, #30032
- M33, *Flowmeters in Water Supply*, #30033
- M36, *Water Audits and Loss Control Programs*, #30036
- M37, *Operational Control of Coagulation and Filtration Processes*, #30037
- M38, *Electrodialysis and Electrodialysis Reversal*, #30038
- M41, *Ductile-Iron Pipe and Fittings*, #30041
- M42, *Steel Water-Storage Tanks*, #30042
- M44, *Distribution Valves: Selection, Installation, Field Testing, and Maintenance*, #30044
- M41, *Fiberglass Pipe Design*, #30045
- M46, *Reverse Osmosis and Nanofiltration*, #30046
- M47, *Capital Project Delivery*, #30047
- M48, *Waterborne Pathogens*, #30048
- M49, *Quarter-Turn Valves: Head Loss, Torque, and Cavitation Analysis*, #30049
- M50, *Water Resources Planning*, #30050
- M51, *Air Valves: Air-Release, Air/Vacuum and Combination*, #30051
- M52, *Water Conservation Programs—A Planning Manual*, #30052
- M53, *Microfiltration and Ultrafiltration Membranes for Drinking Water*, #30053
- M54, *Developing Rates for Small Systems*, #30054
- M55, *PE Pipe—Design and Installation*, #30055
- M56, *Nitrification Prevention and Control in Drinking Water*, #30056
- M57, *Algae: Source to Treatment*, #30057
- M58, *Internal Corrosion Control in Water Distribution Systems*, #30058
- M60, *Drought Preparedness and Response*, #30060
- M61, *Desalination of Seawater*, #30061
- M62, *Membrane Applications for Water Reuse*, #30062
- M63, *Aquifer Storage and Recovery*, #30063
- M65, *On-Site Generation of Hypochlorite*, #30065

M66, *Cylinder and Vane Actuators and Controls—Design and Installation,* #30066

M68, *Water Quality in Distribution Systems,* #30068

M69, *Inland Desalination and Concentrate Management,* #30069

M77, *Condition Assessment of Water Mains,* #30077

M80, *Biological Drinking Water Treatment* #30080

www.ingramcontent.com/pod-product-compliance
Ingram Content Group UK Ltd.
Pitfield, Milton Keynes, MK11 3LW, UK
UKHW052010041026
15590UKWH00008B/2258

9 781647 172329